Reforming Regulatory Impact Analysis

Editors

Winston Harrington,
Lisa Heinzerling, and
Richard D. Morgenstern

Resources for the Future
Washington, DC USA

Printed in the United States of America

Resources for the Future
1616 P Street, NW
Washington, DC 20036–1400

Telephone: 202-328-5000
www.rff.org

Editors:
Elizabeth Stallman Brown
Felicia Day, RFF Managing Editor
Adrienne Foerster, RFF Assistant Editor

Design and production:
Meadows Design Office Inc., Washington, DC
www.mdomedia.com

Cover design by Marc Alain Meadows

ISBN 978-1-933115-75-7

ACRONYMS AND ABBREVIATIONS

ACI activated carbon injection
ACS American Cancer Society
ADHD attention deficit hyperactivity disorder
AEC Atomic Energy Commission
AEI adverse environmental impacts
APA Administrative Procedure Act
BPJ best professional judgment
BPT best practicable control technology currently available
BTA best technology available
CAA Clean Air Act
CAIR Clean Air Interstate Rule
CAMR Clean Air Mercury Rule
CB chronic bronchitis
CBA cost–benefit analysis
CEA cost-effectiveness analysis
CHPAC Children's Health Protection Advisory Committee
C-AVSARTE (National Research Council) Committee on Assessing and Valuing the Services of Aquatic and Related Terrestrial Ecosystems
C-VPRESS (EPA Science Advisory Board) Committee on Valuing the Protection of Ecological Systems and Services
CMAQ community multiscale air quality
CO_2 carbon dioxide
COPD chronic obstructive pulmonary disease
C-R concentration-response
CWA Clean Water Act
CWIS cooling water intake structures
EBA economic and benefits analysis
EDF Environmental Defense Fund
EGU electric generating unit
EIA Energy Information Administration
EIS environmental impact statement
EO executive order
EPA Environmental Protection Agency
EPRI Electric Power Research Institute
GAAP generally accepted accounting principles
GDP gross domestic product
HAP hazardous air pollutant
Hg mercury
HRC habitat replacement cost
I&E impingement and entrainment
ICD International Classification of Diseases
IGEM intertemporal general equilibrium model
IPM Integrated Planning Model
IQ intelligence quotient
kg kilogram(s)
L liter
m^3 cubic meter
MACT maximum achievable control technology
MAF mean annual flow
mg milligram(s)
MMAPS mercury maps
MR mental retardation
MRAD minor restricted activity day
μg microgram(s)
NAAQS National Ambient Air Quality Standards
NEPA National Environmental Policy Act
NHANES National Health and Nutrition Examination Survey
NODA notice of data availability
NO_x nitrogen oxides
NPDES National Pollution Discharge Elimination System
NRC National Research Council
NRDC Natural Resources Defense Council
O&M operations and maintenance
OIRA Office of Information and Regulatory Affairs
OMB Office of Management and Budget
PM particulate matter
$PM_{2.5}$ fine particulate matter
PRA Paperwork Reduction Act
RA regional analysis
RfD reference dose
RFF Resources for the Future
RIA regulatory impact analysis
SAB Science Advisory Board (of EPA)
SIP state implementation plan
SO_2 sulfur dioxide
SO_x sulfur oxides
TB technology-based
TDD Technical Development Document
TRUM transuranic mixed (waste)
TSD technical support document
USDA U.S. Department of Agriculture
UVb ultraviolet b
VSL value of a statistical life
WLD work loss day
WTA willingness to accept
WTP willingness to pay
YPR yield-per-recruit

Contents

PREFACE

The federal rulemaking process will never be the subject of a great novel. While the ponderous analyses and complicated calculations involved in federal rulemaking spark great passion and controversy among a hardy band of scholars and analysts, most citizens have little reason to think about the regulatory processes that affect much of modern commerce. Only when rules impacting their daily lives make news does the public become engaged, especially if the outcomes are perceived as particularly egregious. When the Office of Management and Budget (OMB) announced that it would take into account the fact that life expectancies of older Americans were less than those of younger ones, and would assign lower values to the lives of seniors than those of the young, a major brouhaha followed. The so-called "senior discount" was promptly denounced by the very agency that had developed it. But most controversies over the underpinnings of regulations do not make such news.

Beyond the headlines, important matters of principle and substance are at stake. Two quite divergent groups of scholars find these matters important enough to fight about. One group believes that economic analyses are critical to sound regulatory decisionmaking. The other group questions both the premise that economic considerations should play a prominent role and the particular methods used to develop quantitative estimates of benefits and, to a lesser degree, of costs. Although federal regulators have relied on cost–benefit analyses of regulatory impact analyses (RIAS) for close to four decades, the practice has remained controversial from its inception in the 1970s.

The differences between the proponents and opponents of economic analysis are many and profound, but perhaps the most important are the contrasting attitudes about the value of environmental improvement. In one camp are those, mostly economists, who believe that environmental outcomes can, in principle, be valued just as market goods and services can: by finding what households are willing to pay to improve the quantity, quality, or their own access to the good in question. In the other camp are those who believe that simply asking the willingness-to-pay question lessens the environmental values at stake, and that the answers tend to leave important considerations entirely out of the decisionmaking calculus.

Over the continued and often passionate objections of the cost–benefit opponents, the RIA requirement has become firmly embedded in rulemaking procedures. For their part, the cost-benefit advocates, within both government and academia, have been content to expand the methods and improve the technical content of the analyses, largely ignoring the opponents.

Not surprisingly, much of the debate has been expressed in largely philosophical and rhetorical terms—from both sides. Largely absent has been a practical nuts and bolts approach to the problem, asking quite basic questions: What are current practices, and how can they be improved?

Successful collaborations usually involve individuals with complementary rather than competing ideas. From the very beginning, however, it was clear from the writings of these three editors that significant differences of opinion existed: Harrington and Morgenstern, both economists, were clearly supportive of cost–benefit analysis as an aid to decisionmaking. Heinzerling, a lawyer, has authored a number of critiques of the approach. Yet this report grew out of a belief that, if one assumed that cost–benefit analysis is here to stay, a middle ground could be found—and that if it could be explored, federal rulemaking would be the better for it. For the opponents, this offered an opportunity to identify and suggest changes to some aspects of RIAS. For the advocates, it offered a chance to explore those most objectionable aspects and perhaps sacrifice a little of the things they really care about in order to make the process more acceptable to its critics.

At the outset, we decided to focus on actual RIAS as case studies, and to avoid the philosophical issues as much as possible. We were fortunate to recruit respected scholars from both camps, individuals who had puzzled over the same issues and had some prior familiarity with the particular rules and RIAS being studied. We were also fortunate that these individuals shared our enthusiasm for the overall endeavor and, especially, for the idea of trying to bridge the gap between camps.

We owe many debts in this project, most notably to the chapter authors. We also acknowledge the important contributions of the designated peer reviewers (Frank Ackerman, Tufts University; James Hammitt, Harvard University; and William Pedersen, attorney at law), and to additional experts from the U.S. Environmental Protection Agency's air and policy offices and from OMB (Alexander Cristofaro, Arthur Fraas, Bryan Hubbell, Albert McGartland, and Sam Napolitano)—all of whom attended an authors workshop held at Resources for the Future in June 2008.

As editors, we take sole responsibility for the recommendations for reform contained in the last chapter of the volume. While our intellectual debts to the chapter authors and reviewers are large, they are not responsible in any way for the recommendations. We alone should be held accountable for the results. In this effort, we developed a total of fourteen individual recommendations upon which the three of us agreed. They cover five areas: technical quality of the analyses; relevance to the agency decisionmaking process; transparency of the analyses; treatment of new scientific findings; and balance in both the analyses and the associated processes, including the treatment of distributional consequences. While we make no claim of completeness of these recommendations, we see them as a concrete starting point for discussions about reform of the RIA process.

We owe a special debt of gratitude to our editors, Elizabeth Stallman Brown, Felicia Day, and Adrienne Foerster. Finally, we thank the Smith Richardson Foundation for financial support for the project. We especially appreciate the assistance of Mark Steinmeyer, our project officer, who early on saw both the promise and the peril of this project but persisted in his support and encouragement.

■ ■ ■

CHAPTER 1

Controversies Surrounding Regulatory Impact Analysis

WINSTON HARRINGTON, LISA HEINZERLING, AND RICHARD D. MORGENSTERN

The use of economic methods to evaluate the benefits and costs of new regulations in the areas of health and the environment has expanded dramatically over the past several decades in the United States and is now quite entrenched in the federal regulatory process. In 1981, one of the first actions of the Reagan administration was the issuance of an executive order (EO 12291) that required that "major" regulations—including those with an effect on the economy of $100 million or more—undergo a "regulatory impact analysis" (RIA). Thus, in most federal agencies, a major rule could not be proposed in the *Federal Register* until a cost–benefit analysis (CBA) had been prepared and submitted to the Office of Management and Budget (OMB) for review and approval. Executive orders by subsequent administrations, most notably President Clinton's EO 12866, put greater attention on ensuring an efficient OMB vetting process and a heightened focus on the nonquantitative consequences of major rules. Nevertheless, Clinton's executive order left in place the key components of regulatory benefit and cost estimation and OMB review. This order continues to govern regulatory review today.[1]

Most observers of the regulatory process expect that the outcome of the 2008 presidential election will not affect this general structure. Indeed, some time ago, law professor Cass Sunstein heralded the arrival of the "cost–benefit state" (Sunstein 2002). (Sunstein's views have become all the more important with his nomination to be head of the Office of Regulatory Affairs in the Obama administration.) More recently, New York University School of Law Dean Richard Revesz and attorney Michael Livermore declared that "[c]ost–benefit analysis is here to stay."[2]

At the same time, a growing chorus of scholars and activists have decried what they consider to be an excessive focus on economic analysis and economic efficiency in federal rulemaking. They have argued, for example, that the cost–benefit approach inappropriately values impacts on "priceless" species, habitats, and other important, difficult-to-quantify resources; that the discounting of future regulatory consequences, including human mortality, treats lives unequally and trivializes the future; and that gains and losses to the rich should not be treated the same as those to the poor (Ackerman and Heinzerling 2004). In 2002, leading advocates of alternative approaches to regulatory assessment launched the Center for Progressive Reform, dedicated to the support of regulatory action to protect health, safety, and the environment while "rejecting the conservative view that government's only function is to increase the economic efficiency of private markets."[3]

This report proceeds on two premises. The first is that the opponents of more stringent environmental regulation are not going to get everything they want, nor should they. Clearly, environmental regulation is here to stay, and the American electorate will not tolerate a return to the bad old days when, for example, polluters were free to dump whatever noxious substances they wanted into the air or water.

The second premise is that few want to see a world where every potential environmental risk, no matter how small or fanciful, leads to a new and potentially onerous restriction on product use or manufacture or on corporate or individual behavior.

In sum, both advocates and skeptics of more stringent regulation have been guilty of overreaching in the past, and both sides have paid a price for it. Of course, it is too simplistic to suggest that one could find a happy medium of environmental protection. What is possible from a legislative and regulatory perspective shifts constantly depending on precedent, environmental incidents, the state of technology, the philosophy of the party or individuals in power, and other factors.

The principal focus of this report is not on the lively, and ongoing, philosophical debate between proponents and opponents of the approach as to whether the analytical technique of cost–benefit analysis is necessary, rational, and environmentally protective—or unhelpful, indeterminate, and immoral. Rather, for the purposes of this volume, we embrace the pragmatic view of Sunstein and Revesz that CBA is here to stay.

This volume brings together, for the very first time, distinguished scholars with diverse views in an effort to improve the workings of the basic structure of regulatory impact analysis that we now have. We have asked proponents of CBA to approach fundamental features of current economic analysis with a fresh and skeptical eye. We have asked opponents of CBA, for present purposes at least, to set aside their general objections and to offer constructive possibilities for adjustments to the method. Thus, although a central premise of the volume is that some type of formal economic analysis will be used to support major federal regulations, the design of that analysis and its proper role in the regulatory process are very much at issue.

Rather than considering these matters in the abstract, the report considers the appropriate use of CBA by examining actual RIAS. Case studies of the RIAS for three U.S. Environmental Protection Agency (EPA) rules provide the fodder for the reforms we ultimately propose. The rules are the Clean Air Interstate Rule (CAIR), the Clean Air Mercury Rule (CAMR), and the Cooling Water Intake Structure Rule (Phase II). The case studies help to clarify concrete differences between the sides in the cost–benefit debate and to suggest reforms to the current system for preparing and reviewing RIAS.

Overall, we seek to augment the often philosophical nature of the current debate with a quite pragmatic focus on actual regulatory analyses. We address a number of basic questions: Could particular changes to current practice improve the transparency of RIAS, enhance or modify their content, and increase the acceptability of the resulting studies? Even if the skeptics never fully embrace a cost–benefit framework, could it be made less objectionable?

This first chapter is designed to set the stage for the more detailed assessments that follow. We first present some essential background information on the role of CBA in the regulatory process and on the debates it has spawned. We briefly describe major contentions in this debate—not to answer them, but as part of our effort to improve the current process of regulatory analysis by understanding the perspectives of all sides. We then explain the process followed in engaging the multidisciplinary group of scholars involved in this study. Chapters 2–10 contain the detailed assessments of the three cases, including critiques of each RIA by proponents and opponents of CBA.

The final chapter is an attempt by the editors to seek some common ground on the preferred means of conducting regulatory analysis, including recommendations for improving both the content of RIAS and the process by which they are developed and reviewed.

This report comes at a particularly timely moment, as President Obama, on January 30, 2009, issued a memorandum directed to the heads of executive departments and agencies, asking for their views on how to improve the process of regulatory review. The memorandum directs the head of OMB, "in consultation with representatives of regulatory agencies, to produce within 100 days a set of recommendations for a new Executive Order on regulatory review." The memorandum invites particular attention to "the role of distributional considerations, fairness, and concern for the interests of future generations," and to "the role of the behavioral sciences in formulating regulatory policy."

Background

How CBA Fits into the Regulatory Process

CBA can play several different roles in the regulatory process. First, one environmental law—the Safe Drinking Water Act Amendments of 1996—explicitly calls for formal CBA in deciding on the scope of regulation under the statute. Clearly, under this law, CBA conducted pursuant to the process of regulatory review aligns with the kind of analysis called for by the statute.

Second, quite a large number of health and environmental laws, such as the Toxic Substances Control Act and the Federal Insecticide, Fungicide, and Rodenticide Act, require agencies to conduct a generalized balancing of costs and benefits in coming to their decisions. Here, too, CBA arguably fits comfortably within the statutory framework.

Third, some statutes either dictate a precise regulatory result or forbid altogether the application of CBA in choosing a regulatory approach. In these cases, CBA is not in part of the statutory framework. All versions of the executive orders on regulatory review have provided that, where the law and CBA conflict, the law prevails. Thus, for example, where Congress dictates that a particular performance level must be achieved by regulated sources or rules out the use of CBA in regulatory decisionmaking, congressional directives—rather than the dictates of CB —prevail.

Even in the latter case, proponents of CBA argue that this approach can play a useful role in informing political leaders and the public about the consequences—good and bad—of regulatory decisions. This discussion demonstrates, however, that from the very outset, there may be quite different expectations regarding the analysis conducted pursuant to statutory directives and that undertaken pursuant to the executive orders on CBA. That mismatch can sometimes create legal conflict.

CBA: The Pros

To its proponents, the paramount advantages of a well-done CBA are twofold. First, it forces regulatory designers to think about quantification (that is, the physical effects of regulations they propose) on public health, environmental quality, ecosystem health, and a host of other potentially relevant outcomes. Second, it forces serious consideration of whether and how much those changes matter, and it does so in a particular way. CBA attempts to express the value of those physical changes using, as a metric, a monetary measure of the aggregate change in individual well-being resulting from a policy decision. Individual welfare is assumed to depend on the satisfaction of individual preferences, and monetary measures of welfare change are derived by observing how much individuals are willing to pay or give up in terms of other consumption opportunities. This approach can be applied to nonmarket "public goods" such as environmental quality or environmental risk reduction as well as to market goods and services, although the measurement of nonmarket values is more challenging. When measurement of such nonmarket values is impossible or in some way unacceptable, analysts may resort to cost-effectiveness analysis (CEA), a less ambitious approach in which a policy outcome (for example, a specified reduction in ambient pollution concentration) is taken as a given, and the analysis seeks to identify the least-cost means for achieving the goal, taking into account any ancillary benefits of alternative actions. Every CBA has at least one CEA buried inside.

In addition, the advantages of CBA (and CEA) include the following:

- transparency and the resulting potential for engendering accountability;
- the provision of a framework for consistent data collection and the identification of gaps and uncertainty in knowledge;
- the development of metrics for both the beneficial and adverse consequences of alternative regulatory approaches, allowing those alternatives to be compared to one another; and
- with the use of a monetary metric, the ability to aggregate dissimilar effects (such as those on health, visibility, and crops) into one measure of net benefits.

Most economists would acknowledge that CBA does not incorporate all factors that can and should influence judgments on the social worth of a policy and that individual preference satisfaction is not the only criterion. Nevertheless, most would also argue that rigorous CBA can elucidate for a broader audience how various regulatory choices are supposed to work and who is likely to be affected. At a minimum, then, it can play a useful informational role in the decisionmaking process. CBA also makes a moral argument that private economic activity, as well as regulation, can generate value, and hence that good public policymaking is a balancing process.

From an economist's perspective, the usefulness of CBA is primarily limited by the ability to quantify the effects of regulations and to measure people's willingness to pay for those different health and environmental outcomes. Fortunately, the state of the science of measuring such economic values is quite active. Estimates of the willingness to pay for reductions in mortality and morbidity risks, for avoiding environmental damage to recreation opportunities, and for avoiding visibility degradation are the subjects of much ongoing research. Issues of a higher order stalk the estimation of nonuse values, and a variety of mostly empirical concerns have left material damage poorly understood. Often, estimation of the costs of reducing environmental effects, generally thought to be relatively straightforward, can be as challenging as estimation of the benefits.

The RIAS in which the CBAS are embedded have also undergone considerable changes because so many have been subjected to critical scrutiny, including internal agency and OMB reviews prior to publication of regulatory proposals, commentary from stakeholders, and, in many cases, review by the courts. Nearly 30 years of experience has led to an informal list of best practices—

characteristics and components that, according to various commentators, belong in most if not all RIAS. These include the following:

- the use of clear and consistent baseline assumptions;
- the evaluation of an appropriately broad range of policy options, including alternatives to new regulation;
- transparency in the use of assumptions, data and models, the comparison of alternatives, and the reporting of results;
- appropriate treatment of discounting future benefits and costs and accounting for the cost of risk-bearing;
- the use of probabilistic analyses and other methods to explore the robustness of conclusions;
- the identification of nonmonetizable or nonquantifiable aspects of a policy and the potential incidence of all effects; and
- the use of benefit and cost measures that are grounded in economic theory (measures of willingness to pay and opportunity cost).

Not surprisingly, in practice many RIAS do not contain all these elements. The reasons for the omissions vary: sometimes general resource limitations may be to blame, in other cases the omissions may be more strategic. Overall, the sophistication of RIAS appears to have increased over time. In part, this tracks with the growth in the field of environmental economics. Yet one unfortunate result of this growing sophistication is that RIAS have become much longer and more technically oriented documents, leading perhaps to a certain sacrifice in transparency.

CBA: The Cons

To its critics, CBA is a flawed technique that, among other things, excessively emphasizes the quantification and monetization of risks, trivializes the future through discounting, fails to meaningfully assess the value of avoiding nonmarginal consequences (including environmental catastrophe), and ignores distributional concerns.

Quantification

In many cases, it is currently impossible to quantify all of the important benefits of an environmental regulation. Indeed, it is often impossible to quantify even a substantial portion of them. When the quantified benefits of a rule include only cancer cases averted, yet the rule will also prevent many other illnesses as well as adverse effects on ecosystems, a CBA of that rule will be woefully incomplete (Heinzerling 1998).

Monetization

Even cost–benefit proponents concede that there are no good estimates of the monetary value of many of the benefits of environmental regulation, including the avoidance of many kinds of illnesses and other adverse health conditions and the prevention of harms to species and ecosystems. Where no good estimates of value exist, the benefits will not count for very much in CBA. Even where cost–benefit analysts believe that the estimates are pretty good, cost–benefit critics often disagree.

The value of preventing death is a prime example. It has become standard to measure the health risk reductions associated with proposed regulations as statistical lives. If the risk of dying from a particular cause is reduced by one in one million for one million people, it is said that the regulation would save one statistical life. To measure the value of this statistical life, cost–benefit analysts ask how much individuals are willing to pay to avoid—or how much they are willing to accept to take on—the extra risk of one in one million. If, say, everyone is willing to pay $1 to avoid this risk, then the analysts would say that the value of the statistical life in this case is $1,000,000.

This analysis has many problems, according to cost–benefit critics. Here we cite three. First, in inventing the statistical life, cost–benefit analysts have not escaped the fundamental moral conundrum of valuing life itself; they have merely glossed over it. When a person dies as a result of environmental contamination, she really dies: a real life, not a statistical one, is lost. CBA ignores this fundamental fact.

Second, in assuming that willingness to pay is the measure of value, CBA takes as a given that decisions made in private economic markets will be the same as decisions made by individuals acting as public citizens. But, for many reasons, these settings might produce different decisions, not least among them the fact that in environmental matters, the problem of public goods will press individuals acting alone to devote few or no resources to cleaning up a problem.

Third, valuing life—or health—according to how much people are willing to pay for it invites inequality. Some cost–benefit proponents have advocated, for example, that the rich should be valued more highly than the poor—a position at which many among us would flinch but which is perfectly consistent with the underlying theory of willingness to pay. And, indeed, we see glimmers of this approach in recent EPA analyses. In a preliminary assessment of the social costs of carbon, EPA relied on an estimate that embedded wildly differential values for the lives of people in rich and poor countries. Equally troubling are EPA's fitful efforts to reduce the value of the elderly compared to that of younger people.

Discounting

Much of environmental law aims to protect the future—to protect people living now from illnesses that might befall them in the future, to protect future people from such events, and to preserve ecosystems so that future generations might use and enjoy them much as we do now. Although discounting does account for the costs to present generations of providing these protections, opponents of CBA believe that discounting is not consistent with environmental law's forward-looking premise because the standard technique of constant exponential discounting can have a potentially large adverse effect on the perceived benefits of policies—such as policies to address climate change or policies to protect against long-latency diseases like cancer—that aim to prevent future harms.

Uncertainty

Although the studies increasingly emphasize the incorporation of uncertainty, CBA often assumes stable problems with stable solutions. It works at the margins, but not when the margin is a cliff's edge. Many environmental problems—perhaps the most important example is climate change—involve great uncertainties and potential irreversibility, features ill-suited to the cost–benefit framework.

Distribution

As implied in the above discussion of the value of life, opponents of CBA believe that the willingness to pay criterion has an inherent inequity. There is no corrective to this problem in CBA as it is currently conducted, and indeed CBA is largely blind to the distributional implications of environmental degradation. The kinds of concerns stated above have led numerous cost–benefit critics to call for abandonment of the method—ending it rather than mending it. However, this report assumes that CBA is here to stay. Thus, we focus on the reform rather than the rejection of the approach. Specifically, we aim to improve the cost–benefit method, as currently applied by EPA, without fully engaging the larger debate over whether it is fundamentally flawed. Yet, as we shall see, some of the more fundamental criticisms of the method turn out to be helpful in recommending ways in which current practices might be improved.

Analysis versus Decisionmaking

Throughout this volume, it is useful to distinguish between the analytical and decisionmaking components of rulemaking. Although the two are closely related, they are not one and the same. In fact, some of the differences between the two components were clarified in EO 12866. Specifically, EO 12866 replaced the stipulation contained in EO 12291 that benefits "outweigh" costs with a requirement for "a reasoned determination that the benefits...justify the costs." Further, agencies were mandated to "include both quantifiable measures...and qualitative measures of costs and benefits" and "to select those approaches that maximize net benefits (including potential economic, environmental, public health and safety, and other advantages; distributive impacts; and equity) unless a statute requires another regulatory approach."[4] In effect, EO 12866 embraces social welfare considerations that may not be easily quantified, such as public health and distributional impacts, and rejects the idea that quantified CBA provides a rigid rule for decisionmaking.

Thus, EO 12866 is consistent with the views of most economists, who see CBA as a tool rather than a strict rule for decisionmaking. As Nobel Laureate Kenneth Arrow and others have written:

> *...[In] many cases, benefit–cost analysis cannot be used to prove that the economic benefits of a decision will exceed or fall short of the costs.... [But it] can provide illuminating evidence for a decision, even if precision cannot be achieved because of limitations on time, resources, or the availability of information. (Arrow et al. 1996, 5)*

Arrow et al. (1996) also note that agencies may want to consider other factors in their decisions, such as equity within and across generations, or they may want to place greater weight on particular characteristics of a decision, such as irreversible consequences. They recommend that when the expected costs of regulations far exceed expected benefits, agency heads should be required to present a clear explanation justifying the reasons for their decisions.

Critics of CBA are concerned that even this attenuated process places too much emphasis on CBA and, more generally, on economic efficiency in the decisionmaking process. Despite the language of EO 12866, they also fear that, in its review of new rules, OMB continues to apply the more rigid criteria of EO 12291, namely that benefits outweigh costs.

The true influence of RIAS on regulatory outcomes is not well understood. Indeed, it may be that some of the regulatory "successes" of CBA would have reached a similar outcome even if no CBA had ever been prepared.

Interestingly, the direct effect of a CBA on the regulatory outcome is not the only, and may not be its most important, influence on the regulatory process. Twenty years ago, an EPA report (EPA 1987) listed four specific areas—besides supporting regulatory decisions—where the RIA influenced the development of regulations:

- guiding the development of the regulation;
- adding new alternatives;
- eliminating noncost-effective alternatives; and
- adjusting alternatives to account for differences among industries or industry segments.

The RIA requirement also has been credited with making upper management at EPA and other regulatory agencies more aware of the implications of their decisions.

Our Process

For each of the three RIAS examined in this volume, the first task was to produce a coherent and readable description of the analysis that was conducted by EPA. We assigned this task to two of the volume editors (Harrington and Morgenstern), and to a former Resources for the Future (RFF) colleague who has joined the staff of EPA's Policy Office (David Evans). The assignment for these three authors was to faithfully report the content of the RIA, including the stated justification for the rule; to digest and summarize its technical complexities; and, where appropriate, to highlight potentially controversial issues. However, the authors of these chapters were asked not to take a stand on any of the controversies. Rather, their role is largely reportorial in nature.

The second task, for three scholars on each side of the cost–benefit debate, was to review the RIAS, critique them, and explore what complementary or substitute analyses could have been included. The three authors skeptical of CBA (all law professors) are: Douglas Kysar, Yale Law School; Catherine O'Neill, Seattle University Law School; and Wendy Wagner, University of Texas Law School. The three authors favoring CBA (all economists) are: Nathaniel Keohane, originally Yale University, now Environmental Defense Fund; Alan Krupnick, RFF; and Scott Farrow, University of Maryland. Purposely, we did not specify a precise format for the critiques. The resulting papers cover a wide-ranging set of issues, with the economists generally focusing more on the techniques used in the individual RIAS, and the lawyers often including broader philosophical critiques of CBA.

A small authors workshop was convened in June 2008 at RFF in Washington, DC, to review the papers and consider possible reforms of both the analytical facets of CBA and the process for doing and using this type of analysis. Workshop participants included the paper authors, designated peer reviewers (Frank Ackerman, Tufts University; James Hammitt, Harvard University; and William Pedersen, attorney at law), and selected additional experts from EPA's air and policy offices and from OMB: Alexander Cristofaro, Arthur Fraas, Bryan Hubbell, Albert McGartland, and Sam Napolitano.

Following the workshop, the authors exchanged drafts and had the opportunity to revise their papers on the basis of workshop comments, with the overall goal of ensuring relatively parallel coverage of topics.

Case Study Selection

We make no claim that the RIAS studied in this volume are representative of the 100-plus such analyses prepared by EPA since 1981. However, the regulations selected for inclusion in this project involve issues that are clearly relevant to examining CBA in operation; for example, the valuation of statistical lives, the monetization of ecological and other types of difficult-to-measure benefits, the discounting of future consequences, and the distribution of costs and benefits. All three cases are based on relatively recent rules, thereby ensuring that they reflect current practices in RIA development at EPA. Two of the three rules involve air pollution and one addresses water bodies. The fact that all focus on the electric utility industry is somewhat accidental and certainly not part of any grand design. Arguably, however, an RIA focused on this industry, which has been subject to environmental controls since even before the days of national regulation, is probably among the most sophisticated, thus revealing some of the "best" practices for RIAS. The fact that all three rules have been invalidated by the courts was clearly not part of our design. A description of the three rules follows:

1. The CAIR, issued by EPA in March 2005, aimed at considerable reductions in emissions of sulfur dioxide (SO_2) and nitrogen oxides (NO_x) in the eastern United States. EPA expected that, by 2015, SO_2 emissions would decline by 70 percent from 2003 levels, whereas NO_x emissions would fall by 60 percent. Although the RIA demonstrated that the monetized benefits of the rule—largely in the form of reduced human mortality—exceeded the costs by a considerable margin, thus providing solid justification for the rule, critics contend that the agency's own analysis would have supported even larger emissions reductions with earlier compliance deadlines. Key issues involve both the interpretation and the use of the RIA. The U.S. Court of Appeals in Washington, DC, rejected the CAIR in its entirety, concluding, among other things, that the regional trading program created by the rule paid inadequate attention to the state-specific focus of the relevant statutory provisions.

2. The CAMR, also issued by EPA in March 2005, was designed to reduce mercury emissions from power plants. Although interim emissions reductions of 20 percent from 2003 levels were expected from the installation of controls required under the CAIR, the CAMR itself was expected to result in further reductions in mercury amounting to almost 70 percent reductions by the year 2018. Critics contend that the agency's own analysis supported earlier and deeper cuts of these toxic emissions. Further, they expressed concern that the provision for emissions trading could lead to unacceptably high exposure levels for individuals in certain areas. The U.S. Court of Appeals in Washington, DC, also invalidated this rule, reasoning that EPA had failed to follow proper procedures in taking mercury off the list of pollutants to be regulated under a provision of the Clean Air Act requiring strict technology-based controls for regulated sources.

3. The Cooling Water Intake Structures Rule (Phase II), issued by EPA in June 2006, was designed to minimize the harmful impacts on aquatic life caused by cooling water intake structures at existing power plants. The rule set performance standards (rather than technology requirements) for these plants and also allowed plants to avoid these requirements either through restoration measures or through a site-specific CBA indicating that the costs of meeting the standards were not worth the benefits at a specific plant. EPA's stated reason for the content of its rule was that the CBA did not support a more stringent approach. Critics contend that the agency's RIA failed to

adequately account for certain ecological damages and so understated the net benefits to be achieved from further controls. The U.S. Court of Appeals in New York invalidated the rule, finding that the relevant provision of the Clean Water Act did not permit CBA. The U.S. Supreme Court agreed to hear the case in its 2008–2009 term.

The case studies of these three rules form the backbone of our work, which is to recommend changes, both substantive and procedural, to improve the process of regulatory review.

■ ■ ■

Notes

1. In March 2007 President Bush issued EO 13422, which expanded OMB's jurisdiction to include the review of guidance documents issued by federal agencies. It also required that regulatory agencies provide a written rationale for new regulations and estimates of aggregate annual costs and benefits of all regulatory activities in the agencies' plans. President Obama rescinded EO 13422 on February 4, 2009.

2. Revesz and Livermore 2006, page 11.

3. Center for Progressive Reform (n.d.)

4. EO 12866 1(a), 3 CFR at 638–39 (1995).

References

Ackerman, F., and L. Heinzerling. 2004. *Priceless: On Knowing the Price of Everything and the Value of Nothing.* New York: The New Press.

Arrow, Kenneth, M. Cropper, G. Eads, R. Hahn, L. Lave, R. Noll, P. Portney, M. Russell, R. Schmalensee, K. Smith and R. Stavins. 1996. "Is There a Role for Benefit-Cost Analysis in Environmental, Health, and Safety Regulation?" *Science,* 272:5259 (April 12), 221–222.

Center for Progressive Reform (n.d.), www.progressiveregulation.org.

Heinzerling, L. 1998. Regulatory Costs of Mythic Proportions. *Yale Law Journal* 107: 1981–2070.

Revesz, R.L., and M.A. Livermore. 2008. *Retaking Rationality: How Cost–Benefit Analysis Can Better Protect the Environment and Our Health.* New York: Oxford University Press.

Sunstein, C.R. 2002. *The Cost-Benefit State: The Future of Regulatory Protection,* The American Bar Association (Section of Administrative Law and Regulatory Practice), Chicago.

U.S. Environmental Protection Agency (EPA). 1987. *EPA's Use of Benefit–Cost Analysis, 1981–1986.* USEPA-230-05-87-028. Washington, DC: U.S. EPA.

■ ■ ■

CHAPTER 2

The Clean Air Interstate Rule

RICHARD D. MORGENSTERN

The Clean Air Interstate Rule (CAIR), promulgated by the U.S. Environmental Protection Agency (EPA) in March 2005, mandates reductions in power plant emissions of sulfur dioxide (SO_2) and nitrogen oxides (NO_x) in the eastern United States by 70 percent and 60 percent, respectively, by 2015, with interim reduction targets in 2010. The accompanying regulatory impact analysis (RIA) estimates that the quantified benefits of the CAIR will exceed the quantified costs by $80–100 billion in 2015, with most of the benefits in the form of reduced mortality and improvements in respiratory and cardiovascular health for adults and children. A closely related but distinct action, the Clean Air Mercury Rule (CAMR), mandates that coal-fired electric utilities reduce their emissions of mercury (Hg). Together, the CAIR and the CAMR create a multipollutant strategy to reduce emissions that closely resembles an unsuccessful legislative initiative by the Bush administration known as Clear Skies.

Using an updated version of an analytical framework employed in prior rulemakings, the RIA contrasts a baseline scenario reflecting expected economic and environmental conditions without the CAIR to one anticipated with the rule in place. The SO_2 baseline, for example, reflects the emissions allowed under the current Title IV Acid Rain Trading Program in addition to existing state regulations. The differences between the two economic and environmental scenarios are measures of the costs and benefits, respectively. Of course, these scenarios are underpinned by a complex set of technical and policy judgments involving emissions inventories, air quality modeling, dose–response functions, monetization of benefits, cost estimation, and other elements of the analysis. One very basic judgment concerns the nature and stringency of the options considered in the RIA. Although standard practice calls for the consideration of multiple options, in this case only a single alternative was examined in addition to the status quo. Other judgments involve the extent and type of sensitivity and uncertainty analyses conducted to reasonably portray the uncertainties inherent in the analysis.

Whereas the two subsequent chapters in this volume critique the analytical approach followed by EPA, the aim of this chapter is to consider, in summary form, the RIA developed by the agency and to highlight some of the strategic choices made in the analysis. The organization of this chapter is similar to that of the 400-plus-page RIA itself. Following this brief introduction, the next section presents essential background information used to develop the RIA, including summary data on the emissions and air quality impacts. The following section focuses on the assessment of both physical and monetized benefits and summarizes the qualitative analysis of nonmonetized benefits conducted by the agency. Finally, the last section presents the cost analysis along with the overall assessment of the net benefits of the CAIR, including the uncertainties.

Background

History

Beginning in the mid-1990s, EPA staff began considering options to simplify and rationalize some of the increasingly complex requirements imposed on regulated entities under the Clean Air Act (CAA). Reflecting its size and overall contribution to emissions, the initial focus was on the electric utility industry. Although internal agency discussions were motivated in part by longstanding industry complaints about the burdens of new source performance standards and certain other requirements, these discussions were driven primarily by the recognition that a new round of emissions reductions was on the horizon.

During 1997 and 1998, EPA promulgated new, more stringent National Ambient Air Quality Standards (NAAQS) for particulate matter (PM) and ozone, reflecting the growing evidence regarding health damages associated with these pollutants. Around the same time, the Clinton administration embraced the Kyoto Protocol, which, if submitted to and ratified by the U.S. Senate, would have mandated regulation of carbon dioxide (CO_2) and other greenhouse gases. Overall, about one-third of U.S. CO_2 emissions come from the electric utility sector.

When the Bush administration took office in 2001, calls for reform of the system increased while the interest in imposing mandatory CO_2 controls diminished. Early on, the administration indicated its desire to pull back from the Kyoto Protocol and, for all practical purposes, any other policies involving mandatory CO_2 controls. At the same time, the push for more stringent regulations of SO_2 and NO_x was strengthened by a series of court decisions that supported the agency's new ambient standards promulgated a few years earlier. Concurrent with these developments, the scandals at Enron and other energy-related corporations created additional hurdles for financing new investments in the electric utility industry.

On September 11, 2001, the Bush administration unveiled a straw proposal for reform of environmental regulation of the industry while at the same time achieving significant emissions reductions of sulfur oxides (SO_x), NO_x, and Hg—the so-called 3P approach. Although not a formal legislative initiative, the straw proposal was accompanied by extensive economic and environmental analyses. In many respects, these analyses resembled the RIA that ultimately accompanied the proposed CAIR issued in 2003 when the administration abandoned its legislative strategy and, instead, embraced a regulatory approach.

From 2001 to 2003, extensive public discussion and analysis of the issues raised in the straw legislative proposal took place. Should CO_2 be included in the new program? Should Hg be included? Should the reform elements be explicitly linked to the new round of emissions reductions? Should the program be national or only regional in scope? How large should the reductions be for the different pollutants, and how quickly should they occur? How would different states and different industries be affected by the program?

Rather than addressing the half dozen or more legislative proposals introduced in the Senate during the 2001 to 2003 period—all of which form the backdrop to the CAIR and the CAMR—I focus instead on the regulations themselves. Consistent with the administration's legislative proposal, the regulations do not cover CO_2; rather, they focus exclusively on SO_x, NO_x, and Hg. In contrast to the legislative approach, however, the regulations decouple the reform and emissions reduction elements, and move away from a national program to a regional one. At the same time, the CAIR and the CAMR do represent a commitment to a major new round of emissions reductions that, in turn, involve decisions about the amount and timing of reductions to be achieved for the pollutants. These issues are considered in detail in the RIA and, ultimately, form the central elements of the regulatory decision.

The CAIR

Legal authority for the CAIR derives primarily from Section 110(a)(2)(D) of the CAA, which requires upwind states to control emissions that contribute significantly to downwind nonattainment of the NAAQS. States are required to submit plans to EPA within three years of issuance of revised NAAQS. Among other requirements, these plans must address emissions in the state that contribute significantly to nonattainment downwind.

The CAIR finds that 28 states and the District of Columbia contribute significantly to nonattainment or interfere with maintenance of the NAAQS for fine PM ($PM_{2.5}$) and/or eight-hour ozone in downwind states. Thus, EPA requires these upwind states to revise their state implementation plans (SIPS) to include control measures to reduce emissions of SO_2 and NO_x, which, in turn, will assist the downwind $PM_{2.5}$ and eight-hour ozone nonattainment areas in achieving the NAAQS. In the RIA, EPA argues that this will allow attainment to be achieved in a more equitable, cost-effective manner than if each nonattainment area attempted to achieve attainment by implementing local emissions reductions alone.

Technically, the CAIR does not directly regulate emissions sources. Instead, it requires states to revise their SIPS to include control measures to reduce emissions of NO_x and SO_2. The emissions reduction requirements that would be assigned to the states are based on controls that EPA has modeled as cost-effective for electric generating units, the largest source categories for both SO_2 and NO_x. The CAIR would affect roughly 3,000 fossil fuel–fired units with a nameplate capacity greater than 25 megawatts. Nationwide, these sources accounted for roughly two-thirds of the SO_2 emissions and more than 20 percent of the NO_x emissions in 2003.

A key decision involves the identification of specific policy options to consider in the analysis. In the RIA, EPA argues that the decision to model a specific emissions cap-and-trade program phased in over time—beginning with SO_2 and NO_x caps in 2010 and lowering these emissions caps in 2015—was made based on the time points at which control actions would be needed to help the states in terms of their NAAQS attainment efforts, the feasibility of installing emissions controls, and other factors. The RIA refers to studies conducted by the agency concerning the technical feasibility of producing and installing large amounts of pollution control equipment in a short timeframe. Although the RIA argues that specific caps chosen were derived by determining the amount of SO_2 and NO_x emissions that can be cost-effectively controlled from electric generating units, it is silent on whether the policy option selected actually maximizes net benefits.

The RIA consists of four distinct modeling efforts:

- calculation of the costs and related impacts of the CAIR on electricity generating units assuming a cap-and-trade program based on the national inventory of precursors to PM, specifically NO_x and SO_2;

- air quality modeling for 2010 and 2015 to determine changes in ambient concentrations of ozone and PM, reflecting baseline and postcontrol emissions inventories;
- a benefits analysis to determine the changes in human health and welfare, in terms of both physical effects and monetary value, that result from the projected changes in ambient concentrations of the key pollutants; and
- an uncertainty analysis reflecting data gaps, variability in estimated epidemiological and other relationships, projection errors for population and economic growth, and model misspecifications.

A wide range of human health and welfare effects are linked to the emissions of NO_x and SO_2 and the resulting impact on ambient concentrations of ozone and PM. Potential human health effects associated with $PM_{2.5}$ range from premature mortality to cardiovascular-related symptoms, asthma, and other morbidity effects associated with long-term (chronic) and shorter-term (acute) exposures. Exposure to ozone has also been linked to a variety of respiratory effects, including hospital admissions and illnesses resulting in work and school absences. Although not included in the primary calculation of monetized benefits, the RIA notes that recent research has linked short-term ambient ozone exposure to premature mortality. Welfare effects potentially tied to PM and ozone include materials damage, visibility impacts, and reduced yields of crops and forests. Some of these effects are quantified in the RIA.

The RIA lays out a number of important caveats applicable to the analysis:

- Reductions in ambient concentrations of air pollution generally lower the risk of future adverse health effects by a fairly small amount for a large population.
- The appropriate economic measure is willingness to pay (WTP) for changes in risk prior to the regulation.
- Adoption of WTP as the measure of value implies that the value of environmental quality improvements depends on the individual preferences of the affected population and assumes that the existing distribution of income is appropriate.
- For some health effects for which WTP measures are not available (e.g., hospital admissions), the cost of treating or mitigating the effect is used as the measure of benefits.
- In the absence of direct measurement, three nonmarket methods are used to value endpoints: stated preference or contingent valuation, indirect market (e.g., hedonic wage), and avoided cost methods.
- Benefits transfer, the adaptation of research from similar contexts to estimate benefits for the issues at hand, is widely used in the RIA.
- To account for the effect of future changes in real income on WTP, the estimated health benefits are adjusted upward using category-specific estimates of the income elasticity of demand for the different benefit categories. None of these elasticity estimates is above one.

The RIA recognizes various types of uncertainties in the benefits assessment for PM as well as other endpoints concerning impact functions, PM mortality risk, baseline incidence rates, economic valuation, and the aggregation of monetized benefits. These are displayed in Table 2.1. Some of these uncertainties are explored in the formal uncertainty analysis conducted by the

Table 2.1
Primary Sources of Uncertainty in the Benefits Analysis

Source: EPA Table 4-5, 4-21.

1. Uncertainties associated with impact functions

- The value of the ozone or PM effect estimate in each impact function.
- Application of a single impact function to pollutant changes and populations in all locations.
- Similarity of future-year impact functions to current impact functions.
- Correct functional form of each impact function.
- Extrapolation of effect estimates beyond the range of ozone or PM concentrations observed in the source epidemiological study.
- Application of impact functions only to those subpopulations matching the original study population.

2. Uncertainties associated with ozone and PM concentrations

- Responsiveness of the models to changes in precursor emissions resulting from the control policy.
- Projections of future levels of precursor emissions, especially ammonia and crustal materials.
- Model chemistry for the formation of ambient nitrate concentrations.
- Lack of ozone monitors in rural areas requires extrapolation of observed ozone data from urban to rural areas.
- Use of separate air quality models for ozone and PM does not allow for a fully integrated analysis of pollutants and their interactions.
- Full ozone season air quality distributions are extrapolated from a limited number of simulation days.
- Comparison of model predictions of particulate nitrate with observed rural monitored nitrate levels indicates that the Regional Modeling System for Aerosols and Deposition, or REMSAD, overpredicts nitrate in some parts of the eastern United States.

3. Uncertainties associated with PM mortality risk

- Limited scientific literature supporting a direct biological mechanism for observed epidemiological evidence.
- Direct causal agents within the complex mixture of PM have not been identified.
- The extent to which adverse health effects are associated with low-level exposures that occur many times during the year versus peak exposures.
- The extent to which effects reported in the long-term exposure studies are associated with historically higher levels of PM rather than the levels occurring during the period of study.
- Reliability of the limited ambient $PM_{2.5}$ monitoring data in reflecting actual $PM_{2.5}$ exposures.

4. Uncertainties associated with possible lagged effects

- The portion of the PM-related long-term exposure mortality effects associated with changes in annual PM levels that would occur in a single year is uncertain, as is the portion that might occur in subsequent years.

5. Uncertainties associated with baseline incidence rates

- Some baseline incidence rates (e.g., those taken from studies) are not location specific and therefore may not accurately represent the actual location-specific rates.
- Current baseline incidence rates may not approximate well the baseline incidence rates in 2015.
- Projected population and demographics may not represent well the future-year population and demographics.

6. Uncertainties associated with economic valuation

- Unit dollar values associated with health and welfare endpoints are only estimates of mean WTP and therefore have uncertainty surrounding them.
- Mean WTP (in constant dollars) for each type of risk reduction may differ from current estimates because of differences in income or other factors.

7. Uncertainties associated with aggregation of monetized benefits

- Health and welfare benefits estimates are limited to the available impact functions. Thus, unquantified or nonmonetized benefits are not included.

agency. Others are noted in more qualitative terms. For example, the RIA presents lists of unquantified and nonmonetized benefits and costs (see Table 2.2).[1]

EPA estimates the monetary benefit of reducing premature mortality risk using the value of a statistical life (VSL) approach, which is a summary measure for the value of small changes in mortality risk experienced by a large number of people. The mean value of avoiding one statistical death is assumed to be $5.5 million in 1999 dollars. This represents a central value consistent with the range of values suggested by recent meta-analyses of the wage–risk VSL literature.

The best available estimate of WTP to avoid a case of chronic bronchitis (CB) comes from Viscusi et al.[2] The Viscusi et al. study, however, describes a severe case of CB to the survey respondents; therefore, EPA adjusted the Viscusi et al. estimate of the WTP to avoid a severe case. This is done to account for the likelihood that an average case of pollution-related CB is not severe. The unit values used for economic valuation of premature mortality, CB, and all other endpoints monetized in the RIA are displayed in Table 2.3.

Benefits Analysis—Results

Applying the impact and valuation functions described in the previous section to the estimated changes in ozone and PM yields estimates of the changes in physical damages (e.g., premature mortalities, cases, admissions, and changes in light extinction) and the associated monetary values for those changes. Estimates of physical health impacts are presented in Table 2.4. Monetized values for both health and welfare endpoints are presented in Table 2.5, along with total aggregate monetized benefits. All of the monetary benefits are in constant-year 1999 dollars.

Not all known PM- and ozone-related health and welfare effects could be quantified or monetized. EPA represents the monetized value of these unquantified effects by adding an unknown "B"

Table 2.2 Unquantified and Nonmonetized Effects of the Clean Air Interstate Rule

Source: EPA Table 1-4, 1-10.

Pollutant—effect	Effects not included in primary estimates—changes in:
OZONE: HEALTH	■ Premature mortality
	■ Chronic respiratory damage
	■ Premature aging of the lungs
	■ Nonasthma respiratory emergency room visits
	■ Increased exposure to UVB
OZONE: WELFARE	■ Yields for:
	▪ commercial forests,
	▪ fruits and vegetables, and
	▪ commercial and noncommercial crops
	■ Damage to urban ornamental plants
	■ Recreational demand from damaged forest aesthetics
	■ Ecosystem functions
	■ Increased exposure to UVB
PM: HEALTH	■ Premature mortality: short-term exposures
	■ Low birth weight
	■ Pulmonary function
	■ Chronic respiratory diseases other than chronic bronchitis
	■ Nonasthma respiratory emergency room visits
	■ Exposure to UVB (+/–)
PM: WELFARE	■ Visibility in many Class I areas
	■ Residential and recreational visibility in non-Class I areas
	■ Soiling and materials damage
	■ Ecosystem functions
	■ Exposure to UVB (+/–)
NITROGEN AND SULFATE DEPOSITION: WELFARE	■ Commercial forests resulting from acidic sulfate and nitrate deposition
	■ Commercial freshwater fishing resulting from acidic deposition
	■ Recreation in terrestrial ecosystems resulting from acidic deposition
	■ Existence values for currently healthy ecosystems
	■ Commercial fishing, agriculture, and forests resulting from nitrogen deposition
	■ Recreation in estuarine ecosystems resulting from nitrogen deposition
	■ Ecosystem functions
	■ Passive fertilization resulting from nitrogen deposition
MERCURY: HEALTH	■ Incidence of neurological disorders
	■ Incidence of learning disabilities
	■ Incidence of developmental delays
	■ Potential reproductive effects
	■ Potential cardiovascular effects, including:
	▪ altered blood pressure regulation
	▪ increased heart rate variability
	▪ incidence of myocardial infarction
MERCURY DEPOSITION: WELFARE	■ Impacts on birds and mammals (e.g., reproductive effects)
	■ Impacts to commercial, subsistence, and recreational fishing

Table 2.3 Unit Values Used for Economic Valuation of Health Endpoints (1999$)

Health endpoint	CENTRAL ESTIMATE OF VALUE PER STATISTICAL INCIDENCE		
	1990 income level	**2010 income level**	**2015 income level**
PREMATURE MORTALITY (VALUE OF A STATISTICAL LIFE)	$ 5,500,000	$ 6,000,000	$ 6,400,000
CHRONIC BRONCHITIS	$ 340,000	$ 380,000	$ 400,000
NONFATAL MYOCARDIAL INFARCTION (HEART ATTACK)			
3% DISCOUNT RATE			
Age 0–24	$ 66,902	$ 66,902	$ 66,902
Age 25–44	74,676	74,676	74,676
Age 45–54	78,834	78,834	78,834
Age 55–65	140,649	140,649	140,649
Age 66 and over	66,902	66,902	66,902
7% DISCOUNT RATE			
Age 0–24	$ 65,293	$ 65,293	$ 65,293
Age 25–44	73,149	73,149	73,149
Age 45–54	76,871	76,871	76,871
Age 55–65	132,214	132,214	132,214
Age 66 and over	65,293	65,293	65,293
HOSPITAL ADMISSIONS			
CHRONIC OBSTRUCTIVE PULMONARY DISEASE (COPD) (ICD CODES 490-492, 494-496)	$ 12,378	$ 12,378	$ 12,378
PNEUMONIA (ICD CODES 480-487)	14,693	14,693	14,693
ASTHMA ADMISSIONS	6,634	6,634	6,634
ALL CARDIOVASCULAR (ICD CODES 390-429)	18,387	18,387	18,387
EMERGENCY ROOM VISITS FOR ASTHMA	286	286	286
RESTRICTED ACTIVITY AND WORK/SCHOOL LOSS DAYS			
WORK LOSS DAYS (WLDS)	VARIABLE (NATIONAL MEDIAN=)		
SCHOOL ABSENCE DAYS	75	75	75
WORKER PRODUCTIVITY	$0.95 per worker per 10% change in ozone per day		
MINOR RESTRICTED ACTIVITY DAYS (MRADS)	51	53	54

Note: ICD stands for International Classification of Diseases. *Source: EPA Table 4-11.*

Table 2.4
Clean Air Interstate Rule: Estimated Reduction in Incidence of Adverse Health Effects

Source: EPA Table 4-16, 4-74.

	2010	2015
Health effect	Incidence reduction	
PM-RELATED ENDPOINTS		
Premature mortality		
Adults, age 30 and older	13,000	17,000
Infants, age <1 year	29	36
Chronic bronchitis (adult, age 26 and older)	6,900	8,700
Nonfatal myocardial infarction (adults, age 18 and older)	17,000	22,000
Hospital admissions—respiratory (all ages)	4,300	5,500
Hospital admissions—cardiovascular (adults, older than age 18)	3,800	5,000
Emergency room visits for asthma (age 18 and younger)	10,000	13,000
Acute bronchitis (children, age 8–12)	16,000	19,000
Lower respiratory symptoms (children, age 7–14)	190,000	230,000
Upper respiratory symptoms (asthmatic children, age 9–18)	150,000	180,000
Asthma exacerbation (asthmatic children, age 6–18)	240,000	290,000
Work loss days (adults, age 18–65)	1,400,000	1,700,000
Minor restricted-activity days (adults, age 18–65)	8,100,000	9,900,000
OZONE-RELATED ENDPOINTS		
Hospital admissions—respiratory causes (adults, 65 and older)	610	1,700
Hospital admissions—respiratory causes (children, under 2)	380	1,100
Emergency room visits for asthma (all ages)	100	280
Minor restricted-activity days (adults, age 18–65)	280,000	690,000
School absence days	180,000	510,000

to the aggregate total. The estimate of total monetized health benefits is thus equal to the subset of monetized PM- and ozone-related health and welfare benefits plus B, the sum of the nonmonetized health and welfare benefits.

As noted, total monetized benefits are dominated by the benefits of mortality risk reductions. The primary analysis estimate projects that the final rule will result in 13,000 avoided premature deaths annually in 2010 and 17,000 avoided premature deaths annually in 2015. The increase in annual benefits from 2010 to 2015 reflects additional emissions reductions from the standards, as well as increases in total population and the average age (and thus baseline mortality risk) of the population. Note that unaccounted-for changes in baseline mortality rates over time may lead to reductions in the estimated number of avoided premature mortalities.

Table 2.5
Estimated Monetary Value in Reductions in Incidence of Health and Welfare Effects
(in Millions of 1999$)

Health effect	Pollutant	2010	2015
		Incidence reduction	
Premature mortality			
Adults, age 30 and older			
3% discount rate	$PM_{2.5}$	$ 67,300.00	$ 92,800.00
7% discount rate		56,600.00	78,100.00
Infants, < 1 year		168.00	222.00
Chronic bronchitis (adults, 26 and older)	$PM_{2.5}$	2,520.00	3,340.00
Nonfatal acute myocardial infarctions			
3% discount rate	$PM_{2.5}$	1,420.00	1,850.00
7% discount rate		1,370.00	1,790.00
Hospital admissions for respiratory causes	$PM_{2.5}$, O^3	45.20	78.90
Hospital admissions for cardiovascular causes	$PM_{2.5}$	80.70	105.00
Emergency room visits for asthma	$PM_{2.5}$, O^3	2.84	3.56
Acute bronchitis (children, age 8–12)	$PM_{2.5}$	5.63	7.06
Lower respiratory symptoms (children, age 7–14)	$PM_{2.5}$	2.98	3.74
Upper respiratory symptoms (asthma, age 9–11)	$PM_{2.5}$	3.80	4.77
Asthma exacerbations	$PM_{2.5}$	10.30	12.70
Work loss days	$PM_{2.5}$	180.00	219.00
Minor restricted-activity days	$PM_{2.5}$, O^3	422.00	543.00
School absence days	O^3	12.90	36.40
Worker productivity (outdoor workers, age 18–65)	O^3	7.66	19.90
Recreational visibility, 81 Class I areas	$PM_{2.5}$	1,140.00	1,780.00
MONETIZED TOTAL			
Base estimate:	$PM_{2.5}$, O^3		
3% discount rate		$ 73,300.00 + B	$101,000.00 + B
7% discount rate		$ 62,600.00 + B	$ 86,300.00 + B

Source: EPA Table 4-17, 4-74.

EPA's estimate of total monetized benefits in 2010 for the final rule is $73.3 billion using a 3 percent discount rate and $62.6 billion using a 7 percent discount rate.[3] In 2015, the monetized benefits are about a third higher. Although the magnitude of the unquantified benefits is uncertain, it may be substantial. The monetized benefit associated with reductions in the risk of premature mortality is over 90 percent of total monetized health benefits.

In the aggregate, the ozone benefits associated with the CAIR are positive for the nation as a whole. However, because ozone increases occur during certain hours of the day in some urban areas, a dampening of ozone benefits occurs in both 2010 and 2015. Overall, ozone benefits are low relative to PM benefits for similar endpoint categories because of the increases in ozone concentrations during some hours of some days in certain urban areas.

Probabilistic Analysis of Uncertainty in the Benefits Estimates

A recent National Research Council (NRC) report on estimating the public health benefits of air pollution regulations recommended that EPA begin to move the assessment of uncertainties from its ancillary analyses into its primary analyses by conducting probabilistic, multiple-source uncertainty analyses.[4]

EPA presents two types of probabilistic approaches for the CAIR. The first approach generates a distribution of benefits based on the sampling error and variability in the underlying health and economic valuation studies used in the benefits modeling framework. The second uses the results from a pilot expert-elicitation project designed to characterize key aspects of uncertainty in the ambient $PM_{2.5}$/mortality relationship. Both approaches provide insights into the likelihood of different outcomes and about the state of knowledge regarding the benefits estimates.

The RIA provides likelihood distributions combining the uncertainties from two sources—the concentration-response (C-R) relationship and the valuation—and is estimated with a Monte Carlo method. These estimates should be viewed within the context of the wide range of sources of uncertainty that are not incorporated, including uncertainty in emissions, air quality, and baseline health effect incidence rates.

Because the estimated impact of reductions in premature mortality accounts for such a high proportion of total benefits, it is particularly important to characterize the uncertainties associated with this endpoint. In collaboration with the Office of Management and Budget, EPA conducted a pilot expert elicitation to characterize uncertainties in the relationship between ambient $PM_{2.5}$ and mortality. The pilot project elicited the judgments of five experts in the PM health sciences, all members of at least one of two recent National Academy of Sciences scientific committees focused on PM.

These supplemental analyses yield the following insights:

- Use of statistical error associated with the American Cancer Society (ACS) estimate for the concentration response function for $PM_{2.5}$—premature mortality as well as the statistical error associated with the concentration response functions for each of the other health endpoints to describe the probability distribution of total benefits—yields a distribution in which the 95th percentile is nearly twice the mean ($100 billion in 2015) and the 5th percentile is one-fourth the mean.[5] The overall range from the 5th to the 95th percentile on the total benefits estimate represents one order of magnitude ($26 billion to $210 billion).

- Description of the probability distribution of the concentration response function for $PM_{2.5}$—premature mortality using the results from the pilot expert elicitation (rather than the estimate based on the statistical error associated with the ACS cohort)—yields a larger degree of uncertainty because the elicitation exercise was designed to encompass a broader set of model uncertainties. The mean annual benefits for each expert elicited during the pilot project range from approximately $16 billion to $130 billion in 2015.
- Substitution of the steeper concentration response function for $PM_{2.5}$—premature mortality from the Six Cities study—increases the value of the total benefits from $101 billion to $208 billion in 2015.[6]
- Substitution of the most plausible alternative lag structures has little overall impact on the estimate of total benefits (reductions are on the order of 5 to 15 percent).
- The assessment of alternative assumptions regarding the existence (and level) of a threshold in the $PM_{2.5}$ premature mortality concentration response function highlights the sensitivity of the analysis to this assumption. Only 5 percent of the estimated premature mortality is attributable to changes in exposure above 15 mg/m^3, whereas more than 84 percent of the premature mortality-related benefits are attributable to changes in $PM_{2.5}$ concentrations occurring above 10 $\mu g/m^3$.
- Estimates of premature mortality from ozone exposure may result in an additional 500 premature deaths avoided and an increase in the estimated health benefits of the CAIR by approximately $3 billion annually.

Costs, Net Benefits, and Uncertainties

To evaluate the costs of the CAIR, the RIA uses the Integrated Planning Model, a reasonably transparent modeling tool available to the public. For the affected region, the projected annual incremental private costs of the CAIR to the power industry are $2.36 billion in 2010 and $3.57 billion in 2015. These costs represent the total cost to the electricity generating industry of reducing NO_x and SO_2 emissions to meet the caps set by the rule. Estimates are in 1999 dollars. Costs of the rule are estimated assuming firms make decisions using costs of capital ranging from 5.34 percent to 6.74 percent.

In estimating the net benefits of regulation, the appropriate cost measure is social costs, which represent the welfare costs of the rule to society. These costs do not consider transfer payments (such as taxes) that are simply redistributions of wealth. The social costs of this rule are estimated to be $1.91 billion in 2010 and $2.56 billion in 2015, assuming a 3 percent discount rate.

Retail electricity prices are projected to increase roughly 2.0 to 2.7 percent with the CAIR in the 2010 and 2015 time frame and then drop below 2.0 percent thereafter. The effect of the CAIR on natural gas prices and the power-sector generation mix is also small, with a 1.6 percent or smaller increase in gas prices projected from 2010 to 2020.

EPA estimates continued reliance on coal-fired generation, which is projected to remain at roughly 50 percent of total electricity generated. As demand grows in the future, additional coal-fired generation is projected to be built under the CAIR and the use of coal-fired units will increase. Because of this, coal production is projected to increase from 2003 levels by about 15 percent in 2010 and by 25 percent by 2020.

EPA's cost estimates assume that all states in the CAIR region fully participate in the cap-and-trade programs that reduce SO_2 and NO_x emissions from electricity generating units. The cost projections do not take into account the potential for advancements in the capabilities of pollution control technologies for SO_2 and NO_x removal and other compliance strategies, such as fuel switching or the reductions in their costs over time. EPA projections also do not take into account demand response (i.e., consumer reaction to electricity prices) because the consumer response will probably be relatively small. The RIA notes that costs may be understated because an optimization model was employed and the regulated community may not react in the same manner to comply with the rules. Further, the RIA did not factor in the costs or savings for the government to operate the CAIR program as opposed to other air pollution compliance programs or transactional costs and savings from the CAIR's effects on the labor supply.[7]

Conclusion

The CAIR RIA clearly demonstrates that the benefits of the rule exceed costs by a wide margin. The bulk of the document focuses on the details of the data and methods used to estimate the benefits and, to a lesser extent, the costs, drawing on the well-established approaches used by the agency in prior RIAS. While this particular RIA breaks some new ground, for example, in using formal uncertainty analysis, only quite limited efforts are made to quantify or monetize categories of benefits or costs not previously estimated in prior RIAS. The fact that only a single policy option is examined is a major flaw, as it prevents consideration of alternative emissions reduction goals or timetables, or other approaches to regulating power plant emissions of SO_2 or NO_x.

■ ■ ■

Notes

1. Unquantified costs also include employment shifts as workers are retrained at the same company or re-employed elsewhere in the economy; costs to state and federal governments of running and administering the program; and certain relatively small permitting costs associated with Title IV that new program entrants face.

2. Viscusi, W.K., W.A. Magat, and J. Huber. 1991. "Pricing Environmental Health Risks: Survey Assessments of Risk-Risk and Risk-Dollar Trade-Offs for Chronic Bronchitis." *Journal of Environmental Economics and Management* 21:32–51.

3. The air quality modeling does not include New Jersey and Delaware in the CAIR program. However, EPA's rough estimates suggest that including those states would result in additional reductions of SO_2 and NO_x emissions valued at approximately $1.1 billion in 2010 and $1.5 million in 2015.

4. National Research Council. 2002. *Estimating the Public Health Benefits of Proposed Air Pollution Regulations.* Washington, DC: The National Academies Press.

5. Pope, C.A., III, R.T. Burnett, M.J.Thun, E.E. Calle, D. Krewski, K. Ito, and G.D. Thurston. 2002. "Lung Cancer, Cardiopulmonary Mortality, and Long-term Exposure to Fine Particulate Air Pollution," *Journal of the American Medical Association* 287: 1132–1141.

6. The Six-Cities Study is a longitudinal study of the respiratory effects of air pollution in six northeastern and midwestern U.S. cities. Initial results are reported in Dockery, D.W., Pope C.A. III, X. Xu. 1993. "An Association Between Air Pollution and Mortality in Six U.S. Cities," *New England Journal of Medicine:* 329: 1753–9.

7. *See* note 1 for a list of possible unquantified costs associated with the CAIR program.

CHAPTER 3

The Technocratic and Democratic Functions of the CAIR Regulatory Analysis

NATHANIEL O. KEOHANE

Of the case studies considered in this report, the regulatory impact analysis for the Clean Air Interstate Rule (CAIR) provides particularly fertile ground for analysis and critique (EPA 2005a, henceforth "RIA"). The rule itself was far-reaching, mandating reductions of 60 to 70 percent in the emissions of two major criteria air pollutants—sulfur dioxide (SO_2) and nitrogen oxides (NO_x)—from power plants in the eastern United States. As a result, the stakes were very high. The annual costs of the controls proposed by the CAIR were projected to run into the billions of dollars—and yet they were dwarfed by estimated benefits of nearly $100 billion a year.

In the scale of its economic impact, as well as its proposed reliance on market-based policies, the CAIR was a coda to the immensely successful emissions trading programs for SO_2 and NO_x established by the 1990 Clean Air Act Amendments and the subsequent NO_x Budget Program—programs that produced estimated benefits of $122 billion against costs of just $3 billion (Chestnut and Mills 2005). Since it was announced, the CAIR has been dogged by legal challenges, and its future remains uncertain. But the ambition and potential impact of the program make it a natural case study for the role of regulatory impact analysis in general and cost–benefit studies in particular.

This chapter is divided into two parts, following this introduction. The first part (the bulk of the chapter) offers a technical critique of the CAIR RIA as an exercise in applied cost–benefit analysis. In keeping with the themes identified in the Introduction to this report, I assess the performance of the RIA on a range of dimensions: the scope of alternatives considered; the estimation of costs and benefits, including the expression of benefits in monetary terms using willingness to pay; the consideration of equity and differential impacts among subpopulations; the discounting of delayed effects; and the treatment of uncertainty. The RIA is a more than competent example of cost–benefit analysis on a number of dimensions and should be praised for its innovative approach to considering uncertainty. However, I identify a number of areas where the analysis could have been substantially improved. I close the section with a set of recommendations for improving regulatory cost–benefit analysis.

The second part takes a broader view of the RIA as a public document. The starting point for the discussion is that a purely technical appraisal of the RIA is necessarily incomplete, because such a discussion presumes a formal role for analysis in guiding policy that the CAIR RIA lacked. As several commentators have pointed out, including Wendy Wagner in Chapter 4 of this report, the U.S. Environmental Protection Agency (EPA) did not have statutory authority to base its rulemaking on cost–benefit grounds; indeed, the agency is expressly forbidden to use cost as a criterion in setting or implementing National Ambient Air Quality Standards under the Clean Air Act. Hence the CAIR RIA did not—*could not*—inform policymaking. As a formal matter, the RIA was carried out under Executive Order 12866, which requires agencies to submit an assessment of the costs and benefits of significant regulatory actions to the Office of Information and Regulatory Affairs (OIRA) in the Office of Management and Budget.

If the RIA could not offer guidance to policymakers, nonetheless it represented a potentially powerful means of informing citizens interested in evaluating those decisions after the fact. In contrast to the essentially *technocratic* function usually identified with regulatory analysis, this role suggests a *democratic* function that is often overlooked. The last section of this chapter explores the implications of this function. I argue that an emphasis on informing the public directs attention to order-of-magnitude judgments, estimation of marginal net benefits of stringency at a proposed policy, and above all transparency in presenting information, particularly involving benefits. Nonetheless, the technocratic and democratic roles complement one another, rather than competing with each other. An ideal regulatory analysis should fulfill both functions.

The RIA on Technical Grounds

Here I critique the RIA along several dimensions: scope, estimation of costs and benefits, equity, time discounting, and treatment of uncertainty. I close with an overall assessment, along with concrete recommendations for how this analysis, or any similar applied cost–benefit analysis, could be improved.

Scope

One of the most striking characteristics of the CAIR RIA is how narrow it is. The analysis evaluates the costs and benefits associated with only one emissions target, considers only one policy instrument, and compares the policy to a single baseline case representing one possible business-as-usual outcome.

First, the CAIR RIA is essentially an up-or-down assessment of the final rule versus the status quo. No other policy alternatives were considered in detail. Although the RIA conclusively demonstrates that the CAIR was preferable to the status quo on cost–benefit terms, it fails to show that the CAIR was better than an alternative policy. And there certainly were alternatives; what is more, EPA already had much of the information necessary to evaluate them. As Richard Morgenstern notes in Chpater 2, the rule itself was preceded by vigorous debate in Congress over several legislative proposals to reduce air pollution from power plants. These proposals ranged from the Bush administration's own Clear Skies legislation, introduced by Senators Inhofe (R-OK) and Voinovich (R-OH), to the more ambitious proposal by Senator Jeffords (I-VT). The cost to EPA of expanding the scope of the analysis, in terms of time and resources, would have been modest; as it was, the agency performed several runs of its Integrated Planning Model of the electric power sector and could have added more. Indeed, much of the necessary work was already done because EPA staff themselves had considered a range of policy scenarios in their assessment of the administration's

Clear Skies legislation.[1] The CAIR RIA, therefore, should have considered a range of alternatives to the policy that was eventually chosen.[2]

Second, the agency assumed that the states covered by the rule would implement a cap-and-trade program for SO_2 and NO_x emissions from electric generating units (EGUS), rather than using more prescriptive (and costly) approaches such as performance standards. But the states were free to regulate emissions in whatever manner and from whatever sources they chose. In defense of its approach, EPA argued that states could be expected to adopt a trading program for EGUS because it would be the least-cost approach. But the decision to focus solely on an emissions trading system effectively limited EPA's analysis to a *best-case scenario*—and one it lacked authority to impose.

At the same time, EPA failed to demonstrate the cost savings that could be expected from using a market-based approach. By focusing on a single policy instrument, EPA ruled out a cost-effectiveness analysis—that is, a cost-based comparison among several policies that would achieve the same reduction in emissions. Such an analysis, however, would have been directly germane to EPA's task under the Clean Air Act. As it turned out, costs were more relevant than benefits in issuing guidelines for state implementation plans (SIPS)—an ironic outcome, given the Clean Air Act's prohibition on using costs to set ambient air quality standards. In particular, Section 110(a) of the Clean Air Act—more precisely, Section 110(a)(2)(D)(i)(I)—requires SIPS to prohibit emissions in upwind states that "contribute significantly to nonattainment" of ambient air quality standards in downwind states. By anchoring EPA's authority in the binary distinction between attainment and nonattainment in downwind states, the statute made the quantification of benefits largely irrelevant.[3] On the other hand, in creating an emissions trading program for NO_x in 1998, EPA had used cost-effectiveness as a criterion for defining the "significant contribution" of upwind states—a position that was upheld by the U.S. Court of Appeals in *Michigan v. EPA*.[4]

EPA was well aware of the importance of the cost-effectiveness criterion: the preamble to the official announcement of the CAIR in the *Federal Register* repeats the phrase "highly cost-effective" like a mantra (EPA 2005b). Yet the preamble offers a somewhat languid defense of the CAIR's emissions targets as "highly cost-effective," based on a comparison of the expected per-ton costs under the CAIR with cost estimates from a smattering of other air pollution controls, along with an ad hoc "knee of the curve" analysis that attempts to discern the level of abatement at which diminishing returns set in. Notably absent is any formal demonstration or evidence that the costs of an emissions trading program for EGUS would be substantially less than the costs of implementing more direct controls (e.g., technology standards requiring scrubbers or low-NO_x combustion technologies).

The third and final dimension of scope concerns the choice of baseline scenario. Baseline scenarios embed a host of assumptions just as surely as policy scenarios do, even if they tend to be less obvious: outcomes under business as usual, and therefore the incremental costs and benefits of a particular policy, can be highly sensitive to assumptions about economic growth, electricity demand, technology, energy prices, and so on. Throughout almost all of its analysis of the CAIR, however, EPA compares the costs and benefits of its rule against a single business-as-usual scenario. The agency did run a few simulations assuming higher energy prices, but these are mentioned only occasionally and appear almost as an afterthought.

Estimation of Costs

The CAIR RIA makes two basic distinctions among types of costs: direct versus indirect costs, and private versus social costs. *Direct costs* refer to the actual costs of complying with the rule, such as installing the emissions control equipment (flue-gas desulfurization to reduce SO_2 and selective catalytic reduction to abate NO_x). These were estimated using EPA's model of the electric power sector. *Indirect costs,* on the other hand, are the "ripple effects" in the economy as a whole that ultimately result from the regulation. For example, an increase in electricity rates—as the compliance costs are passed on to utility customers—will translate into higher production costs for businesses and lower real wages for workers, which can in turn lead to adjustments and changes in manufacturing output, investment, and labor supply. Gauging these indirect costs requires a general equilibrium model of the economy.

For the CAIR RIA, EPA estimated the general equilibrium (indirect) costs using the well-regarded Intertemporal General Equilibrium Model (IGEM), but these turned out to be negligible. For example, the projected decrease in gross domestic product (GDP) was only 0.03 of a percent relative to the business-as-usual case; even in energy-intensive industries, such as chemical manufacturing, impacts on output were well under 0.1 percent. As a result, EPA essentially, and quite reasonably, ignored the indirect costs and focused entirely on the direct costs in the bulk of its analysis.

Those direct costs, in turn, can be expressed either as *private* or *social* costs. In this case, the term *social costs* does not refer to the full social costs, taking into account the externality from pollution, which in this case would be a negative cost (i.e., a benefit). Rather, the wedge between private and social costs in this analysis is a much narrower one, including only taxes, which appear as costs to the firm but are transfers from the perspective of society, and the difference between private and social discount rates (with the private rate, representing the cost of capital, higher than the social rate). Appropriately, the agency used social costs as the basic measure of costs, although the presentation was somewhat confusing: only private costs appear in the summary table in the chapter of the RIA that discusses costs, whereas social costs appear in the tables in the executive summary.

A number of concerns arise regarding how EPA estimated the direct costs. Some aspects of EPA's methodology understate the costs of the regulation. Not only did EPA assume that states would employ the recommended cap-and-trade system (which would achieve the emissions targets at lower cost than traditional command-and-control regulation), but in modeling the cost of that approach, EPA also essentially assumed a frictionless market with perfectly cost-minimizing utilities. This is far from an accurate description of reality. For example, ex post analyses of the SO_2 trading program under the 1990 Clean Air Act Amendments have demonstrated that the costs of that program, although lower than they would have been under more prescriptive regulations, are substantially higher than what would have prevailed in a perfectly efficient market (Carlson et al. 2000; Keohane 2006).

On the other hand, several factors tend to overstate the costs of the regulation. For example, EPA's model of the electric power sector uses a static representation of abatement technologies. Because technology would be expected to improve between 2005 (the date of the analysis) and 2015 (when the final emissions targets were in place)—especially given the additional spur provided by the regulation in question—ignoring technological change amounts to overstating the costs of compliance. This effect is reinforced by the use of conservative capital cost figures that, by EPA's own admission, did not even reflect the most recent data available at the time of the analysis (EPA 2005a, 7-19).

The cost analysis also fails to consider adjustments on the demand side. Higher electricity prices would dampen electricity demand, which in turn would bring compliance costs down because the lower electricity generation would translate into decreased emissions and less need for abatement. Again, EPA acknowledges this flaw in the RIA and goes out of its way to say that this might be significant (EPA 2005a, 7-20). On balance, the cost estimates in the RIA are probably overstated, as EPA itself acknowledges. Indeed, this is a common theme in discussions of regulatory cost estimates: Harrington, Morgenstern, and Nelson (2000) found that ex ante estimates of the cost of regulation tend to overstate the actual realized costs.

Nonetheless, it is hard to find too much fault with the agency. An ideal cost–benefit analysis would use up-to-the-minute cost data, incorporate technological change, take demand response into account, and model the frictions and transactions costs of real-world markets. But those things—especially the second and fourth items—are all at the frontier of economic research. Moreover, EPA is admirably up-front about the caveats and limitations of its analysis; indeed, all of the concerns listed above are raised by EPA itself. As I argue below, transparency is next to accuracy in cost–benefit analysis; indeed, it may be even more essential.

Estimation of Benefits

Despite the pitfalls associated with estimating the costs of environmental regulation, that exercise is widely viewed as straightforward compared with evaluating the benefits. The CAIR RIA is no exception to this rule.

In line with standard economic practice, EPA strives not only to quantify but also to monetize the benefits from reduced air pollution. It uses willingness to pay (WTP) as the basis of this valuation as far as possible (although the analysis resorts to other techniques in a few cases, such as cost-of-illness measures reflecting direct hospital costs and lost wages), and it accords more consideration to revealed preference methods of estimating WTP (such as using the variation in wages to estimate the value of a statistical life [VSL]) than to stated preference methods based on surveys. The fundamental justifications for these approaches are well known, even if they are hotly contested by critics: expressing benefits in dollar terms provides a convenient and consistent yardstick with which they can be compared to costs; WTP is an appropriate measure of economic value (and it is economics, after all, that provides the framework for analyzing and comparing costs and benefits); and revealed preference methods are generally considered more reliable measures of benefits than survey-based methods, which are prone to a number of biases.

A discussion of the merits or flaws of this basic approach is well beyond the scope of this chapter.[5] For present purposes, I posit the appropriateness of the basic economic valuation paradigm and focus on three particular issues: the use of WTP in evaluating public decisions, the scope of the benefits estimation, and the transparency of the results. The first concerns the applicability to the public sphere of willingness-to-pay measures based on private decisions. It is not at all evident that people's willingness to trade off higher wages for greater risk in the workplace should be used to infer their valuation of reduced risk from premature mortality resulting from air pollution, but this is precisely the logic behind wage-based VSL estimates. Individuals may regard risks that they willingly accept in the workplace as fundamentally different from risks over which they have no control—such as the consequences of breathing air pollution caused by power plants hundreds of miles away. Moreover, even if those risks are considered to be commensurate, observed wage premiums offer an imperfect basis for estimating willingness to pay in general. The people who accept risky jobs have (almost by definition) preferences that are systematically different from those of the general population: in particular, they are likely to have a higher tolerance for risk.

Relying on stated preference measures cannot address these basic concerns. As several authors have argued, there is an important difference between people's valuation for amenities as *consumers* and their valuation as *citizens*.[6] Indeed, experimental evidence suggests that social context matters: people value public goods differently when asked to state their valuation publicly, versus recording it privately (List et al. 2004). For all these reasons, determining a socially based *citizen valuation* for reduced risk would be a valuable topic for further research.[7]

A second and fundamental issue to consider in the benefits estimation is its scope: what gets measured and monetized, and what does not. The most striking aspect of the CAIR benefit analysis is how much it leaves out. As Richard Morgenstern discusses in Chapter 2 of this report, the list of unquantified and nonmonetized benefits is long and includes a variety of health benefits from lower air pollution (e.g., a reduction in premature mortality from ozone pollution or short-term exposure to particulates and a reduction in respiratory problems other than asthma) as well as virtually all nonhealth benefits, from impacts on commercial agriculture and fisheries to ecosystem functions.

Given that some benefits are expressed in dollar terms, the demands of completeness and consistency suggest that the largest possible fraction of benefits should be monetized. However, the difficulties inherent in monetizing benefits mean that some omissions are inevitable. Starting from that premise, the relevant questions are as follows: How should analysts determine which benefits to monetize? What should they do with impacts for which they lack willingness-to-pay estimates? How should they present their results?

To EPA's credit, it captures what appears to be the most important benefit category, by several orders of magnitude: namely, the reduced premature mortality risk from ambient concentrations of particulate matter (PM). EPA estimates that the CAIR would prevent 17,000 premature mortalities each year when fully implemented, corresponding to annual benefits of $80 billion to $90 billion. By comparison, the next highest category of monetized benefits (reduced chronic bronchitis) is valued at just $3 billion. On the other hand, EPA also devotes a great deal of attention to benefit categories that amount to much less than rounding error: two pages of the RIA are given up to the details of assessing the impact of school absence days ($36 million), and other benefit categories include emergency room visits for asthma ($3.6 million) and lower and upper respiratory symptoms in children ($4 million and $5 million, respectively). It may be unfair to criticize EPA too much for analyzing these ultimately inconsequential (in an order-of-magnitude sense) effects of the regulation: perhaps the only way to determine how relatively small those benefits were was to conduct the analysis. Nonetheless, given the potentially significant benefit categories that were not monetized, EPA does not appear to have gotten the biggest "bang for its buck."

Of the large number of benefit categories omitted from the estimate of monetized benefits, three stand out. The first of these is reduced premature mortality from ozone pollution. The benefits from reduced ozone-related mortality do not appear in the RIA's estimate of total benefits, apparently because of doubts on the part of EPA's Science Advisory Board (SAB) that the ozone effect could be distinguished from the mortality effects of PM. Nonetheless, on the basis of recent research on the subject, EPA suggests that reduced premature mortality from ozone could contribute an additional $3 billion annually to the estimated benefits—making it the second-largest single category of estimated benefits (though it would still be dwarfed by the benefits of lower particulate pollution). This estimate is based on meta-analyses of the literature on ozone pollution that were in press, but not yet published, at the time of the RIA. Thus, the quantification of ozone pollution appears to have been a case of the SAB's recommendation lagging behind the literature. It is not too great a feat of inference to conclude that EPA staff disagreed with the SAB's recommendation and would have strongly preferred to include the estimate of ozone-related benefits in their official total.

The facts of the case are similar in the second major missing category of benefits: reduced acidification of lakes and streams in the northeastern United States. Again, a study was available (Banzhaf et al. 2004) that would have allowed estimation of the monetary benefits from reduced acidification. In contrast to the case of ozone, however, EPA chose not to apply the study to the CAIR RIA on the grounds that the study was still undergoing peer review.

A third category of missing benefits concerns visibility. Here EPA's action appears to have been more arbitrary and much less well explained. The agency states that it is able to quantify visibility impacts throughout the area affected by the new regulation; but it provides monetized benefit estimates only for Class I areas in the southeastern United States, without explaining why it limits the geographic scope of its analysis. As it is, EPA relies on a study of WTP for visibility improvements in the southwestern United States and extrapolates those results to the southeastern states and hence is already obliged to use benefits-transfer methods to extrapolate. If transferring benefits in this way is valid, it ought to be equally valid for the other areas where the CAIR would improve visibility.

These omitted benefit categories raise a methodological concern about the criteria used to include or exclude relevant information from the scientific and economic literatures. In particular, EPA (or the SAB) seems too rigid in its distinction between acceptable and unacceptable studies. Given the problems and uncertainties inherent in any applied valuation analysis, this either-or approach is essentially arbitrary, especially when it means the difference between having some number and having no number at all. At the very least, such a bright-line approach leaves the results of a regulatory impact study subject to the whims of peer review timing and publication schedules. Meanwhile, efforts to get around the application of a strict rule may lead to asymmetric treatment of essentially similar cases, as in the example of ozone pollution versus acidification benefits.

An alternative and preferable approach would be both more flexible and internally consistent. EPA should present a range of estimates based on studies of varying degrees of authority. At a minimum, the range would include two estimates of benefits: one using only peer-reviewed studies with high confidence, and a second using the best available estimates even where they have not yet been published or where other concerns pertain (e.g., the age of the study). The former number, based on peer-reviewed studies, could still serve as the official or preferred estimate of benefits. The latter would provide a sense of the range of estimates. A "best available estimate" would also focus attention on EPA's choice of which benefit categories to include and which to leave out, thereby helping to identify the categories of benefits that are most deserving of further research and study.[8]

Even with this more flexible approach, however, a complete accounting of monetized benefits will never be feasible. A third crucial issue to consider in benefits estimation, therefore, is transparency. What are the precise reasons for including or excluding a benefit category? How much does the decision not to include a benefit category matter for the overall results?

To help address these questions, a comprehensive list of all impacts should be presented, along with an indication of whether they were quantified or monetized, and a brief explanation for this choice (e.g., no willingness-to-pay information available; studies available but deemed unreliable; or quantification of impacts unavailable). The CAIR RIA takes a good first step in this direction by providing an exhaustive list of nonmonetized impacts. Even so, the analysis is inconsistent in explaining why these impacts were not monetized and what their magnitude might have been. In the case of ozone pollution, EPA (apparently chomping at the bit against the SAB's restrictive recommendations) presents its own "informal" estimate and emphasizes its significance. In the other two cases, however, the agency is much vaguer about the potential benefits; the potential magnitude of missing visibility benefits is mentioned almost as an afterthought, and the potential benefits from reduced acidification are described as "substantial," without any further elaboration. Yet strong evidence suggests that each of the missing benefits could be several hundred million dollars or more annually—at least an order of magnitude greater than several of the health-related benefit categories to which EPA devotes much more attention.

The lack of transparency is even more problematic in the case of other missing benefit categories. Regarding the remaining nonmonetized health benefits related to PM (involving low birth weight, pulmonary function, and other effects), EPA simply asserts that "we feel these benefits may be small relative to those categories we were able to quantify and monetize," without any mention of the evidence on which the agency bases that judgment.

A further step toward improving transparency is explicitly acknowledging that the dollar-valued-benefit estimate is incomplete. The CAIR RIA scores well on this dimension, collecting nonmonetized impacts into a term "B," which is then carried throughout the cost–benefit analysis. Although some observers criticize this approach as effectively ignoring a range of benefits by collapsing them into a single unknown parameter, acknowledging the missing impacts explicitly—putting them "on one side of the ledger"—is certainly preferable to the default alternative of assigning those impacts a zero value.

Distributional Incidence

Geography

One would expect that a cost–benefit analysis of an environmental regulation focused on the interstate transport of air pollutants would consider how the consequences of the policy were likely to vary with location. Indeed, the uneven incidence of the costs and benefits of pollution control provides the central rationale for the CAIR itself: upwind states should be held accountable for the impacts of their emissions on air quality in downwind states.

EPA's analysis of costs largely reflects the central importance of geography, presenting region-specific estimates of the projected impacts on coal production and retail electricity prices (EPA 2005a, Tables 7-7 and 7-9). On the benefit side, however, EPA's performance is more mixed. The RIA estimates the qualitative impacts on lakes and streams in three regions (EPA 2005a, Table 5-1), projects visibility impacts for each of 29 individual Class I areas (Table 3-10), and maps the expected percentage reductions in sulfur and nitrogen deposition (Figures 5-1 and 5-2). Similarly, the Notice of Final Rulemaking preamble (EPA 2005b) presents projections of precisely which counties will be in nonattainment for PM and ozone in 2010 and 2015 under both the base case and the CAIR, along with estimated ambient pollution concentrations.

Regarding the geographic distribution of monetized benefits, however, the RIA is silent. This must have been a conscious omission, although it is unacknowledged. EPA already estimates changes in air quality at the county level, and data on population density and hence exposure are used implicitly to translate those air quality changes into benefits. Thus, no new work would have been required to discuss how the estimated benefits are distributed geographically—but the resulting analysis would have been of considerable interest in understanding the impacts of the policy. Given that benefits are presented in monetary terms at the aggregate level, it is difficult to see why they should not be presented that way at local and regional levels as well.

Income

Any air quality regulation as sweeping as the CAIR can be expected to have disparate impacts across different income groups. A given rise in electricity rates has very different implications for rich and poor households, and may have regressive effects if not countered by other policy measures. The benefits are likely to be unevenly distributed as well. Air quality improvements may disproportionately benefit low-income households, to the extent that they are concentrated in urban areas or in places with poor initial air quality. On the other hand, visibility benefits are likely to accrue disproportionately to richer households, who are more likely to visit places such as national parks where visibility is most valuable.

The CAIR RIA, however, ignores distributional incidence across income groups. As in the case of geographic distribution of benefits, this would have required little extra work: EPA would only have had to match its county-level estimates of air quality improvements to similarly disaggregated data on average household income. Greater information on the distributional incidence of costs, meanwhile, could be gleaned by comparing estimated increases in electricity prices with average expenditures on electricity across households of different income levels (data already collected by the Energy Information Administration). The payoff from employing these approaches would have been a good deal of insight into how the consequences of the regulation fell on different groups.

The reluctance to report distributional effects may stem from a justified concern about the role of income in determining WTP. Strictly speaking, WTP depends on ability to pay and increases with income. According to this logic, benefits to richer people should be more highly valued. Although such an approach may be consistent with a narrow application of economic theory, it violates basic principles of fairness.[9] Quite appropriately, EPA elects not to scale its measure of monetized benefits on the basis of cross-sectional variation in the income of the affected population.

At the same time, EPA's decision not to adjust benefits for cross-sectional variation in income is apparently contradicted by its use of income adjustments over time. To account for economic growth, EPA assigns higher value to improved air quality in future years (when incomes will be higher in real terms), relying on estimates of income elasticities of WTP drawn from the economics literature. At first blush, this seems logically inconsistent with EPA's decision not to take cross-sectional income variation into account. Indeed, the RIA itself directs attention to this problem, by characterizing income disparities across subpopulations and income growth over time as two manifestations of how income differences can affect WTP. Having thus lumped the two sources of variation together, EPA leaves itself little room to explain its decision to adjust WTP over time but not across income groups. As justification, the agency cites a statement by the SAB highlighting the "sensitivity of making such distinctions [among income groups], and because of insufficient evidence available at present" (EPA 2005a, 4-16). The clear implication is that if EPA (or the SAB) were more insulated from political "sensitivities," or had greater evidence about the income elasticity of WTP, then it would be justified in adjusting WTP across populations.

This apparent contradiction can be resolved—but only by taking a conceptual step that is missing from EPA's analysis. The measure of value that EPA desires (appropriately) to estimate, in assessing the benefits from its proposed policy, is not the willingness to pay of the actual affected population (which would depend on income, as well as age, education status, and so on), but rather the WTP of a representative U.S. population.[10] In this conception, whether smog settles over a poor neighborhood or a rich suburb does not affect society's estimation of the damages caused, or the benefits of better air. Such an approach would seem to be a fundamental tenet of true environmental justice, and consistent with basic concerns of fairness and equity. Moreover, such an approach resolves the logical contradiction. Estimating the WTP of a representative population is perfectly consistent with making an adjustment for economic growth over time—as the United States as a whole gets wealthier, so does a representative population.

Discounting

The use of discounting to express future costs and benefits in present-value terms is at once one of the most standard and one of the most controversial approaches in applied cost–benefit analysis. The thorniest issues arise when comparing costs and benefits across long periods of time (e.g., decades or centuries), because discounting then carries with it an implicit judgment about intergenerational welfare comparisons. In the case of the CAIR, the benefits and costs were examined over a much shorter time horizon, in 2010 and 2015. Hence, no intergenerational comparisons are implicated. If the only use of discounting were to provide a common yardstick for costs incurred in 2010 (for example) with benefits realized in 2015, there would be little need for comment.

However, discounting still enters into the analysis in a fundamental way because of the lag time involved in the health consequences of exposure to air pollution. Following the recommendations of the SAB, EPA uses a segmented lag structure that allocates 30 percent of the PM-related mortality reductions to the first year, 50 percent to years 2 through 5, and the remaining 20 percent to years 6 through 20 (EPA 2005a, 4-45). To express the benefits from reduced mortality in present-value terms, therefore, EPA spreads the estimated reductions in mortality over 20 years and applies the standard Office of Management and Budget 3 percent and 7 percent discount rates to the resulting time profile of benefits.

This discounted lag approach has serious flaws. To begin with, as EPA notes, the lag structure is essentially arbitrary: because there is no "specific scientific evidence of the existence or structure of a PM effects lag," the segmented approach is simply "intended to reflect the combination of short-term exposures in the first year, cardiopulmonary deaths in the 2- to 5-year period, and long-term lung disease and lung cancer" in the later years (EPA 2005a, 4-45–44-6). But when later effects are discounted, the choice of lag structure (as well as the choice of discount rate) matters considerably. Although EPA rightly performs a sensitivity analysis of the impact of the lag structure, that analysis underscores the problem: the estimated health benefits when no lag structure is applied are 250 percent greater than when an exaggerated (15-year) lag structure is applied and benefits are discounted at 7 percent.

What makes the choice of lag structure matter, of course, is the decision to discount future benefits. This may appear at first to be entirely unobjectionable—at least if one grants the appropriateness of discounting costs and benefits in principle. Upon reflection, however, very little justification seems to exist for discounting the reduced mortality from air pollution. The measure of benefits is the VSL. The underlying damage at issue—whose reduction is being quantified as a benefit and then discounted—is not an actual death, but rather an increase in risk. The benefit of cleaner air is an immediate reduction in the statistical likelihood of death—whether that death is imminent or lies in the distant future. Because the benefit is realized at the time that air quality improves, it should not be discounted.[11]

It may be that people are willing to pay more to reduce the chance that they will die within a year than to reduce their chance of dying within two decades. But this is not evident or deducible from first principles: some might well prefer a sudden heart attack to a drawn-out struggle with a chronic and debilitating disease. It is an empirical question. What is needed to resolve it is not simply better information about the lag structure, but also—and crucially—better information about how people value reductions in the risks of different kinds of deaths. Applying a discount rate, as EPA does, is a crude approach that imposes a particular and arbitrary assumption about how people value reductions in the risk of future death. Curiously, EPA appears largely oblivious to these considerations. The RIA is concerned only with whether the lag structure is correctly determined, but does not acknowledge that there is a more fundamental question about valuation at stake.[12]

As the discussion in the CAIR RIA makes clear, the issue of lagged health effects is a crucial one that will apply to many future regulatory analyses. A high priority for research, therefore, should be to gather better empirical estimates of willingness to pay for reductions in different kinds of mortality risk—at a minimum, distinguishing near-term impacts from chronic ones. Once such evidence is available, the proper approach will be to apply the appropriate measure of VSL *at the time when the reduction in risk takes place*—in other words, when the air becomes cleaner, not when the eventual mortality would have occurred. Until then, EPA should acknowledge the fundamental problem with discounting in this context and should include the case of zero lag (which is of course equivalent to an arbitrary lag structure with no discounting) as one of its core benefit estimates, rather than relegating it to a sensitivity analysis.

Treatment of Uncertainty

The CAIR RIA addresses uncertainty in three ways: through conventional sensitivity analysis (using parameter values chosen to represent plausible alternative assumptions), through Monte Carlo analysis using estimated distributions for dose–response parameters and health endpoints, and through Monte Carlo analysis using distributions of PM-related mortality impacts drawn from an expert elicitation process.

The results of all three approaches underscore the importance and pervasiveness of uncertainty in the case of the CAIR. As the sensitivity analysis shows, basing the estimated PM-related mortality impacts on results from the Harvard Six Cities study (Dockery et al. 1993), rather than the American Cancer Society study (Pope et al. 1995) used in the base case, more than doubles the estimated benefits of the CAIR. At the other extreme, assuming a fairly high (but still plausible) threshold for the effects of PM cuts the estimated benefits by 96 percent to less than $1 billion. The choice of income elasticity also has a sizeable effect because it interacts multiplicatively with the PM-related mortality effects that make up, by far, the greatest share of the benefits.

The Monte Carlo analyses demonstrate the extent of uncertainty even more clearly. Using estimated standard errors from the underlying studies that provided the basis for the health impacts, the Monte Carlo analysis finds a 90 percent confidence interval spanning an order of magnitude, from $26 billion to $210 billion (with a mean of $100 billion). When the Monte Carlo is based on the results of expert elicitation rather than estimated standard errors, the 90 percent confidence interval balloons, extending from $3 billion to $240 billion (around a mean of $74 billion).

This is an admirably varied, complete, and even innovative approach to assessing uncertainty. The three techniques employed by EPA represent a substantial and sophisticated effort to account for uncertainty, and they complement each other well. The conventional sensitivity analysis facilitates focused consideration of particular discrete and often qualitative alternatives; it answers such questions as (a) What happens if we ignore the lag structure for reductions in mortality risk? (b) What happens if we value all cases of chronic bronchitis? (c) What happens if we assume that willingness to pay is more or less sensitive to increases in income? The "classical" Monte Carlo approach allows for simultaneous consideration of multiple sources of uncertainty. It answers the question (conditional on the specified parameter distributions, of course), What is the central range or most likely magnitude of benefits? Finally, the expert elicitation approach combines the flexibility and scope of Monte Carlo analysis with a fuller and more nuanced treatment of uncertainty on a particular dimension—in this case, the dose–response curve for PM-related mortality impacts.

While one can quibble with how the techniques were applied in the case of the CAIR RIA, for the most part these concerns are minor. For example, the classical Monte Carlo analysis is applied only to health effects; impacts on ecosystems and visibility are treated as constants in the analysis. In practice, however, this omission probably matters little, in part because some of those same effects (in particular, visibility) *are* explored in the conventional sensitivity analysis; this is another example of complementarities among the three approaches. Similarly, the expert elicitation procedure was imperfect in a number of ways: only five experts were consulted, limited review appears to have taken place beforehand, no pre-elicitation workshop was held, and so on. But these criticisms hardly seem fair when one considers that the CAIR RIA represents the pilot phase of expert elicitation: it was explicitly designed as a trial run, and many of its deficiencies were remedied in subsequent applications. Indeed, the formal use of expert judgment to evaluate uncertainty represents an important innovation.

For all its merits, however, the uncertainty analysis in the CAIR RIA—like the selection of policy scenarios and the choice of which benefits to monetize—is too narrow in scope. Faced with pervasive uncertainty, the RIA considers only a subset of the sources of that uncertainty. Estimating benefits in dollar terms requires performing several independent analyses in sequence, with each link in the chain subject to uncertainty. Because the regulation proposed an emissions-trading system rather than mandates on individual power plants, a model of the electric power sector is required to translate the overall emissions targets into plant-level emissions estimates. Because the emissions affected by the regulation (SO_2 and NO_x) contribute to air pollution hundreds of miles away, and in different chemical forms (e.g., NO_x combining with volatile organic compounds to produce ground-level ozone), analysts must employ necessarily imperfect models of pollution dispersion and atmospheric chemistry to translate emissions into pollution concentrations. In turn, those concentrations must be translated into effects on human health and ecosystems, using often poorly understood dose–response relationships. And finally, the physical impacts (*x* number of premature deaths among human populations, or *y* percent of lakes and streams affected by acid deposition) must be expressed in monetary terms using often scant or incomplete measures of value, some of which depend on uncertain projections of population or income growth.

The problem is not that EPA fails to acknowledge these sources of uncertainty; indeed, in its discussion of benefits estimation the agency goes to great lengths to enumerate the uncertainties (EPA 2005a, Table 4-5). Rather, the problem is that the RIA addresses only a subset of the sources of uncertainty head-on. The sensitivity and Monte Carlo analyses explore the relationships between pollution concentrations and physical impacts, and between physical impacts and monetary values. But they give short shrift to the first two links in the causal chain outlined above—that is, the uncertainties in the distributions of pollution emissions and of ambient concentrations.

When the RIA does address the air transport models used to derive ambient concentrations, it does so in a self-referential way—evaluating their performance by comparison to the performance of other models or of other analyses. Thus, the model runs performed for the CAIR analysis are deemed "appropriate" simply because they are no worse than prior model runs. Although some statistics are provided on the predictive abilities of the models used relative to actual measured conditions, they are given without any context for the level of fractional error, for example, that might be deemed "good" or "bad" in an absolute sense. And no evidence is provided on the correlation between predicted and actual changes in air quality, even though the accuracy of predicted changes in air quality resulting from policy-induced changes in emissions is of central importance in the reliability of the model.

Meanwhile, the RIA includes essentially no discussion of the uncertainty in the spatial pattern of predicted emissions based on the projected outcome of emissions trading using EPA's model of the electric power sector. Indeed, one is hard-pressed to find any recognition at all in the RIA that these projected emissions might be a major source of uncertainty in the analysis as a whole.[13] This omission is especially glaring in light of the July 2008 *vacatur* ruling overturning the CAIR: one of the reasons cited by the court for its ruling was EPA's failure to conclusively demonstrate any connection between its chosen regulatory approach (emissions trading) and the likely reductions in the contributions by sources in specific upwind states to downwind air quality.

EPA's failure to sufficiently explore explicitly these two major sources of uncertainty—the impacts of regulations on emissions and of emissions on concentrations—constitutes a major gap in its analysis. This omission is all the more striking given how straightforward it would be, at least conceptually, to integrate these sources into its formal modeling of uncertainty—in particular, its Monte Carlo analyses. The air transport models, which are based on Gaussian plumes (i.e., probabilistic analyses of air movements), ought to be readily amenable to Monte Carlo analysis. Although more work would probably be required to formally model uncertainty in the spatial distribution of emissions under trading, that could be done as well—for example, by explicitly modeling the cost of pollution abatement at each individual EGU as a draw from a distribution rather than as a point estimate.

Recommendations

The CAIR RIA is, in many ways, an admirably comprehensive account of the benefits and costs associated with a particular regulation. EPA staff brought to bear a huge amount of relevant information. They carefully described how they conducted the analysis and, for the most part, explained why they made the choices they did. They presented many estimates in natural units as well as in dollar terms, and employed a set of sophisticated and complementary techniques to assess uncertainty.

Nonetheless, like any such document, the CAIR RIA had a number of weaknesses and blind spots that help to highlight ways that regulatory impact analysis could be improved. Distilling the preceding discussion yields a number of recommendations for regulatory impact analyses—some, but not of all, of these were followed in the case of the CAIR.

Scope of analysis. An RIA should consider multiple policy alternatives and, if possible, multiple policy baselines. When the policy instrument is not mandated by regulation—as in the case of the CAIR, which could suggest but not require an emissions trading program—the scope of policy alternatives considered should include other policy instruments as well as other targets.

Use of willingness-to-pay measures. Because the chosen value for the VSL plays a central role in the analysis, high priority should be placed on further research into appropriate VSL measures, particularly measures that explicitly capture public or social values rather than being derived purely from private risk-taking behavior.

Choice of primary sources. Although it is appropriate to base the "main" estimate of benefits only on studies meeting a well-defined and rigorous set of criteria (e.g., peer-reviewed articles published within a certain period of time), at least one additional benefit estimate should be presented that incorporates a wider set of studies, especially where doing so can expand the set of benefit categories considered. For example, in the CAIR RIA, a second estimate should have been presented that drew on still-unpublished but leading-edge research into the benefits from ozone-related mortality reductions and from visibility improvements.

Monetization of benefits estimates. If any benefits are expressed in monetary terms, then as many as possible should be expressed this way, and the reasons for not doing so should be clearly and fully explained on a case-by-case basis. Moreover, the existence of nonmonetized benefits should be explicitly acknowledged in the presentation of results—for example, through a generic term labeled "B."

Distributional incidence. The distributional incidence of costs and benefits should be presented in depth—in particular, by geographic region and household income.

Use of a "representative population" for estimating benefits. Estimates of WTP should be defined with respect to a representative U.S. population. Benefits should be adjusted for income growth over time, but not for income disparities across subpopulations.

Discounting and lag structures for health effects. If health effects lag behind exposure, separate VSLs for different types of mortality (e.g., acute versus chronic impact) should be used rather than an arbitrary discounting approach. Because the benefit from the policy is a reduction in risk, that benefit should not be discounted—regardless of how far off in the future the death is likely to occur.

Uncertainty analysis. Multiple analyses of uncertainty should be used, including conventional sensitivity analysis, Monte Carlo analysis, and (if feasible) expert elicitation. To the extent possible, all sources of uncertainty should be explicitly accounted for, including uncertainty in modeling emissions and air quality, rather than just dose–response relationships and valuation.

The RIA as a Guide to Policymaking or a Source of Information

Having delved into the details of methodology and scope, we now pull back to a loftier vantage point and asks more fundamental questions: What can a regulatory impact analysis like the CAIR RIA achieve? What roles does it serve? Here, I identify two main functions, which I term the technocratic and democratic functions. Focusing solely or primarily on cost–benefit methodology as a technical input to policymaking (as in the above section) ignores the equally critical but often overlooked democratic function. Moreover, the two roles are complementary rather than mutually exclusive. Improving the performance of the RIA in informing and educating the public (the democratic function) cannot help but improve its usefulness to policymakers.

The Technocratic Function: The RIA as a Guide to Policymaking

The conventional view among efficiency-minded economists is that an RIA, and particularly the cost–benefit analysis at its heart, should guide policymakers in designing policy—setting the stringency of the emissions reductions required, selecting the appropriate policy instrument, and so on. According to this view, the RIA logically precedes the choice of policy. It should explore a range of relevant policy alternatives, illuminating the trade-offs among them and determining which would yield the greatest net benefits. Of course, most advocates of this view hasten to add that a cost–benefit analysis need not be narrowly determinative: economic efficiency need not be a necessary or sufficient criterion for sound policymaking (Arrow et al. 1996). Nonetheless, the role of the cost–benefit analysis in this framework is prescriptive: to provide the policymaker with a definitive assessment of the relative efficiency of various policy options.

As the section critiquing the RIA on technical grounds makes clear, the CAIR RIA performs well on many technical aspects of cost and benefit estimation. And yet the RIA fails to meet the most basic requirement of sound economic policy analysis: namely, the consideration of multiple alternatives. A document that considers the costs and benefits of the proposed policy only relative to the status quo cannot possibly have been used to design that policy.

Ironically, the CAIR RIA's outward embrace of the technocratic ideal partly explains its failures. The benefits and costs are patiently catalogued, summed up, and presented to three significant digits as if decimal points will lead to better policy. Indeed, the RIA is almost compulsive in its precision—as illustrated by its patient exploration of categories of impacts (such as school absence days and asthma events) that do not even amount to rounding error, being measured in the tens of millions relative to total benefits in the tens of billions. For all its impressive features, the CAIR

RIA is a document consumed by relatively minor details, providing little guidance or rationale for how the policy itself was chosen.

To be sure, the recommendations presented in this chapter's section critiquing RIA on technical grounds (as well as similar recommendations from other authors, including those represented in this report) offer suggestions for improving the RIA as a technical document. All the same, criticizing the RIA on the grounds that it did not provide the basis for informed policymaking misunderstands the statutory context. As noted in the introduction to this chapter, the Clean Air Act itself expressly forbids the EPA administrator to consider costs in setting ambient air quality standards. Indeed, EPA itself, in announcement of the CAIR (EPA 2005b), takes pains to explain that the required emissions reductions were based not on the RIA but rather on the agency's judgment of what was "highly cost-effective" and necessary to reduce the contribution of the states affected by the rule to ambient air pollution in downwind states. Far from playing a central role, the RIA is summarized cursorily on pages 144 to 151 of the 155-page document.

The Democratic Function: The RIA as a Source of Public Information

If the RIA was not actually used to guide policy, what is its purpose? In Chapter 4 of this report, Wagner characterizes the CAIR RIA as a "litigation support document." In the tradition of public choice analysis, she treats the RIA as a tool designed by EPA to support its rulemaking against legal challenge and evaluates it on that dimension.

But there is another alternative, and indeed one that is suggested by EPA itself in its introductions of recent RIAS: "to inform the public and states about the potential costs and benefits of implementing these important air quality standards." What happens if we take seriously the proposition that a central aim of an RIA is (or ought to be) informing the public (and policymakers) ex post about the consequences of a decision—even if the decision, as in the case of CAIR, was made on other grounds? Three implications stand out.

Order-of-magnitude judgments

First, from this different vantage point, precision in estimating benefits and costs takes on less importance. Rather, the most useful and relevant pieces of information concern order-of-magnitude judgments: Are benefits likely to be greater than costs? With what degree of confidence?

Emphasizing order-of-magnitude impacts rather than precise figures communicates a more honest realization of the deep uncertainties involved in assessing regulatory policies as far-reaching as the CAIR. As noted above, the process of estimating the regulation's costs, and especially its benefits, involve a chain of reasoning—from regulation to emissions, emissions to concentrations, concentrations to physical impacts, and impacts to monetary values—in which each step involves great difficulty and fundamental uncertainty. The impression of staggering complexity is confirmed by the sensitivity of the results to key assumptions, as well as the extraordinarily large 90 percent confidence intervals found in the Monte Carlo studies.

Moreover, the measured sources of uncertainty just described amount to the known unknowns—the gaps in knowledge that can be catalogued, assessed, and assigned standard errors or other measures of uncertainty. On top of these lie what might be called the unknown unknowns.[14] Several major categories of benefits are not monetized at all; as a result, they are absent from the discussions of standard errors or Monte Carlo analysis. And there may well be whole categories of benefits that are not accounted for at all. The impetus behind the SO_2 trading system in the 1990 Clean Air Act Amendments, for example, was the problem of acid deposition in forests, lakes, and streams. Only later did the contribution of SO_2 to ambient concentrations of PM—now considered to be far and away the most important benefit from reducing SO_2 emissions—become known. This pervasive uncertainty suggests caution in drawing precise conclusions from the CAIR RIA or from any assessment of similarly complex regulations.

In the case of the CAIR, is fairly certain that benefits are greater than costs; after all, the regulation would pass a simple cost–benefit test even assuming the lower bound of the 90 percent confidence interval using expert elicitation ($3 billion). Indeed, one may reasonably conclude that benefits are much greater than costs—by at least an order of magnitude. Beyond these order-of-magnitude statements, however, it is hard to pin down any number with confidence. In a sterling example of mistaking precision for accuracy, the CAIR RIA presents results to three significant digits without regard to the considerable error bounds surrounding its estimates. A more honest approach would be to replace the precise numbers presented in the executive summary with a simple conclusion: "Based on the best available evidence, the benefits from the CAIR are at least an order of magnitude greater than the costs, with net benefits measured in the tens of billions of dollars annually."

This is not to say that analysts should not seek appropriate precision in their estimates—but rather to argue that in presenting their main results, analysts should abstract from the precise numbers and offer conclusions that are consonant with the underlying uncertainties. Presenting results in order-of-magnitude terms would also direct attention toward a fruitful set of questions that are unanswered in the CAIR RIA. What are the most important sources of uncertainty in driving the results? How wide is the range of plausible benefits and costs? Perhaps most importantly, what would have to change to alter the basic conclusion from the analysis (i.e., reverse the sign of net benefits)?

Estimation of marginal net benefits of the proposed policy

Second, if the primary objective of the analysis is to describe the impacts rather than prescribe an outcome, identifying and considering a particular set of alternative policies may not be as crucial. Instead, it becomes more useful to estimate the marginal impacts of increasing or decreasing the stringency of the policy. Such an approach can help answer a question of considerable interest: How does the proposed policy compare to the economically efficient one?

Note that the performance of a given policy relative to the efficient one cannot be ascertained simply by computing its total benefit and cost. In the case of the CAIR, for example, EPA found projected net benefits of $83 billion to $99 billion a year at full implementation in 2015, even without considering a range of nonmonetized benefits; benefits outweighed costs by roughly 30 to 1. Such figures might seem to provide prima facie grounds for more stringent action. But, in fact, that conclusion does not follow at all: in principle, net benefits could have been even larger for a weaker policy (although this did not prove to be the case in the CAIR, as I discuss below). Simply calculating total benefits and costs does not shed light on marginal benefits and costs, which—as any economics student knows—must be equated to satisfy efficiency.

Nor is the mere consideration of a few alternative policies sufficient to determine how a given policy compares to the efficient one. Suppose policy A represents the status quo, policy B is a proposed regulation, and policy C is a more stringent alternative. An analysis finds that total net benefits are greatest under B, but greater under A than under C. Comparing A and C might suggest that the efficient policy must be less stringent than B; but in fact nothing rules out the possibility that the efficient outcome actually lies between B and C—meaning that the proposed policy is not stringent enough, rather than too stringent.

By comparing sufficiently many alternatives, of course, one could determine the efficient policy. In the limit, the analyst would estimate the marginal benefit and cost schedules (for all possible policies), with the efficient point lying at their intersection. In the real world, estimating these entire schedules may be infeasible. Nonetheless, it may be possible to estimate the marginal benefit and marginal cost of the proposed policy. Whether the proposed policy is more or less stringent than the efficient one can then be determined by computing marginal net benefit (marginal benefit minus marginal cost): if it is positive, the policy is too lax from an efficiency perspective; if negative, the policy is too stringent.

This approach was feasible in the case of the CAIR. EPA itself estimated the marginal cost of its proposed rule; other analysts had little trouble estimating marginal benefit. The advocacy group Environmental Defense Fund, using a methodology employed by EPA in prior analyses, made a back-of-the-envelope estimate that the benefits of SO_2 reductions amounted to $15,000 per ton—more than an order of magnitude greater than EPA's estimate of the marginal abatement cost (Shore et al. 2004).[15] Researchers at Resources for the Future estimated that marginal benefits were between $1,800 and $4,700 per ton of SO_2 reductions, well above EPA's estimated marginal cost of $700 to $1,400, and $700 to $1,200 per ton of NO_x reduction, somewhat less than EPA's estimate marginal cost of $1,300 to $1,600 per ton (Banzhaf et al. 2004). On the basis of these estimates, the SO_2 reductions—despite their impressive net benefits—proved to be too small from an efficiency perspective, while the NO_x reductions may have been slightly too stringent.[16]

Transparent and accessible presentation of benefits and costs

Third, an eye toward informing the public rather than shaping policy also suggests that much more attention should be given to how costs and especially benefits are presented in the analysis.

A key step is to quantify the consequences of the policy in natural units (i.e., physical impacts) as well as in monetary terms. The monetization of costs and especially benefits is one of the most common targets for critics of cost–benefit analysis—and with some justification, given the large number of assumptions involved. On the other hand, proponents of cost–benefit analysis point out (with equal justification) that expressing impacts in dollars—or any other common metric—is a necessary step in aggregating disparate benefits and comparing them with costs in a consistent fashion.

The primary justification for boiling everything down to net benefits is to ensure that decisionmakers rely on some "objective" measure of value rather than substituting their own personal preferences. In this context, WTP is best seen as one possible system of weights, among many, to use in comparing disparate impacts. An estimate of net benefits is essentially a summary statistic; like all summary statistics, in compressing a great deal of information it leaves much out. If only a single set of weights is to be used (necessarily the case in deriving a single "societal" estimate for net benefits), then economic theory provides a strong argument that for all its faults, WTP based on revealed-preference measures is the best set of weights.[17] Rather than viewing the calculation of net benefits as a central goal (the implicit approach taken in the CAIR RIA), however, the estimation of net benefits should be regarded as one means to a larger end: informing the public about the consequences of a proposed regulatory policy, and the trade-offs involved.

When the function of a regulatory analysis is defined as informing the public, the personal preferences of individuals take center stage. From this perspective, the goal of the analysis should be to give the reader enough information to answer the question, would I vote for this? This means supplying enough information that an individual reader can substitute her own weights on various policy outcomes—that is, her own preferences—rather than relying solely on the "objective" weights provided by willingness-to-pay measures.

Wherever possible, therefore, benefits ought to be quantified and presented in natural units (e.g., a reduction in the incidence of premature mortality or the increase in visibility in natural parks). The presentation of quantified benefits in natural units is a complement to, rather than a substitute for, the presentation of monetized benefits: the two sets of numbers convey distinct information. To its credit, EPA does present a variety of health impacts in natural units (*see* EPA 2005a, tables 4-16 and 4-17.

Benefits (and costs) could also be made more informative by conveying them in both total and per-capita terms. For example, health effects could be presented in terms of incidence rates as well as in totals. After all, what is being measured (and valued) is a reduction in risk, not a reduction in specific deaths.[18] In addition to presenting the PM-related health impacts as 13,000 fewer deaths per year, therefore, EPA should present them as such-and-such a reduction in the risk of premature death (e.g., per 100,000 individuals). The total number conveys a sense of the magnitude of the program's impact (and could be compared to the total cost of the program); the reduction in risk is what is relevant for an individual (and could be compared to, say, the cost of the program per household).

Recognizing that individuals may have difficulty identifying their own preferences over unfamiliar nonmarket goods, a well-designed RIA could provide contextual clues to help readers make those assessments. A reader may wonder how much she, as an individual, ought to value the reduced risk associated with air pollution that this policy will achieve. To inform this contemplative process, an RIA could describe the underlying trade-offs by transposing them to other, more familiar spheres of decisionmaking. For example, in the case of the CAIR, the reduced mortality risk from air pollution might be described as follows:

> *This proposed regulation is estimated to cost roughly $2 billion annually and to prevent 13,000 premature deaths each year. The implied cost per avoided premature death is therefore $150,000. If one were to apply this same trade-off to other, more familiar decisions, it would be equivalent to an individual paying $15 per year to reduce his or her annual risk of dying by 1 in 10,000—equivalent to the risk from [smoking X cigarettes per day] [rock climbing at X elevation], and so on.*

Similarly, results from revealed-preference studies could explicitly serve as guides to personal reflection, rather than being offered as "true" or "objective" measures of value. Continuing the example above, the RIA might point out that empirical studies of wage premiums suggest that workers earn, say, roughly $600 more in a year for every 1-in-10,000 increase in the risk of death.

Finally, the same considerations also suggest that RIAs should explicitly include discussions of other regulatory policies. In the conventional view of an RIA as a guide to policy, discussion of other policies is essentially irrelevant; the costs and benefits of the policy at hand are what matters. The CAIR RIA, for example, never mentions any other environmental regulation; it is as if the analysis takes place in a vacuum. In contrast, if the primary goal of the RIA is to inform the public, then a discussion of other regulatory policies—such as their costs and consequences—can provide crucial context.

Conclusion

In the last section I emphasized the democratic function of regulatory analysis—partly because, in the case of the CAIR, the technocratic function was essentially made moot by the Clean Air Act's prohibition on using costs in setting air quality standards. In general, however, an effective regulatory impact analysis should fulfill both the technocratic and the democratic functions—guiding policy makers ex ante and informing citizens ex post.

At a practical level, the technocratic function is already well enshrined in the regulatory review process. Elevating the democratic function implies a change of approach. In particular, the team at OIRA responsible for reviewing and commenting on draft RIAS should include not only the typical assortment of economists, engineers, and scientists—i.e., technical experts—but also at least one member charged with assessing the transparency and informativeness of the review. One can imagine a new office of "OIRA ombudsman" serving in this new role.

If the technocratic and democratic functions would place somewhat different emphasis on different aspects of analysis, their combined effect would be complementary. A technically sound cost–benefit analysis is an obvious prerequisite for an informative one. Less apparent, but equally the case, is that a truly democratic regulatory analysis—one that is transparent and accessible to the lay reader, and designed to inform the public—will also provide better guidance to policymakers. In part, this is because the ultimate consumer of a cost–benefit analysis is often little more than an educated layperson, at least relative to the technical experts, steeped in the nuances of a particular regulation, who author the analyses. Policymakers need plain language, transparent presentation of results, and order-of-magnitude conclusions just as surely as the public does. Moreover, the process of making a regulatory analysis transparent and informative can only improve the clarity of thought and quality of reasoning that go into the document itself. Finally, from a dynamic perspective, the democratic function is integrally important to good policymaking in the long run: each successive regulation offers a chance to inform and educate the public, and thereby strengthen popular support for sound and well-designed regulations.

■ ■ ■

Notes

1. Personal communication from Sam Napolitano, Director of EPA's Clean Air Markets Division, June 6, 2008.

2. The narrow scope of the CAIR RIA also runs counter to the language of Executive Order 12866, which mandated it. Section 6(C)(iii) of the executive order requires "An assessment, including the underlying analysis, of costs and benefits of potentially effective and reasonably feasible alternatives to the planned regulation... and an explanation why the planned regulatory action is preferable to the identified potential alternatives."

3. In other words, the binary distinction does not recognize the *magnitude* of the benefits from moving from nonattainment to attainment. A large improvement in air quality in a very polluted county could still fail to bring that county into attainment, while a much smaller improvement in a marginal county could cross the threshold.

4. *State of Michigan, Michigan Department of Environmental Quality and State of West Virginia, Division of Environmental Protection, Petitioners v. U.S. Environmental Protection Agency, Respondent, New England Council, Inc., et al., Intervenors,* 213 F.3d 663 (D.C. Cir. 2000).

5. Later in this chapter, I probe into how a valuation approach based on willingness to pay could be supplemented with other information.

6. *See,* e.g., Sen (1995) and Sunstein (1997, Ch. 2).

7. Although there are principled reasons to question the use of a wage-based VSL for air pollution, the practical implications for cost–benefit analysis are less clear-cut, at least in the case of the CAIR. First, few alternatives exist. Indeed, EPA itself acknowledges the limitations of using wage–hedonic estimates of VSL, but defers to its own Scientific Advisory Board in continuing to rely on such estimates for valuation. Second, the number value EPA uses—$6 million in the year 2010—is on the high end of available estimates; for comparison, the cost–benefit cost analysis of the CAIR by Banzhaf et al. (2004) used a value of $2.25 million. While that comparison does not reveal whether $6 million is "high" or $2.25 million is "low," it does insulate EPA from the charge that it chose a low-end estimate. Third, the estimated benefits from the CAIR are already far greater than the costs. Selecting a higher value for the VSL feels a little like running up the score.

8. Note that the proposal to base estimates of benefits (and costs, which should be treated symmetrically) on a wider range of studies is distinct from the treatment of uncertainty discussed later in this chapter. The point here is not to explore the consequences of this or that assumption, but rather to show the total impact on estimated benefits and costs from incorporating the very latest research, even if that research has not yet completed its journey through the publication process. The importance of taking leading-edge research into account is especially great in areas of active research, as in the case of the CAIR.

9. An alternative approach, equally consistent with economic theory, would be to base valuation on *willingness to accept* (WTA; also known as "equivalent variation" in welfare economics) rather than WTP (also known as "compensating variation"). While WTP is more commonly used in applied settings, there is no theoretical ground for preferring it, and under standard conditions it is weakly smaller than WTA.

10. *See* Revesz (1999, 967), who advocates on equity grounds that a uniform VSL be applied across all environmental programs on the basis of a representative population of the United States.

11. *See* Heinzerling (2000, 204–5) for a similar argument.

12. The only hint of these deeper issues comes in the context of EPA's discussion of "uncertainties" surrounding the valuation of premature mortality, when the agency discusses the theoretically attractive but practically infeasible "survival curve" approach, which would account for the effect of improved environmental quality on the probability of survival as a function of age, health status, and so on (EPA 2005a, 4-58).

13. Although EPA performed a sensitivity analysis to gauge the effects of alternative assumptions about energy prices and electricity demand, it appears to have considered the consequences only for estimated costs—not for the spatial pattern of emissions.

14. Although former Secretary of Defense Donald Rumsfeld captured the idea in his memorable phrase "unknown unknowns," economists generally credit Frank Knight (1921) with distinguishing between risk, involving a number of possible events whose probabilities can be known in advance (i.e., known unknowns), and uncertainty, involving events whose likelihood is unknown and unmeasurable (i.e., unknown unknowns).

15. The author, although currently on the staff of Environmental Defense Fund, was not at the organization at the time that the CAIR was finalized or at the time of the comments.

16. It is worth noting that, in generating these benefit estimates, Banzhaf et al. used a much lower estimate for the value of a statistical life than EPA used in its analysis: $2.25 million versus $5.5 million.

17. That statement would surely be challenged by opponents of cost–benefit analysis. Nonetheless, as a conditional statement it seems hard to challenge. *Conditional* on the need to use a single set of weights, economic theory provides a strong basis for using WTP. Scholars uncomfortable with cost–benefit analysis have a stronger case in challenging the need to use a single set of weights, as I discuss in the text.

18. For a differing perspective, *see* Heinzerling (2000), who challenges the notion of a "statistical life." But while Heinzerling argues that risk and death are distinct harms, the distinction makes little sense from an analytic perspective. What is relevant for the policy analyst is the incremental effect of the policy being considered. Consider an analogy to cigarette smoking. Smoking raises the risk of death from lung cancer, but many smokers never get cancer, and many nonsmokers do. An antismoking campaign could be evaluated by the resulting total reduction in the number of deaths from lung cancer (controlling for other factors) *or* by the change in the risk, i.e., the incidence of lung cancer in the affected population. Those two measures correspond to the same underlying effect (the latter is simply the former divided by the total population) and hence are equivalent: they do *not* represent distinct harms and should not both be counted.

References

Arrow, Kenneth J., Maureen L. Cropper, George C. Eads, Robert W. Hahn, Lester B. Lave, Roger G. Noll, Paul R. Portney, Milton Russell, Richard Schmalensee, V. Kerry Smith, and Robert N. Stavins. 1996. "Is There a Role for Benefit-Cost Analysis in Environmental, Health, and Safety Regulation?" *Science* 272 (12 April): 221–222.

Banzhaf, H. Spencer, Dallas Burtraw, and Karen Palmer. 2004. "Efficient Emission Fees in the U.S. Electricity Sector." *Resource and Energy Economics* 26:317–341.

Carlson, Curtis, Dallas Burtraw, Maureen Cropper, and Karen L. Palmer. 2000. "Sulfur Dioxide Control by Electric Utilities: What Are the Gains from Trade?" *Journal of Political Economy* 108:1292–1326.

Chestnut, Lauraine G., and David M. Mills. 2005. "A Fresh Look at the Benefits and Costs of the U.S. Acid Rain Program." *Journal of Environmental Management* 77:252–266.

Dockery, D.W., C.A. Pope, X.P. Xu, J.D. Spengler, J.H. Ware, M.E. Fay, B.G. Ferris, and F.E. Speizer. 1993. "An Association between Air Pollution and Mortality in Six U.S. Cities." *New England Journal of Medicine* 329 (24): 1753–1759.

Harrington, Winston, Richard D. Morgenstern, and Peter Nelson. 2000. "On the Accuracy of Regulatory Cost Estimates." *Journal of Policy Analysis and Management* 19 (2):297–322.

Heinzerling, Lisa. 2000. "The Rights of Statistical People." *Harvard Environmental Law Review* 24:204–205.

Keohane, Nathaniel O. 2006. "Cost Savings from Allowance Trading in the 1990 Clean Air Act: Estimates from a Choice-Based Model." In Jody Freeman and Charles D. Kolstad, eds., *Moving to Markets in Environmental Regulation: Lessons from Thirty Years of Experience* (New York: Oxford University Press), 194–229.

Knight, Frank H. 1921. *Risk, Uncertainty, and Profit.* Boston: Hart, Schaffner & Marx.

List, John A., Robert P. Berrens, Alok K. Bohara, and Joe Kerkvliet. 2004. "Examining the Role of Social Isolation on Stated Preferences." *American Economic Review* 94:741–752.

Pope, C.A., III, M.J. Thun, M.M. Namboodiri, D.W. Dockery, J.S. Evans, F.E. Speizer, and C.W. Heath, Jr. 1995. "Particulate Air Pollution as a Predictor of Mortality in a Prospective Study of U.S. Adults." *American Journal of Respiratory Critical Care Medicine* 151:669–674.

Revesz, Richard L. 1999. "Environmental Regulation, Cost–benefit Analysis, and the Discounting of Human Lives." *Columbia Law Review* 99:941.

Sen, Amartya. 1995. "Environmental Evaluation and Social Choice: Contingent Valuation and the Market Analogy." *Japanese Economic Review* 46 (1):23–37.

Shore, Michael, Amber Munger, Cindy Copeland, and Vickie Patton. 2004. *Stop Blowing Smoke in the Heartland: Protecting Human Health from Air Pollution.* New York: Environmental Defense Fund. www.edf.org/documents/ 3887_blowingsmoke.pdf.

Sunstein, Cass R. 1997. *Free Markets and Social Justice.* New York: Oxford University Press.

U.S. Environmental Protection Agency (EPA). 2005a. *Regulatory Impact Analysis for the Final Clean Air Interstate Rule.* EPA-452/R-05-002 (March).

———. 2005b. "Rule To Reduce Interstate Transport of Fine Particulate Matter and Ozone (Clean Air Interstate Rule); Revisions to Acid Rain Program; Revisions to the NO_x SIP Call." *Federal Register* 70 (91): 25162–25405.

■ ■ ■

CHAPTER 4

The CAIR RIA: Advocacy Dressed Up as Policy Analysis

WENDY E. WAGNER

Critics allege that regulatory impact analyses (RIAS) serve as little more than a fig leaf to hide the contributions of controversial participants in the rulemaking process and provide an illicit entry point for the White House to tinker with agency decisions when an RIA indicates that the agency's proposal is too costly. Others argue that RIAS provide an unaccountable forum for conservative-leaning economists to hijack or at least delay agency policies by requiring analyses that use inherently under protective economic methods and assumptions. The Clean Air Interstate Rule (CAIR) RIA now provides strong evidence for a third, somewhat overlapping source of concern: namely, that RIAS may serve primarily as a mechanism for promoting agency decisions rather than scrutinizing them.

While the role of the RIA as a post hoc rationalization document is surprisingly ignored in the large and growing literature on cost–benefit analysis, it is safe to say that at least in the case of CAIR, this role is indisputable. The CAIR RIA is remarkable for providing almost no information about alternative policies, while at the same time touting the wisdom of the U.S. Environmental Protection Agency's (EPA) preferred program. This key move undoubtedly saved EPA added headaches that would have resulted from a more candid and comprehensive policy analysis, particularly given the unique vulnerability of the CAIR to legal challenge and political opposition. Indeed, by employing economically conservative assumptions, high discount rates, and quantification choices, EPA crafted an RIA peculiarly designed to protect its rule from devastating criticism, at least with respect to the CAIR's aggregate economic impact.[1]

From the standpoint of bureaucratic rationality, in fact, it is quite sensible for agencies to use RIAS as propaganda documents rather than self-critical policy analyses. For "significant rules," which are the subset of rules required to undergo an RIA, the agency faces a very high probability of being sued as well as having its rules criticized by the media, Congress, and even the White House. Moreover, the RIA is generally prepared at the very end of a rulemaking, after a policy decision has solidified. Under these circumstances, few agency attorneys or appointed officials would allow the agency to undertake an honest, searching public analysis of the costs and benefits of various alternatives relative to the agency's rule.

In this chapter, I argue—based on the CAIR case study—that the real challenge to improving RIAS may lie less in perfecting the methodology, and more in overcoming the multiple, entrenched institutional forces that discourage agencies from engaging in public self-evaluation. I advance this argument in four sections. In the first two, I evaluate the CAIR RIA from two different vantage points—first, as an instrument to help insulate the agency from inevitable legal and political attack, and second as a policy instrument. After concluding that the RIA does a superb job at the former but a poor job at the latter, I consider in the third section whether the two goals are mutually exclusive. If they are, then of course that complicates the project for those who would like RIAS to serve as important vehicles for open-minded policy analysis. Although I suspect that the RIA process is too badly broken on an institutional level to be salvaged, I conclude in the final section with some tentative recommendations for reform that may counteract at least some of the pressures for agencies to transform their policy analyses into litigation support documents.

Before proceeding, there are a few caveats about how the CAIR case study might extrapolate to other rulemakings. First, because the CAIR was extremely contentious and was inevitably going to lead to litigation, it may illustrate the worst case rather than the mean in terms of defensive RIAS. Without a threat of significant litigation, an agency might feel freer to conduct honest policy analysis rather than engage in post hoc rationalization crafted to support its rule in the courts. Second, and perhaps most unique to the CAIR, because of the high level of controversy surrounding it, EPA may have had few real options. Legislative attempts to address the problem had largely failed, putting EPA on the hot seat for coming up with an immediate solution. When the alternatives have been largely exhausted and little time remains for meaningful analysis, the benefits of rigorous policy analysis tend to decline, and an agency's failure to engage in candid policy analysis is much more justified institutionally and analytically.

The RIA as Evidence of Bureaucratic Rationality

Based almost exclusively on adverse health effects, particularly mortality from particulates, the CAIR rulemaking is justified by a benefit-to-cost ratio of at least 25 to 1 (EPA 2005c, 25166). Even more encouraging, EPA suggests that in the end the benefits will probably be considerably larger, although how much larger EPA cannot say (EPA 2005a, 1-9). EPA also concedes that the costs to utilities are likely to be overstated, making the proposal even more justified in terms of a benefit-to-cost comparison than its executive summary suggests (EPA 2005a, 7-19–7-21).

The RIA supports this upbeat conclusion with an analysis that is 240 pages in length with a separate, 180-page appendix. The analysis contains dozens of interlocking assumptions, some of which are understandable only to a small group of social scientists or natural scientists. The RIA includes more than four dozen tables and figures that provide the raw numbers to support EPA's quantification and monetization calculations. It also provides the results of two independent studies EPA conducted to quantify the uncertainties surrounding its health benefits quantifications.

From the standpoint of bureaucratic rationality, the RIA is impeccable. Under Executive Order 12866, EPA is required to provide the Office of Management and Budget (OMB) with a cost–benefit analysis for all significant rulemakings. EPA does this—perhaps with the assistance of OMB itself—in ways carefully crafted to support the viability of a highly controversial rule in both the media and the courts. It also puts its strongest opponents on the defensive by adopting most of their preferred methods and assumptions to support its conclusion that the CAIR is an indisputable social bargain, costing $1 for every $25 in returns.

Before describing the specific ways in which the RIA is bureaucratically brilliant, it is important first to underscore the contentious environment from which the CAIR emerged. As the first chapter explains, a long history preceded the CAIR and influenced its development. Congressional and presidential efforts to design new legislation to address the problem of nonattainment of the National Ambient Air Quality Standards (NAAQS) and other, related problems from utilities had failed or appeared likely to fail. The only viable option left amidst the rubble was EPA's own proposal, the CAIR, as well as two partner rules, the Clean Air Mercury Rule and the Clean Air Visibility Rule. Few would view any of these rules as the "best" solution; each constituted an uneasy compromise among competing interests. But these rules did offer one virtue lacking in the legislative proposals—the prospect of implementation.

The CAIR not only was an uneasy compromise solution to the problem, it was also a somewhat heroic regulatory intervention because EPA's explicit legislative authority to address interstate pollution problems under the Clean Air Act (CAA) was quite constrained. The CAIR was not of the ordinary type of rulemaking that responds to a discrete, deadline-driven command by Congress. Instead, in the CAIR, EPA was attempting to address—on its own initiative and without a congressional requirement that it do so—the extensive nonattainment problems in more than 28 states with regard to fine particulates and ozone (EPA 2004, 4580–81). In these states, at least part of the continuing nonattainment problems is the result of interstate transport of pollutants. For example, EPA observes that "the ozone levels floating into Maryland . . . actually exceed the new 8-hour ozone standard before any Maryland emissions are added" (McGuffey and Sheehan 2005, 67). For such interstate pollution problems, a national or regional approach is warranted, and EPA was the ideal agency to develop such a large-scale strategy.

Because the states have the ultimate authority under the CAA to determine how to attain the NAAQS, however, EPA needed to propose its regional solution in a firm but noncoercive way. EPA's primary vehicle was its somewhat obscure State Implementation Plan (SIP) Call authority, which allows it to require certain states to make emissions reductions when they are significantly affecting attainment in a downwind state (42 U.S.C. § 7410(a)(2)(D)). In fact, the CAIR SIP Call requires states to cut emissions of sulfur dioxide (SO_2) and nitrogen oxides (NO_x) roughly in half (EPA 2005c, 25172–73).

One option EPA offers states—the most obvious and probably the least expensive—is the adoption of its cap-and-trade market for electric utilities, which automatically satisfies its reduction requirements. Electric generating units (EGUS) are an especially good target in this regard because their emissions are very high and arguably excessive given the capabilities of existing control technologies (Ibid.).They are also already trading in a national SO_2 market, so some of the basic features of the cap-and-trade program are in place and time-tested. If states choose not to join these regional markets, they must then create their own state cap-and-trade market or otherwise determine how best to meet the reduction requirements by EPA's deadlines and must adjust their SIPS accordingly (Ibid., 25167).

EPA's rule is legally precarious, however. To make the case for a SIP Call, EPA must establish, among other things, that each state subject to the Call (in this case, 28 states) is "contribut[ing] significantly" to the problematic pollutants (ozone precursors and fine particulates) in downwind states (42 U.S.C. § 7410(a)(2)(D)), and that EPA's SIP Call solution utilizes "highly cost-effective" controls to mitigate these interstate impacts (EPA 1998, 57363). This first showing—that there is a connection between the "significant" contributions of an upwind state to a downwind state—is largely outside of the reach of a cost-benefit analysis. Indeed, it was this feature of the CAIR that led to its reversal by the DC Circuit (*State of North Carolina v. US EPA* 2008a).

The CAIR RIA focuses its firepower instead on supporting the latter requirement—namely, that EPA's interstate trading program will result in "highly cost-effective" controls for problem pollutants—and accomplishes this by showcasing the CAIR's large benefit-to-cost ratio. Yet in touting these economic virtues of its program, EPA also shrewdly crafted its RIA in a way that dodges and even attempts to conceal fundamental assumptions inherent in the CAIR which leaves it vulnerable to attack. For example, EPA studiously avoided conceding the possibility that trading could lead to hot spots or areas that did not improve with respect to emissions (EPA 2005a, 8-21–8-22). Such evidence would have only resurfaced in the briefs filed by those challenging the CAIR. By instead circling its analytical wagons around the CAIR and side-stepping a meaningful alternatives comparison, EPA avoided making itself unnecessarily vulnerable to a number of attacks that lay just on the horizon (*State of North Carolina v. US EPA* 2008a; cf. *Motor Vehicle Manufacturers Ass'n v. State Farm* 1983).

EPA's strategic use of the RIA as an advocacy document is evident not only from the RIA's bottom line, but also from a number of individual, methodological decisions that are discussed in more detail below.

Consider only one policy alternative and make sure it produces a whopping benefit-to-cost ratio in support of the rule.

Perhaps most dazzling from the standpoint of blatant agency advocacy, the RIA only considers EPA's proposed option against the status quo. A robust and insightful policy analysis, as discussed below, would of course consider at least a handful of alternatives in relation to one another. But in the CAIR RIA, by examining only one policy option, EPA positions its final rule as a legal and political no-brainer. The resulting benefit-to-cost ratio of as much as 40 to 1 makes it difficult for the administration to reject the rule as bad public policy and complicates the ability of regulated parties, who shoulder the costs of the CAIR, to complain publicly about the rule. Even better, by presenting a regulatory option that produces such a high benefit-to-cost ratio, EPA helps buttress its legal case that its proposed reductions are "highly cost-effective."[2]

Not only does EPA set up a comparison deliberately designed to make its option look fabulous from a political and legal standpoint, but by providing such a limited glimpse of the policy alternatives, EPA reduces the risk of being hung up in litigation about the viability of close competitor approaches. Acknowledging and, worse yet, providing rigorous documentation of alternative approaches would only provide fodder for its many critics and exacerbate its vulnerability in the litigation that was inevitably to follow. In fact, in light of such an overwhelmingly strong benefit-to-cost justification, industry opponents will be quick to do the math and realize that EPA could have justified additional reductions and still supported its rule with a very favorable benefit-to-cost ratio, but declined to do so, presumably in part to protect the interests of industry (Graham 2007, 183; Graham 2008, 473–74; Shore and Patton 2004, 4). Utilities, which are the parties most affected economically by the rule, may thus find EPA's analysis of alternatives to be woefully incomplete, but only in ways that accrue to their benefit.

Thus, although EPA's one-option RIA violates the rules of robust policy analysis, it satisfies the general counsel's aim of limiting litigation risks. By supporting its rule with extensive, flattering analysis and studiously avoiding conceding and documenting other equally appealing alternative approaches, EPA transforms this bureaucratic speed bump into a rulemaking asset.

Use opponent-friendly assumptions to minimize the rule's vulnerability to attack as not "highly cost-effective."

To shore up the vulnerable features of its favorable benefit-to-cost calculation, EPA uses relatively conservative (utility-friendly) assumptions in monetizing the social benefits of the CAIR. Although the utilities take issue with some of the scientific assumptions about the relationship between sulfur dioxide (SO_2), nitrogen oxides (NO_x), and health effects, these same opponents to the CAIR are noticeably silent about the monetization-related decisions EPA makes for the social benefits (EPA 2005b, 58–90, 128–29, 922–26). This silence occurs for good reason: EPA's methods uniformly and sometimes heavily tilt in favor of the utilities' narrow interests. For example, EPA acknowledges a long list of very substantial social benefits that will result from decreased emissions of SO_2 and NO_x at the levels specified in the CAIR, but only a subset—perhaps fewer than half of these benefits—are quantified. The rest are only referenced qualitatively as "+ B" on the benefits side of the monetized benefit-to-cost ratio used to justify the rule (EPA 2005a, 1-10). If these benefits could be quantified and monetized, however, the true ratio of benefits to costs would be greater, perhaps as much as double (EPA 2005b, 232–66, 929–30).

EPA's actual monetization of the health benefits it does quantify similarly uses assumptions that tilt in industry's favor. EPA relies on standard willingness-to-pay (WTP) estimates for death and chronic bronchitis, but for most of the remaining health harms, EPA explicitly underestimates their value, relying only on the costs of treatment and lost wages to calculate monetary values (EPA 2005a, 4-14). Finally, EPA conservatively monetizes this subset of social benefits that it is able to quantify and discounts the resulting benefits at both the 3 percent and 7 percent discount rates for 2010 and 2015, thus appeasing even the most vigorous proponents of discounting (Ibid., 4-52).

EPA is also conservative in calculating the costs of the CAIR to utilities, making assumptions that generally overestimate compliance costs (Ibid., 7-19). In so doing, EPA acknowledges that past experience with markets and industry compliance reveals that its compliance cost estimates are likely to be overstated, perhaps significantly (Ibid., 7-19–7-21). EPA also assumes that no innovations or unexpected cost reductions in pollution control technology will occur, even though one could credibly make those adjustments if one were interested in the most accurate (mid range) estimate of compliance costs (Ibid., 1-12).

The result is an explicitly inflated estimate of compliance costs associated with the CAIR set against an explicit undervaluation of social benefits. Those who tout economic analysis as an important regulatory tool and view the rule from the perspective of hostile utilities are forced—by their own arguments and philosophies—to be mollified by EPA's conservative approach to evaluating the costs from the CAIR.

Reinforce the most vulnerable features of the decision with rigorous uncertainty analyses.

Health benefits were an important part of justifying the rule, constituting more than 90 percent of the monetized benefits, but they were scientifically difficult to establish. Determining first how

SO_2 and NO_x reductions from utilities translate into reduced ambient levels of particulates and ozone, and then predicting how these reductions will lead to fewer deaths and health harms requires a number of fragile and uncertain scientific assumptions. On this score, however, adopting utility-friendly assumptions about the relevant science would diverge too far from mainstream scientific views (some utilities appear to contend that neither of these links is supported by the available science) and might force EPA to contradict some of its earlier scientific conclusions (EPA 2005b, 58–90, 128–29, 922–26).

Therefore, to make its rule less vulnerable to this anticipated line of attack against the causal linkages it felt compelled to make, EPA used both Monte Carlo probabilistic analyses and expert elicitation to identify the extent of uncertainty surrounding its quantification of this limited set of health benefits (EPA 2005a, 4-77–4-83). These uncertainty analyses were performed only on the quantified and ultimately monetized health–benefits, thus capturing only a slice of the total social benefits at issue in the CAIR (Ibid., 1-6).

Fortunately for EPA, both of these uncertainty analyses confirm that its quantifications are well within the middle of the range of only moderate uncertainty (Ibid., 1-6–1-8). Moreover, the clever use of both expert elicitation and cutting-edge Monte Carlo analyses provides a redundant measure of the residual uncertainties that effectively shifts the burden to the challengers to identify concrete problems with EPA's scientific analysis. Although Keohane points out in Chapter 3 that EPA's uncertainty analyses could have been still better, EPA's elaborate use of uncertainty analysis to protect its rulemaking soft spot is nevertheless a telling indication of its strategic use of the RIA process.

Don't sweat the stuff that won't be used against you.

At least two executive orders, governing children's health and environmental justice, are arguably triggered by the CAIR, but they are skillfully interpreted by EPA to be inapplicable (Executive Orders 13045 and 12898). EPA's move is skillful because these requirements are not of concern to the industry challengers and addressing them could begin to unravel EPA's preferred option. For example, the possibility of hot spots arising from the hoarding of allowances by some utilities seems to raise, at least in the abstract, the possibility of both environmental justice problems and undue health impacts on subsets of children (Ackerman and Heinzerling 2003, 142–45; Drury et al. 1999). EPA concludes, however, that because its required emissions reductions will result in a net improvement in public health, there is simply nothing to analyze with respect to these two executive orders: everybody will be better off (EPA 2005a, 8-21–8-22).

In summary, the CAIR RIA reflects EPA's excellent strategic skills: EPA followed the maxim "know thine enemies" and built its RIA to please them. It thus seems ironic that despite EPA's determination to develop an analysis that supported the wisdom of the CAIR, its rule was ultimately dragged through the court system and reversed by the DC Circuit (*State of North Carolina v. US EPA* 2008a). Does this suggest that using the RIA as an advocacy document failed or was unwise? Definitely not. Such a conclusion is far too simplistic and presumes a one-issue character for a highly complex and multi-faceted regulatory proposal. In invalidating the CAIR, the court found a number of legal errors related to EPA's reliance on an interstate trading scheme to solve nonattainment problems in individual, downwind states, errors that cumulatively made the program "fundamentally flawed" in the court's view (*State of North Carolina v. US EPA* 2008a, 58–59). Yet none of these statutory lapses were even remotely connected to whether the CAIR would require controls that were "highly cost-effective," the central argument advanced by the RIA.

Indeed, the United States' argument for rehearing and the court's subsequent decision to allow the CAIR to proceed while EPA revises its rule may be due in part to the fine advocacy work embodied in the RIA (DOJ and EPA 2008; *State of North Carolina v. US EPA* 2008b). The court ultimately decided that vacating CAIR was not in the best interests of the country given the tremendous health and related benefits associated with the program's promised (and highly cost-effective) reductions in criteria pollutants (*State of North Carolina v. US EPA* 2008b). By contrast, if the litigation had turned on a finding that the compliance costs were not justified by societal benefits or that the controls were not "highly cost-effective" (issues that the RIA arguably took out of contention), then vacating the CAIR may have been legally necessary since those prejudiced by it would be the utilities and others bearing the brunt of the expense of the program.

The CAIR RIA as Policy Analysis

The CAIR RIA's value in advancing policy analysis is much less impressive. Indeed, the "A" or "A+" that EPA earned in the previous, strategic bureaucratic category drops to an "F" in the policy analysis category. And, consistent with the potentially mutually exclusive nature of the litigation support and policy analysis goals, virtually all of the assets or positive features of the RIA from the standpoint of bureaucratic rationality are mirrored by weaknesses in terms of policy analysis.

The best gauge of the RIA's success as a policy instrument is to assess whether it meets the general objectives set for it by cost–benefit analysts. According to adherents of cost–benefit analysis, or CBA, the advantages of CBA (and cost effectiveness analysis, or CEA) include:

- transparency and the resulting potential for engendering accountability;
- the provision of a framework for consistent data collection and identification of gaps and uncertainty in knowledge;
- the development of metrics for both the beneficial and adverse consequences of alternative regulatory approaches, allowing those alternatives to be compared to one another (CEA); and
- with the use of a monetary metric, the ability to aggregate dissimilar effects (such as those on health, visibility, and crops) into one measure of net benefits.

Compared against this list, the CAIR RIA is a major disappointment. Indeed, it provides almost a textbook example of how not to do cost–benefit analysis. At least a few of the most egregious problems are outlined below.

Considering Only One Alternative

The biggest strength of the RIA in terms of bureaucratic survival is also its primary weakness as a matter of policy analysis. From this standpoint, it is essentially disqualifying to consider only one alternative and, in fact, a one-alternative approach to policy analysis is flatly rejected in the RIA's analog, the environmental impact statement (EIS) requirement of the National Environmental Policy Act (NEPA) (42 U.S.C. § 4332(C)(iii)) statement; in the OMB's RIA guidance (OMB 2003

16); in the mainstream policy analysis literature (Keeney 1996); and in the adherents' objectives for cost–benefit analysis quoted above.

Moreover, EPA simply has no excuse for its decision to consider only one alternative. It would not have been that difficult to design a few easy-to-analyze alternatives in its RIA. For example, EPA could have considered alternative caps and deadlines, as some of the commenters (including Environmental Defense Fund (EDF)) did in their comments (Shore and Patton 2004, 10–19; EPA 2005b, 232–66, 385–94), as well as considering other sources, in addition to electric generating utilities, as potential participants in the interstate market, leading to even more ambitious reduction targets (Shore and Patton 2004, 20–24; EPA 2005b, 152–76). Ideally, EPA could have also considered alternative strategies to its model market that did not involve pollutant trading, such as strict emissions reduction requirements for certain sectors of industry and for transportation.

Indeed, subsequent reports on the CAIR reveal that this alternatives analysis was undertaken and vigorously debated within the executive branch; yet there is scarcely a trace of these critical choices in the RIA itself. For example, the administrator of the Office of Information and Regulatory Affairs (OIRA) at OMB during the CAIR RIA, Dr. John Graham, reports that EPA and OIRA advocated a more stringent reduction target (90 percent) for sulfur emissions, but the White House rejected their proposal and prevailed on the CAIR (Graham 2007, 183; Graham 2008, 469–70). Yet this 90 percent reduction target, the preferred alternative for both EPA and OIRA, is not mentioned, much less analyzed, in the RIA.

Similarly, OIRA advocated that additional sources, like industrial boilers, be included in the rule to accomplish more reductions, and OIRA even went so far as to meet with some of these sources to discuss the proposal (Graham 2008, 473). The RIA, however, makes no mention of the possibility of including these sources, nor does it consider how their inclusion would affect the costs and benefits of the rule. There was also considerable internal executive branch analysis about the extent to which the CAIR would reduce mercury emissions, with internal estimates that CAIR would "reduce mercury emissions to 34 tons by 2020" (Graham 2007, 184). Again, mercury reductions are listed only cursorily in the RIA and these reductions are never quantified or analyzed (EPA 2005a, 1-9).

The distinct possibility that EPA was deliberately holding back on sharing its extensive analysis of alternatives in the RIA is further evidenced by EPA's publication of a very elaborate alternatives analysis of the CAIR as part of a legislative briefing only seven months after the RIA was finalized (EPA 2005d). This EPA analysis compares the CAIR against five separate legislative proposals and analyzes alternate emissions reductions of NO_x and SO_2; alternate attainment rates for fine particulates and ozone; and the effects of the six alternatives on coal production, greenhouse gas emissions, mercury emissions, and renewable generating capacity.

An honest consideration of alternative approaches in the RIA might have also helped EPA develop a rule with a much closer fit to efficiency, if that was the ultimate goal. EDF conducted an analysis on marginal costs and benefits that revealed that EPA may not have selected the most efficient alternative (EDF 2004, 7–8), a problem that also concerned OIRA and EPA staff (Graham 2007, 183; Graham 2008, 473–74). Intuitively, in fact, with a benefit-to-cost ratio of about 25 to 1, there should be some room for further reductions in emissions to reach an efficient policy endpoint (Shore and Patton 2004, 4). Given the CAA's strong legislative commitment to public health protection, moreover, it is conceivable that Congress (and the public) would expect an outcome in which the quantified marginal benefits at least equal the quantified marginal costs, particularly if the benefits side is the one that is only partially quantified.

The disconnect between EPA's approach and that suggested by rigorous policy analysis again provides support for the observation that the RIA is intended only to support the final CAIR rule and insulate it from opposition from the courts, the White House, other agencies, and Congress. The RIA makes no pretense of seriously analyzing EPA's approach for efficiency—or any other social goal, for that matter. EPA concedes as much in its conclusory response to commenters who criticized it for its one-alternative approach:

The EPA conducted extensive analyses to determine highly cost-effective control levels, and the optimal criteria for significant contribution determinations. The CAIR will result in significant air quality improvements, reductions in the unhealthy levels of $PM_{2.5}$ [fine particulate matter] and for many areas of the CAIR region, and is highly beneficial to society (EPA 2005b, 929).

Potentially Huge Unquantified Benefits

EPA admits that the long list of unquantified social benefits could overwhelm the benefits it does quantify and then monetize, but it still ignores them in conducting its net cost–benefit calculation (EPA 2005a, 1-9, 4-22, 4-24). Ecological benefits are perhaps the most significant in this lengthy list of unquantified benefits (Ibid., 1-10). Fine particulates, ozone, nitrates, and sulfates introduce stressors that can lead to a variety of known and untraced adverse consequences for ecosystem health and productivity (Ibid., 4-70–4-73). In cases where these pollutants are reduced substantially—more than halved in the case of the CAIR—agricultural and ecological benefits could be substantial because the affected ecosystems support great expanses of crops, forests, water supplies, and fisheries. A number of those who submitted comments, in fact, underscored the significance of these unquantified ecological benefits in their comments on the CAIR (EPA 2005b, 918–22, 926–27).

Ultimately, however, because of the methodological difficulties that afflict measurements of these ecological benefits, EPA determined that they could not be quantified reliably. (EPA was also unable to quantify all health benefits, such as premature mortality from short-term ozone exposures [EPA 2005a, 4-26].) Thus, although EPA acknowledged that "[t]he net effect of excluding benefit and disbenefit categories from the estimate of total benefits depends on the relative magnitude of the effects" that remain unknown but are potentially significant, the agency perceived it had no choice in this deterministic monetization exercise other than to simply assign this huge set of unknowns a placeholder value of "+B." Consistent with the nature of the cost–benefit assignment, moreover, these unquantified variables inevitably dropped out of the monetized cost–benefit comparisons (Ibid., 4-22).

From the standpoint of inevitable litigation, EPA's methodological approach—bracketing and then effectively ignoring these large and uncertain categories of unquantified social benefits—puts a heavy thumb on the side of industry in the RIA analysis that should only help, or at least not hurt in subsequent opposition by that sector. Indeed, by virtue of its approach of considering only one option against the status quo, these added unquantified benefits are effectively irrelevant in any event. Once EPA has justified its proposal as clearly better on the basis of the benefit-to-cost ratio, the fact that even more benefits would accrue than were initially imagined is gravy.

With regard to policy analysis, however, EPA's decision to ignore a large portion of the benefits because they cannot be quantified is untenable. To identify the most efficient policy option, as noted above, several alternatives need to be considered; this, in turn, means putting all of the benefits on the table to identify the appropriate point at which the marginal benefits and marginal costs are in equipoise. If a large, or potentially large, portion of the benefits cannot be quantified and are removed from consideration, then the analysis is badly incomplete and will not yield the correct outcome. In fact, when a good portion of the benefits are unquantifiable, prominent economists maintain that cost–benefit analysis is no longer appropriate (Morgenstern and Landy 1997, 455, 465, 472, 476).

The goals of policy analysis are also slighted by EPA's decision to simply mark the unquantified benefits with a placeholder (+B) and ignore them, without proposing how or whether to reduce these uncertainties in the future. If EPA had been serious about policy analysis, it would have recommended ways to collect information on these large sources of uncertainty. The opportunity to underscore important areas of future research and data collection would have been a major benefit of conducting such an analysis. This future information collection could involve, for example, linking measures of agricultural or ecosystem productivity to changes in the concentrations of ozone and fine particulates over time.

Excessive Quantification of the Remaining Subset of Benefits

It follows from the previous weakness that if EPA cannot even be sure it has quantified the bulk of the benefits, subsequent monetization of the remaining quantified benefits becomes practically useless. If $(x+y) =$ *social benefits,* and y is unknown but is potentially large and perhaps even greater than x, then excessive efforts at monetization of x is not going to move the ball forward in finding the efficient balance point where marginal benefits meet marginal costs. This is not meant to suggest that the appropriate remedy is for EPA to simply put more resources into quantification of y (or in EPA's terminology, +B), however. EPA persuasively made a case that the ecological benefits were so difficult to predict, both qualitatively and quantitatively, that any estimation would amount to an unverifiable guess. The appropriate response to these quantitative problems is to acknowledge them and abort efforts to arrive at aggregate, monetized costs and benefits.

Indeed, to nevertheless persist with incomplete quantification in such circumstances is not only analytically corrupt, but also, undercuts statutory commands in the Clean Air Act that EPA err on the side of protecting the public health and welfare (42 U.S.C. § 7409(b)). As Doug Kysar points out in Chapter 10, an inflexible commitment to monetization in cost–benefit analysis, particularly when quantification cannot be accomplished, causes the innumerable scientific unknowns arising with respect to health and environmental harms to be zeroed out and chalked up against public health and environmental protection, despite the fact that in most cases the authorizing statute (and the public) has adopted the opposite value choice.

Nevertheless, in its RIA, EPA engages in this "fifth-significant-digit" sort of analysis for only a section of the health benefits and then provides an excruciatingly detailed monetization of that subset of benefits to compare against the industry costs. Except for the promotional benefits achieved by boasting of a 25 to 1 ratio for the CAIR, it is difficult to find any analytical value in this added monetization exercise. In fact, a good argument could be made that monetizing half of the benefits (or some unknown portion) only makes the analysis that much more misleading and confused.

EPA's obsession with the precise quantification of a subset of benefits begins by mounting a long succession of scientific hurdles to produce a final, error-prone quantitative estimate of a portion of the health benefits. In fact, EPA acknowledges that this final estimate of a subset of health benefits is scientifically precarious:

[M]any inputs were used to derive the final estimate of [a subset of health] benefits, including emission inventories, air quality models (with their associated parameters and inputs), epidemiological health effect estimates, estimates of values (both from WTP [willingness-to-pay] and COI [cost-of-illness] studies), population estimates, income estimates, and estimates of the future state of the world (i.e., regulations, technology, and human behavior). Each of these inputs may be uncertain and, depending on its role in the benefits analysis, may have a disproportionately large impact on final estimates of total [quantified health] benefits. For example, emissions estimates are used in the first stage of the analysis. As such, any uncertainty in emissions estimates will be propagated through the entire analysis. When compounded with uncertainty in later stages, small uncertainties in emission levels can lead to large impacts on total [quantified] benefits (EPA 2005a, 4-19).

In describing these sources of error, however, EPA's analysis is highly technical, almost always opaque, and at times contradictory. For example, although the agency seems to concede significant sources of error in its predictions of ambient concentrations (Ibid., 3-1, 3-5, 3-11, 3-20) and further concedes a number of contestable assumptions in predicting mortality from particulates (Ibid., 4-11), EPA still boasts in the executive summary of relatively precise estimates of a range of health effects, which contain no error bars (Ibid., Ch. 1 and 9). The agency also lapses into passages intended to illuminate these uncertainties, which are effectively indecipherable:

The procedures for determining the RFFs (relative reduction factors) are similar to those in EPA's draft guidance for modeling the $PM_{2.5}$ standard (EPA, 2000). This guidance has undergone extensive peer review and is anticipated to be finalized this year. The guidance recommends that model predictions be used in a relative sense to estimate changes expected to occur in each major $PM_{2.5}$ species. The procedure for calculating future-year $PM_{2.5}$ design values is called the "Speciated Modeled Attainment Test (SMAT)." EPA used this procedure to estimate the ambient impacts of the CAIR NPR [Notice of Proposed Rulemaking] emissions controls. The SMAT procedures for the No Further Remediation (NFR) have been revised. Full documentation of the revised SMAT methodology is contained in the Air Quality Modeling TSD [Technical Support Document] (Ibid., 3-12).

Although these impenetrable passages may alienate casual readers, they are unlikely to pose a barrier to resourceful utilities. And to defend against these more vigorous stakeholders' criticisms, EPA then takes its obsessive quantification of a subset of benefits one step further by commissioning not one, but two extravagant uncertainty analyses to shore up its estimates of a subset of health benefits. Monte Carlo probabilistic analyses and a separate expert elicitation ultimately confirm that EPA's quantification of a subset of health benefits is in the ballpark, at least within one order of magnitude (Ibid., 4-77–4-83). Yet although these analyses help protect EPA's CAIR from criticism that it is too costly in light of the benefits, from the standpoint of policy analysis they do little to nothing. Understanding the range of uncertainty surrounding x does little to clarify the uncertainty surrounding the larger set of social benefits $(x+y)$ or how to factor these social benefits $(x+y)$ into an evaluation of competing policy options (Ibid., 5-17).

Even assuming that it turns out (presumably through the agency's omniscience) that EPA's quantification of x is exactly right and that $y=0$, EPA's next quantification step—its monetization of x—consistently underestimates the dollar value of these remaining benefits, perhaps dramatically. Because EPA lacks WTP values for many of the health harms,[3] the agency values most illnesses tied to air pollution events, such as emergency room visits, heart attacks, and pneumonia, based only on generic estimates of lost wages and medical care costs (Ibid., 4-52 to 4-53). An emergency room visit for asthma is thus valued at a flat rate of $286 (Ibid., 4-53). Intangible pain and suffering, inconvenience, nonwage opportunity cost, and loss of consortium associated with these harms are all valued at $0 (Ibid., 4-14). As a result, a number of EPA's monetized calculations are not only inconsistent with one another, but are socially offensive. For example, according to EPA's RIA, a heart attack is worth more than twice as much if it occurs in a person aged 55–65 as compared with a child or young adult (aged 0 to 24) or a senior (over 65) (Ibid., 4-62). This is because youth and the elderly are generally not employed and there are no wage-related opportunity costs; only the medical costs of treatment. Equally disconcerting is EPA's monetization of the costs for a child who misses one day of school because of air pollution-caused illness. In this case, the only monetized costs arise for those children whose caretakers work. If one parent is unemployed, the value assigned to a missed day of school is $0 (Ibid., 4-65).

In a rigorous policy analysis (and putting aside the $x+y$ problem, which arguably makes such efforts to monetize x a nonissue in the first place), EPA presumably would have noted or even highlighted these inconsistencies in its valuation of different health harms and tried to redress them. Yet in the RIA, these disparities are noted only in passing and treated as largely irrelevant to the exercise, which they of course are if one assumes that this is a bureaucratic survival document prepared to protect the CAIR from opposition by the utilities.

To add insult to injury, EPA then proceeds to discount this subset of monetized health harms at two separate levels without offering any nondiscounted calculation or noting the legitimate disagreements about discounting nonmarket goods, especially in the intergenerational context (Heinzerling 1999a, 1999b, 1999c, Revesz 1999). The resulting discounted health benefits are presented in the executive summary at 3 percent and 7 percent discount rates with three significant digits and no error bars (EPA 2005a, 1-2).

Behavior of Market Participants

EPA makes note of the fact that cap-and-trade markets may not function exactly as planned because of unpredictabilities in market behavior, but it does not explain or explore what these sources of unpredictability might be or how they could affect its proposal. Instead, EPA flatly assumes in its analysis, without elaboration, "that all States in the CAIR region fully participate in the cap and trade programs" (EPA 2005a, 1-12).

A rigorous policy analysis would dedicate at least several pages or even a chapter to the viability of this assumption, particularly because a cap-and-trade approach is only one of several competing approaches available to EPA to recommend to states in the SIP Call and because unpredictabilities in trading behavior could affect the states' achievement of the reduction targets. In fact, as EPA well knows, one problem with its cap-and-trade proposal from the standpoint of meeting the health targets is the now-familiar "hot spot" problem, which occurs if a utility purchases large amounts of allowances and produces excessive emissions, thus creating dangerous concentrations of pollutants in localized areas.

Because the CAIR permits banking allowances, including the banking of pre-2010 SO_2 allowances, the hoarding of SO_2 allowances may also allow for high emissions over time in ways that may not have been adequately explored by EPA (EPA 2005a, 7-4). Such concerns about the effectiveness of the cap-and-trade market present tremendous legal and political vulnerabilities for the CAIR, however. As a result, EPA did not even acknowledge these potential sources of market slippage in its RIA, much less analyze their implications.

Costs to Utilities

EPA's estimates of the utilities' compliance costs are based on overly pessimistic projections that do not incorporate the insights gained from retrospective studies of compliance costs in general and SO_2 markets in particular (EPA 2005a, 7-19). The agency in fact notes that its compliance cost estimates in the past were overestimated by as much as 80 percent and concedes that all of the errors in its RIA tend in the overestimation direction (EPA 2005a, 7-19; McGarity and Ruttenberg 2002). It further acknowledges that changes in scrubber technology and in demand for electricity are both projected in ways that may cause the compliance cost estimates to be too high, perhaps significantly so (Ibid. 7-19–7-21). Yet EPA never explains why it makes sense to err in the direction of overestimating compliance costs, particularly when it appears that the overestimates may be more than double the expected costs based on the literature.

The Bigger Picture

Utilities contribute not only ozone precursors and fine particulates, but also mercury and greenhouse gases to the airshed; neither of the latter two important pollutants is analyzed in the RIA.[4] Unlike some of EPA's other decisions, this limitation in the scope of the RIA is more defensible: one must find a stopping point, and limiting the analysis to the two NAAQS-related pollutants seems reasonable.

On the other hand, from the standpoint of policy analysis, at least some mention of the important relationships among utility-generated pollutants is warranted. For example, in the RIA, EPA predicts that coal-fired plants will become more important in the future as a source of energy production (EPA 2005a, 1-11), despite growing recognition of their disproportionately high contribution of greenhouse gas pollutants as well as mercury. Might the multiple pollutants produced by utilities caution against cost–benefit solutions that are tipped so heavily in favor of the continued low-cost operation of coal-fired power plants? Do technological solutions exist that might not be much more costly but could begin to reduce the mercury or carbon dioxide (CO_2) emissions simultaneously with the control of SO_2 and NO_x emissions? Would it even be in the utilities' long-term interest to address these pollutants all at once in a predictable way?

In fact, some of the legislation that EPA compared with the CAIR in an economics briefing addressed these issues and required more renewables and considerably less dependency on coal (EPA 2005d, 15). This type of more unified cost–benefit analysis explaining how the proposal affects all four of these significant pollutants—or explaining how the other two pollutants intersect with the proposal—would seem to be far preferable in an RIA from the standpoint of useful policy analysis (EPA 2005b, 106–27, 422–40). A more comprehensive analysis would also help readers assess how well the president was meeting his pledge to reduce all four pollutants under the Clean Air Act (*see* Graham 2008, 470).

Equity

EPA shrugs off concerns about equitable disparities arising from its proposal by reassuring readers that everybody's health will improve as a result of the CAIR reductions (EPA 2005c, 8-21–22). However, this simplistic assurance begs the question of whether the improvements will be equitable. Will poor people in some areas pay a 2 percent increase in their electricity bills and receive in return very negligible health benefits, or even a reduction in health benefits, because they live in predicted hot spots surrounding some of the dirtiest participating utilities? At the same time, perhaps a number of wealthy communities might pay a similar 2 percent increase in their energy bills, but experience dramatically better air quality—maybe as much as a 10 percent improvement—because they are located in less utility-polluted areas.

The agency also fails to outline other demographic characteristics of those who are most likely to benefit from the reductions. Presumably the biggest beneficiaries are the elderly, children, and African Americans, who are particularly susceptible to asthma, but EPA never suggests that this is the case (Ibid. 4-50). It instead concedes only that the demographic characteristics are important and that it used "projections based on economic forecasting models developed by Woods and Pole, Inc.," without further elaboration of what these projections revealed (Ibid. 4-24).

Transparency

Among the touted virtues of cost–benefit analyses are the added transparency and political accountability they provide for submerged agency policy choices as well as their value in positioning the agency's preferred policy against credible alternatives. As the preceding discussions make clear, this noble goal of transparency has also been forsaken in EPA's RIA. The numerous blind spots and limitations of the agency's analysis—for example, considering only a single alternative—makes the resulting analysis quite opaque because it is completely unclear why EPA made many of the decisions it did. Moreover, even if its analysis were more complete, the 240-pages of technocratic explanations are generally accessible only to experts. Arguably, in fact, even these experts will be unable to evaluate the RIA without assistance from interdisciplinary expert teams because the RIA uses complex models from several disciplines, including the natural sciences, statistics, and economics.

One could argue that the crisp executive summary in the RIA nevertheless makes an important advance in communicating the policy bottom line of the CAIR in an understandable sound bite. In fact, the RIA does make it clear that EPA's CAIR is, in terms of benefits versus costs, better than doing nothing if one takes all the uncertainties in favor of the affected industry (Ibid, 1-1). But that is about all the RIA does say; it offers no insights about whether EPA's CAIR is the best policy among alternatives. Thus, to the extent that readers are left feeling as though EPA has struck the right balance in its rule after reading this short summary, they have been badly misled.

The Goal of the Analysis

The RIA is obviously framed around Executive Order 12866's requirement that agencies conduct cost–benefit analyses on influential rulemakings. However, in a high-quality policy analysis, this singular goal would be supplemented with analyses keyed to other policy objectives more relevant to the regulatory task at hand.

According to some policy analysts, in fact, the most important feature of policy analysis is the goal it selects to evaluate policy alternatives (Keeney 1996, viii, 1, 22). Professors Shapiro and Schroeder in fact persuasively document the legal irrelevancy of cost–benefit endpoints for most environmental mandates (Shapiro and Schroeder 2008, 436). If policy analysis should be done, they argue, it should be based, at least in part, on the goals of the statute (Ibid. 471).

Thus, if EPA's mandate for SIP Calls actually did require it to identify "highly cost effective" controls in upwind states, then its analysis should focus on evaluating alternatives based on this criterion. Social benefits in such an analysis would arguably be irrelevant; instead, the evaluation would consider only the capabilities of competing pollution control technologies and programs over a broad range of total emitters.

Yet although an extensive cost-effectiveness analysis is included in the preamble of the final rule (EPA 2005c, 25195–229), no cost-effectiveness comparison is provided in the RIA despite the centrality of this assessment to the statutory mandate and to economic analysis. EPA may simply have been out of time and unable to include it in the RIA. Or perhaps EPA decided that adding a chapter in the RIA on cost-efficiency and identifying the knee of the curve would only reveal the existence of equally cost-effective alternatives and distract from its otherwise clear and compelling message that the CAIR was justified by a benefit-to-cost ratio of at least 25 to 1.

At an even broader level, to the extent that efficiency—and Kaldor Hicks efficiency, to be precise—is the benchmark against which rulemakings are measured, acknowledging the stark limitations of this goal, particularly with regard to ensuring equitable outcomes, is essential (Ackerman and Heinzerling 2003; Markovits 2008; Sagoff 1981). Indeed, even if equity can still be qualitatively factored into the analysis, the limitations of efficiency as an end goal are too important not to at least acknowledge in the analysis.

Can EPA Navigate Its Rules through the Executive Branch and the Courts and Conduct Rigorous Policy Analysis at the Same Time?

Rigorous policy analysis and bureaucratic survival are not only two very different goals, but they may work at cross purposes. Before developing reforms for regulatory impact analyses, the first order of business is to consider whether the current agency incentive structure creates a viable environment for meaningful policy analysis under any circumstances. If, even with the best tools and guidelines, agencies will be strongly inclined to transform RIAs into advocacy documents, then reforms that focus exclusively on fine-tuning the agency's methods for conducting cost–benefit analyses will miss the target by a considerable margin.

Consideration not only of the CAIR RIA case study, but also of the literature more generally reveals multiple reinforcing reasons, both in theory and experience, to expect that a large number of RIAs will be prepared as self-serving, post hoc rationales rather than open, honest policy analyses. The best analog to the RIA requirement is the National Environmental Policy Act (NEPA), which requires agencies to analyze alternatives to their proposed projects (42 U.S.C. § 4332(c)(iii)). Almost 40 years of experience with NEPA reveals that, although its analytical requirements may help eliminate some of the very worst projects, much of NEPA's promise of probing policy analysis and agency transparency has given way to agencies that now "act as if the detailed statement called for in the statute is an end in itself, rather than a tool to enhance and improve decision-making," and turn the environmental impact statement (EIS) into a "litigation-proof" document that does not adequately raise or consider alternatives (CEQ 1997).

Yet in comparison to RIAS, NEPA analyses should be more complete and probing. Agencies conducting them may at least be sued for arbitrary fact-finding or the inadequate consideration of alternatives (*Calvert Cliffs v. Atomic Energy Commission* 1971).

When agencies conduct an RIA, however, the absence of judicial review means that OMB is the lone analytical police, at least in ensuring the RIA meets minimal standards of rigor and transparency. In some cases, obviously the CAIR RIA being one of them, this permits agencies to prepare RIAS that would not survive judicial review under NEPA because of fundamental analytical flaws, such as failing to consider more than one alternative in the analysis (OMB 2003, 16). This is not to suggest that judicial review should be required for RIAS, for the net benefit of judicial review under NEPA itself is highly contestable and uncertain, but only that experience with NEPA reveals that agency-conducted policy analyses are often not done in rigorous or honest ways.

A similar phenomenon has been observed in agency responses to the Administrative Procedure Act's (APA) notice and comment process. Several scholars have argued that, because notice and comment periods often allow interest groups to have a field day, agencies work behind the scenes to perfect their rules before they are published as proposed rulemakings (West 2004). The result is a great deal of backroom policymaking, including potentially extensive, unrecorded meetings with interest groups that fall outside the protections of the APA because they are done before the proposed rule is published (Ibid.). These APA-generated pressures, which include the prospect of judicial review, may be so great that in some cases agencies may avoid informal rulemaking altogether, particularly when they can make policy another way, through guidances or adjudications for example (Mashaw and Harfst 1987; Pierce 1991).

The possibility that RIAS might follow this same pattern of nontransparent, defensive policymaking observed under NEPA and the APA seems plausible, and the CAIR RIA only serves to reinforce this possibility in practice. In the CAIR, the RIA was published at about the same time that EPA issued its final rule. The RIA was thus finalized 2 years after the proposed rulemaking and 7 to 10 years into the larger policy exercise. The possibility that the RIA was going to provide an honest exposition of alternatives at this stage of the rulemaking seems not only naïve but fantastical, at least from a legal perspective. Under current circumstances, then, bad policy analysis may make for better policy outcomes, all things considered (Graham 2008, 182–83; Graham 2008, 469–74).

Yet the effect of these institutional forces on the RIA process may be even more perverse. Beyond providing agencies with rational incentives to transform their economic analyses into advocacy documents, institutional pressures may actually create a recurring substantive bias in favor of regulated parties and against environmental and health protection in these analyses. This arises from two overlapping features of the hostile institutional environment in which RIAS are produced. First, empirical research reveals that environmental and health rules tend to be challenged throughout their life cycle much more heavily by regulated parties than those representing the diffuse public who benefit from the protections (Yackee and Yackee 2006). These empirical observations are supported by theory; in the often arcane and highly technical area of environmental and public health regulation, regulated parties have the greatest stake in the rulemaking outcomes and also have the most resources to participate. Pluralistic processes in such a setting will be imbalanced just by virtue of the resource and stake equation (Gormley 1986; Komesar 1995).[5]

In such a lopsided, adversarial environment, agencies are more likely to adopt assumptions that routinely tip in the direction of industry or other powerful stakeholders that enjoy disproportionate influence in the political process. In the case of the CAIR, for example, OIRA demanded that EPA provide "less optimistic" alternative estimates of the benefits of sulfur reduction in its RIA. While it is possible that the conservative tilt of these resulting assumptions may have had nothing to do with efforts to appease industry and related skeptics, the "less optimistic" assumptions led to benefit estimates that were ten times lower than EPA's original estimates and triggered strenuous objections from some EPA staff (Graham 2007, 183; Graham 2008, 471).

The conservative tilt of the economic analyses is further exacerbated by a second feature: the litigation-driven inclination of agencies to quantify as much as possible in their RIAS to reach a flattering cost–benefit justification. Strong incentives for quantification help to insulate a rule from judicial and political controversy, but for environmental and public health rulemakings, this quantification bias will also cause agencies to bracket and effectively ignore unquantifiable features (as EPA did in the CAIR), almost all of which are most likely to arise on the social benefit side of the equation (Ackerman and Heinzerling 2003, 108–10). Again, the environmental and health benefits will tend to be undervalued in these litigation-motivated quantifications.

It does ultimately appear that RIAS on balance tend, as an institutional matter, to expose rulemakings to challenge by the most aggressive and resourceful stakeholders. If in response to this pressure agencies tend to tip their assumptions in a direction that protects them as much possible from these same future litigants, then the result is not simply an unhelpful policy analysis, but one that may be routinely biased against the public interest.

RIA Reform

In light of these perverse institutional forces, the most obvious and perhaps the only viable corrective is to devise ways to separate the RIA process as much as possible from judicial review and related political pressures on the agency. Ideally, agencies would be rewarded for conducting searching policy analyses or, at the very least, not penalized for self-critical and transparent consideration of alternatives.

If this type of safe analytical space cannot be created within the existing administrative structure, it may be counterproductive to recommend that agencies engage in analyses that only serve to make their already fragile regulatory decisions even more vulnerable to litigation and tipped to favor the most litigious parties. A tool intended for honest policy analysis will be transformed into yet another legal lever that the wealthiest stakeholders can use to pin the agency's rule to their preferred policy outcomes (Schmidt 2002).

This reform section begins with several tentative recommendations for agencies to engage in honest policy analysis without dooming their rulemakings in the process. I remain dubious that these reforms would be sufficient to coax the agency out into the analytical sunshine, however. The second part of the reform section then offers several tentative recommendations for substantive reform of RIA methodology that should accompany these procedural adjustments. Again, these recommendations are only a start and must be supplemented with other, creative reforms if RIAS are ultimately to become useful instruments for encouraging and assessing good policy.

Creating a Safe Administrative Space for Probing Policy Analysis

The first and most vital step for RIA reform is to correct the institutional disincentives—those emerging both from judicial review and the political process—that discourage agencies from conducting honest, searching policy analysis. With regard to the perverse incentives created by judicial review, one reform possibility is to reward an agency that conducts a rigorous RIA with superdeference (that is, a clearly erroneous standard) for its substantive policy choices. For example, in the case of the CAIR, if EPA had conducted an RIA that followed rigorous policy analysis guidelines or survived review by an expert advisory board, the court would attach a strong presumption in favor of the policy choices made in the rulemaking, well beyond the arbitrary standard. Only in cases where the agency does not conduct a rigorous RIA (as evaluated against a respected and established benchmark) would the agency's fact-finding and ultimate proposal receive a harder look with regard to the underlying assumptions and the record as a whole.

This judicial review safety zone would not only provide some reward to agencies that engage in a rigorous policy analysis, but also would protect them from having that analysis used against them. The courts have periodically taken a "hard look" at agency actions and have remanded rules when an agency offers an inadequate explanation for its selection of a particular alternative (among a larger set). This appears even more likely when the agency's own record suggests equally compelling alternatives (*Motor Vehicle Manufacturers Ass'n v. State Farm* 1983).

Under the proposal here, the agency would be freed of this type of judicial oversight if it provides a high-quality analysis of alternatives. In contrast to current practices of judicial review, then, the adequacy of the agency's actual explanation (or reasoned analysis) for selecting one option over others would receive great deference and would not be subjected to a hard look or even arbitrary and capricious review. Because explaining one choice over others involves primary political considerations, it is the area least amenable to judicial oversight in any event (*Chevron v. NRDC* 1984). Agency accountability under this recommendation is instead provided by the agency's rigorous and accessible evaluation of the available alternatives, including alternatives that on paper might appear more attractive to some stakeholders than others.

To enjoy the reward of superdeference, however, the agencies would have to first establish that they followed a set of respected guidelines for conducting policy analysis. Guidelines for these policy analyses could be issued by the National Academy of Sciences or some other neutral and respected expert body. The items discussed in the next section provide a start on the substance for such guidelines.

Even with this legal safe harbor, however, the political costs of coming out in the open with controversial policy choices may still lead agencies to shy away from conducting a truly searching policy analysis. As a result, agencies must also receive greater political insulation, which could be accomplished in part by requiring that the RIA or policy analysis process occur at a much earlier point in the rulemaking. An early RIA process would obviously focus the agency on alternatives analysis at a point in its rulemaking when it is still open to such options. Because the agency's RIA would be more preliminary, moreover, it would not need to provide rigorous cost–benefit justifications for each alternative, but could instead provide only crude approximations that are mostly qualitative. Even if the agency does attempt to quantify the alternatives, the time between the RIA and the final rule—the publication of a proposed rule, notice and comment, and final rule—will provide an opportunity for refinement and thus help distance the analysis from the final decision.

If it is also true that agencies are already conducting some of their policy analyses behind closed doors at an early stage in the process, then requiring an RIA at an earlier point may offer a more open and transparent forum for conducting these analyses. In the case of the CAIR, such an early RIA may have emboldened EPA to share its deliberations on alternative reduction targets, additional sources, and additional pollutants in a way that was more insulated from subsequent, strategic litigation decisions. Obviously, however, in situations where the White House or other political officials strike a deal at the outset regardless of a cost–benefit analysis, earlier opportunities for an RIA are unlikely to lead to more honest analyses. In situations where the deal has not been struck until later in the process, however, requiring an early RIA might give the agency a running start in analyzing alternatives in an open-minded way. Such a public analysis could also inhibit subsequent dealmaking, or at least limit the credible options available to political negotiators.

Substantive Adjustments to the RIA Process

Once agencies have the freedom to conduct more searching policy analyses, they may be able to develop creative approaches that not only meet, but exceed most analysts' wildest expectations. In this "let 1,000 analyses bloom" world of RIAs, the need for formulaic guidelines and rigid itemized lists of characteristics might become unnecessary and inadvertently chill creative and probing analysis.

Assuming, however, that some substantive direction is still needed—at the very least to identify when an agency's RIA should be entitled to superdeference—there are at least three substantive features of cost–benefit analysis (present at least in the CAIR RIA) that would benefit from reform. They are as follows:

Limit unreliable quantifications and eliminate monetizations of nonmarket goods to make the analysis more comprehensive and technically accurate.

Reform begins with an explicit rule that places a premium on presenting an analysis that provides a comprehensive snapshot of all costs and benefits and alternatives, leaving no significant costs or benefits off the table in the final decision.[6] Quantified benefits should not dominate when no information suggests that they, in fact, compose the most significant benefits. Charts such as those found in the CAIR RIA in the executive summary that add a "+B" to account for uncertain benefits would thus be strongly discouraged because they omit from the final analysis a potentially significant portion of the benefits.

In this setting of great uncertainty regarding both the quantification of social harm and its monetization, efforts to put only that subset of benefits that can be quantified into monetized units will result in an incomplete and potentially skewed factual basis for decisionmakers and could lead them down the wrong path. The powerful and familiar metric of dollars, for example, tends to anchor thinking in ways that are biased against remaining uncertainties. Partial monetization is also likely to fall prey to the "availability" heuristic: simple, but badly incomplete information—taking the form of cost–benefit price tags—overshadows the uncertainties and related complexities of the underlying tradeoffs (cf. Tversky and Kahneman 1974). Indeed, even if it is true that the incomplete, monetized benefit-to-cost ratio of the CAIR was crucial in changing skeptics' minds (Graham 2008, 472), the CAIR RIA reveals that in closer cases (which could be the lion's share of rules), unsound policies could result from the natural tendency of decisionmakers to focus primarily only on those benefits and costs that have been monetized.

Even if the quantification of all significant types of social harms could be accomplished reliably, however, monetization of the social harms remains highly problematic in a policy analysis setting because the monetization assumptions are inescapably value-laden, and at the same time are multilayered and difficult to coordinate with or explain to policymakers. As a result, the multiple assumptions used by EPA economic analysts, however well-meaning, can diverge significantly from what a policymaker might believe is appropriate. In the case of the CAIR, for example, EPA provides no indication in the RIA that an estimate of $0 for lost school days or a cost-of-illness approach for most health harms (which gives lower valuations to youth and the elderly) comports with the policy judgments held by decisionmakers or the public at large or is otherwise the best monetized choice among the large range of alternative valuations. Although including WTP for other health estimates at least provides some accounting of public valuations for these intangible costs, the available WTP studies, as well as their extrapolation to a particular question, are similarly value-laden and contestable, with a wide range of equally credible estimates.

At a more conceptual level, the legitimacy of using the market to value nonmarket goods has been vigorously challenged by a number of political scientists, ethicists, lawyers, and even some economists (Shabman and Stephenson 2000). Because the goods are not bought and sold on the market, identifying market-analog prices is highly artificial and also ignores the civic reasons that motivate citizens to protect the public goods or the general public welfare in the first place (Sagoff 1981; Shabman and Stephenson 2000). Moreover, using market analogs (such as revealed preferences or WTP) to isolate market prices for nonmarket goods is based on a number of rosy assumptions about consumer behavior and consumer valuations that are empirically tenuous at best and arguably refuted at worst (Shabman and Stephenson 2000, 383–84).

Set against these multiple problems with the monetization of nonmarket goods is the fact that more transparent and informative means can be used to educate policymakers about the comprehensive costs and benefits of policy options that do not require monetization as a prerequisite. Natural units can be used in place of monetized benefits and costs in ways that provide a much less assumption-laden picture of the societal consequences of a policy. Rather than monetize all of the various health benefits, these benefits are simply listed and set against the costs an individual consumer would be charged in their electricity bills. Empirical work in fact suggests that decisionmakers themselves prefer the more direct, natural unit measurements of the implications of policy alternatives. In the hydropower arena, for example, analysts have found that stakeholders actually prefer qualitative, nonmonetized projections as the basis for negotiation and decisionmaking and tend to ignore monetized valuations (Gowan et al. 2006).

In the CAIR, EPA could have taken at least two, complementary approaches to its economic analysis following these recommendations. The first and most straightforward approach for assessing whether EPA's proposed controls were "highly cost-effective" against alternatives would have been to prepare a cost-effectiveness analysis that considers only the cost per ton of emissions reductions in alternative scenarios (Figure 4.1). In this quantitative comparison, the analysts' main job is simply to quantify alternative compliance costs and their associated pollutant reductions;

Figure 4.1
Marginal Cost Curve of Abatement for SO_2 Emissions from EGUs in 2010
(NO_x Emissions at 1.5 million tons)

Reprinted from EPA Rule 2005c, 25204.

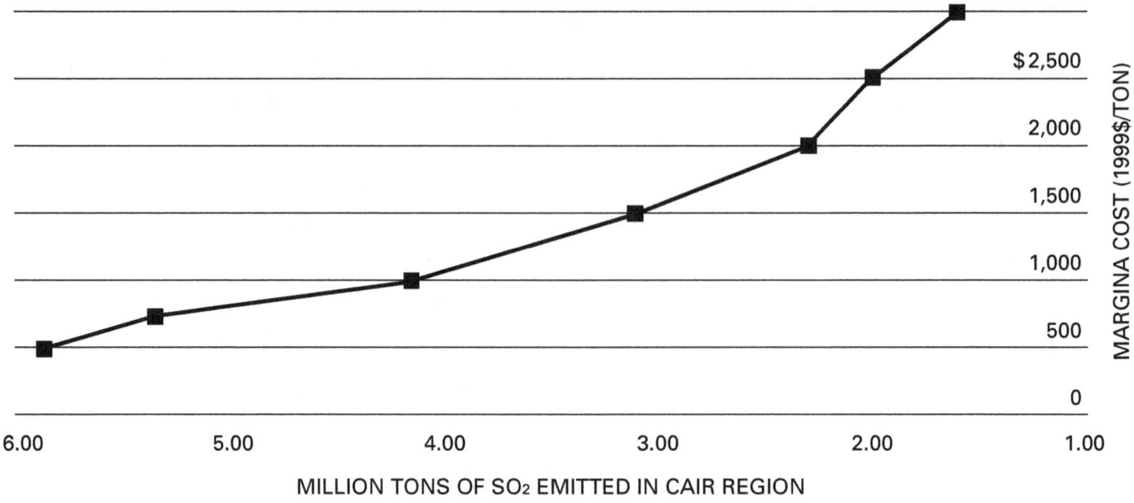

there is no need for quantification of social benefits, monetization of social benefits, or discounting. To provide a broader range of alternatives, EPA would also ideally consider sources in addition to the electric utilities, as well as other regulatory methods for accomplishing reductions, beyond markets.

An argument can be made, particularly in the wake of the CAIR litigation, that EPA's assignment was broader and included, but did not stop with, this assurance of cost-effectiveness. For situations when an aggregate analysis of all of the benefits in relation to the costs is deemed necessary, it is still possible to develop mixed quantitative–qualitative cost–benefit analyses that are much improved in technical accuracy, comprehensiveness, and transparency compared with the incomplete monetized assessment in the CAIR RIA executive summary. In this situation, EPA would summarize its rule using an alternatives table that lists the aggregated costs—ideally to the consumer—on one side compared against the significant quantified and unquantified (but not monetized) benefits, presented on the other side. See Table 4.1, which serves as an illustration (many of the numbers are fictitious; others are drawn from analyses done by Shore and Patton 2004, 14–15). Because this approach avoids the problems associated with monetization, it is substantially less error-prone and assumption-laden.

Indeed, this table (or something like it) actually provides the equivalent of a WTP matrix specifically tailored to the rulemaking that sidesteps all of the multiple, assumption-laden steps involved in monetizing the nonmarket goods and considers policy options more holistically (Ackerman and Heinzerling 2003, 210–12). Unlike with WTP surveys, moreover, policymakers would take direct responsibility for choosing the preferred option from the alternatives. The table could even be translated by economists and social scientists into surveys that provide more direct information about the general public's preferences for specific types of trade-offs. To provide more direct public input, the table of alternatives could also be submitted to citizen advisory groups or vetted in public hearings or through notice and comment. Even without these focused sources of public input, the increased transparency should create sufficient political and media pressure on policymakers to make the best choice.

Table 4.1
Alternative Scenarios in Natural Units

Cost to consumers and society at large	Benefits or harms averted
0.5% INCREASE IN ELECTRICITY BILLS OVER 10 YEARS	■ Health 5,000 (± 2,500) premature deaths 7,000 (± 3,500) nonfatal heart attacks 4,000 (± 2,000) hospital admissions 7,000 (± 3,500) acute bronchitis, children
	■ Ecosystem 3 million fewer tons of SO_2 into ecosystems 0.05 million fewer tons of NO_x into ecosystems
	■ Visibility 2% improvement
2% INCREASE IN ELECTRICITY BILLS OVER 10 YEARS	■ Health 13,000 (± 6,000) premature deaths 18,000 (± 9,000) nonfatal heart attacks 11,000 (± 5,000) hospital admissions 16,000 (± 8,000) acute bronchitis, children
	■ Ecosystem 5.5 million fewer tons of SO_2 into ecosystems 1.9 million fewer tons of NO_x into ecosystems
	■ Visibility 8% improvement
3.5% INCREASE IN ELECTRICITY BILLS OVER 10 YEARS	■ Health 19,408 (± 9,500) premature deaths 27,000 (± 13,000) nonfatal heart attacks 16,000 (± 8,000) hospital admissions 24,000 (± 13,000) acute bronchitis, children
	■ Ecosystem 6.5 million fewer tons of SO_2 into ecosystems 2.2 million fewer tons of NO_x into ecosystems
	■ Visibility 12% improvement
5% INCREASE IN ELECTRICITY BILLS OVER 10 YEARS	■ Health 21,000 (± 10,000) premature deaths 29,000 (± 15,000) nonfatal heart attacks 18,000 (± 8,000) hospital admissions 26,000 (± 13,000) acute bronchitis, children
	■ Ecosystem 7 million fewer tons of SO_2 into ecosystems 2.6 million fewer tons of NO_x into ecosystems
	■ Visibility 14% improvement

Note: Ecosystem benefits from reduced pollutant loading include benefits to commercial forestry, fishing, recreation, and ecosystem function.

Distributional impacts

If they are likely to be different for vulnerable subpopulations like the poor, the elderly, children, or minorities, then the demographics of those experiencing health effects (and benefiting from the reductions) at each option should also be included more fully in the analysis as well as in the summary tables. The distributional features of each alternative could be graphed as a pie chart of winners and losers in demographic categories or might be flagged only when the distribution of gains and losses, even in a localized area, is likely to vary among alternatives. However they are determined, the distributional implications of alternatives are critical to policy analysis and are missing in the standard methodology of aggregating costs and benefits (Graham 2008, 420–22, 516–24).

If these distributional impacts cannot be estimated quantitatively, then qualitative statements should be provided, both in summary form and in an expanded version, which explains how different vulnerable populations might be affected by the alternatives, not only in absolute terms, but also relative to one another. For example, if middle- and upper-class communities will experience the greatest health benefits under a preferred option relative to the poor, then this feature needs to be communicated. Conversely, if the benefits are likely to accrue primarily to asthmatics and sensitive subgroups, such as children and the elderly, then this important feature of the proposal deserves mention.

Adaptive learning

A number of commentators have criticized administrative processes for inadvertently discouraging agencies from collecting information over time, after a policy is in place (Blais and Wagner 2008; Doremus 2001; McGarity 1992). If RIAS are to accomplish their full potential, agencies must identify the types of technical information that would benefit from further study or collection and propose how this research might be accomplished. Unfortunately, and as discussed earlier, the CAIR RIA simply inserts what is known and moves on. Incorporating adaptive learning into the analysis will not only produce better short-term analyses, but also better policies and information in the long term.

Conclusion

The CAIR case study reveals that EPA has become quite adept at using the RIA to reinforce the wisdom of its rulemaking. That the RIA offers nothing to policy analysis is, in fact, precisely the point; in other words, the point is to protect the rulemaking, not to open it up to attack.

Reform of the RIA process must address this institutional reality. RIAS will never be produced under ideal conditions of academic solitude away from the pressures of litigation and political interference. Instead, a safe harbor for meaningful agency policy analysis needs to be created that insulates the agency from at least some of these institutional pressures. Until then, recommendations for methodological refinements of the RIA process are likely to fall on rationally deaf, bureaucratic ears.

■ ■ ■

Notes

1. I bracket the interesting but currently unanswerable question of who the primary proponent of the CAIR actually was. The evidence suggests, in fact, that the ultimate decision to select the CAIR over alternatives may have been an edict from the White House (Graham 2008). Because this project is focused on the analysis contained in the RIA itself, however, I do not attempt to speculate about the CAIR's origins.

2. In the CAIR, EPA considers the highly cost-effective controls to be those at the "knee of the curve." (EPA 2005c, 25201). Undoubtedly, controls with a 20 to 1 benefit-to-cost ratio are at or below this knee.

3. There may also be disagreements about EPA's values of death and chronic bronchitis that at least included some intangible, pain and suffering factor through WTP surveys. For example, EPA assumes life is worth about $5.5 million per life (in 1990$) (EPA 2005a, 4-52). Because these estimates at least include an intangible component, they are presumably less contestable on monetization grounds than many of the other harms discussed in the text.

4. The operation of coal-fired power plants also leads to an increase in worker deaths and health harms as a result of coal mining (EPA 2005a, 4-13) (alluding to worker safety, but not analyzing it).

5. High-salience rulemakings like CAIR and NAAQS-related rules are likely to generate interest from the full range of affected stakeholders, however, and thus these rules may ultimately be exceptions to this general rule of skewed participation by regulated parties (Blais and Wagner 2008).

6. I am assuming that economic analysis and efficiency remain the goal and bracket my own concerns about the wisdom of those objectives.

References

Clean Air Act, § 109(b), 42 U.S.C. § 7409(b).

Clean Air Act, § 110(a)(2)(D), 42 U.S.C. § 7410(a)(2)(D).

National Environmental Policy Act, § 102(C)(iii), 42 U.S.C. § 4332(C)(iii)

Calvert Cliffs v. U.S. AEC, 449 F.2d 1109 (D.C. Cir. 1971).

Chevron v. NRDC. 1984. 467 U.S. 837 (S.Ct.).

Motor Vehicle Manufacturers Ass'n v. State Farm Mutual Automobile Insurance Co. 1983. 463 U.S. 29 (S. Ct.)

State of North Carolina v. EPA. 2008a. slip opinion, to be published at 531 F.3d 896 (D.C.Cir.).

State of North Carolina v. EPA, 2008b. 2008 WL 5335481 (D.C. Cir.).

Executive Order 12866.

Executive Order 12898.

Executive Order 13045.

Ackerman, Frank and Lisa Heinzerling. 2003. *Priceless: On Knowing the Price of Everything and the Value of Nothing.* New York: The New Press.

Blais, Lynn and Wendy Wagner. 2008. Emerging Science, Adaptive Regulation, and the Problem of Rulemaking Ruts. *Texas Law Review* 86:1701.

Council on Environmental Quality (CEQ). 1997. *The National Environmental Policy Act: A Study of Its Effectiveness After Twenty-five Years.* Washington, DC: CEQ.

United States Department of Justice and U.S. Environmental Protection Agency (DOJ and EPA). 2008. Brief in support of Petition for Rehearing or Rehearing En Banc for *State of North Carolina v. U.S. EPA*. September 24. Washington, DC: DOJ.

Doremus, Holly. 2001. Adaptive Management, the Endangered Species Act, and the Institutional Challenges of "New Age" Environmental Protection. *Washburn L.J.* 41:50.

Drury, Richard Toshiyuki et al. 1999. Pollution Trading and Environmental Injustice: Los Angeles' Failed Experiment in Air Quality Policy. *Duke Environmental Law & Policy Forum* 9:231.

Environmental Defense Fund (EDF). 2004. Stop Blowing Smoke in the Heartland, available at www.edf.org/documents/3887_blowingsmoke.pdf

U.S. Environmental Protection Agency (EPA). 1998. Finding of Significant Contribution and Rulemaking for Certain States in the Ozone Transport Assessment Group Region for Purposes of Reducing Regional Transport of Ozone. *Federal Register* 63:57356.

———. 2004. Proposed Rule to Reduce Interstate Transport of Fine Particulate Matter and Ozone (Interstate Air Quality Rule). *Federal Register* 69:4566.

———. 2005a. *Regulatory Impact Analysis for the Final Clean Air Interstate Rule.* EPA-452/R-05-002. Washingon, DC: U.S. EPA, available at http://yosemite1.epa.gov/ee/epa/ria.nsf/vwRef/A.2005.4+A?OpenDocument.

———. 2005b. *Corrected Response to Significant Public Comments on the Proposed Clean Air Interstate Rule, Docket Number OAR-2003-0053,* March; corrected April. Washington, DC: EPA.

———. 2005c. Final Rule to Reduce Interstate Transport of Fine Particulate Matter and Ozone (Clean Air Interstate Rule); Revisions to Acid Rain Program; Revisions to the NO_x SIP Call; Final Rule, *Federal Register* 70:25162.

———. 2005d. *Multi-Pollutant Legislative Analysis: Comparison Briefing.* Washington, DC: U.S. EPA

Gormley, Jr., William. 1986. Regulatory Issue Networks in a Federal System. *Polity* 18:595.

Gowan, Charles, Kurt Stephenson, and Leonard Shabman. 2006. The Role of Ecosystem Valuation in Environmental Decision Making: Hydropower Relicensing and Dam Removal on the Elwha River. *Ecological Economics* 56:508.

Graham, John D. 2007. The Evolving Role of the U.S. Office of Management and Budget. *Review of Environmental Economics and Policy* 1: 171.

———. 2008. Saving Law through Administrative Law and Economics. *University of Pennsylvania Law Review* 157: 395.

Heinzerling, Lisa. 1999a. Discounting Life. *Yale Law Journal.* 108:1911.

———. 1999b. Discounting Our Future. *Land & Water Law Review* 34:39.

———. 1999c. Environmental Law and the Present Future. *Georgetown Law Journal* 87:2025.

Keeney, Ralph L. 1996. *Value-Focused Thinking: A Path to Creative Decision-making.* Cambridge, MA: Harvard University Press.

Komesar, Neil. 1995. *Imperfect Alternatives: Choosing Institutions in Law.* Chicago: University of Chicago Press.

Markovits, Richard. 2008. *Truth or Economics: On the Definition, Prediction, and Relevance of Economic Efficiency.* New Haven, CT: Yale Press.

Mashaw, Jerry L. and David L. Harfst. 1987. Regulation and Legal Culture: The Case of Motor Vehicle Safety. *Yale Journal on Regulation* 4:257.

McGarity, Thomas O. 1992. Some Thoughts on "Deossifying" the Rulemaking Process. *Duke Law Journal* 41:1385.

McGarity, Thomas O. and Ruth Ruttenberg. 2002. Counting the Cost of Health, Safety, and Environmental Regulation, *Texas Law Review* 80:1997.

McGuffey, Mark and Gary R. Sheehan. 2005. Taking Care of CAIR. *Natural Resources and Environment.* Summer 2005: 67.

Morgenstern, Richard D. and Marc K. Landy. 1997. Economic Analysis: Benefits, Costs, Implications. In *Economic Analyses at EPA: Assessing Regulatory Impact.* Edited by Richard D. Morgenstern. Washington, DC: RFF Press.

U.S. Office of Management and Budget (OMB). 2003. Circular A-4. Washingon, DC: OMB.

Pierce, Jr., Richard J. 1991. The Unintended Effects of Judicial Review of Agency Rules: How Federal Courts Have Contributed to the Electricity Crisis of the 1990s. *Administrative Law Review* 43:7.

Revesz, Richard. 1999. Environmental Regulation, Cost-Benefit Analysis, and the Discounting of Human Lives. *Colorado Law Review* 99:941.

Sagoff, Mark. 1981. At the Shrine of Our Lady of Fatima or Why Political Questions Are Not All Economic. *Arizona Law Review* 23:1283.

Schmidt, Patrick. 2002. Pursuing Regulatory Relief: Strategic Participation and Litigation in U.S. OSHA Rulemaking. *Business and Politics* 4:71.

Shabman, Leonard and Kurt Stephenson. 2000. Environmental Valuation and Its Economic Critics. *Journal of Water Resources Planning and Management* Nov./Dec. 2000: 382.

Shapiro, Sidney A. and Christopher H. Schroeder. 2008. Beyond Cost–Benefit Analysis: A Pragmatic Reorientation. *Harvard Environmental Law Review* 32:433.

Shore, Michael and Vickie Patton. 2004. Environmental Defense Fund, Comments on Proposed CAIR Rule to U.S. EPA. March 30 (available through regulations.gov for Docket OAR 2003-0053)

Tversky, Amos and Daniel Kahneman. 1974. Judgments under Uncertainty: Heuristics and Biases. *Science* 185:1124.

West, William F. 2004. Formal Procedures, Informal Processes, Accountability, and Responsiveness in Bureaucratic Policy Making: An Institutional Policy Analysis. *Public Administrative Review* 64:66.

Yackee, Jason W. and Susan W. Yackee. 2006. A Bias toward Business? Assessing Interest Group Influence on the Bureaucracy. *Journal of Politics* 68:128.

■ ■ ■

CHAPTER 5

The Clean Air Mercury Rule

DAVID A. EVANS[1]

In May 2005, the U.S. Environmental Protection Agency (EPA) adopted the Clean Air Mercury Rule (CAMR). The rule established a national cap-and-trade program for mercury from coal-burning electricity generating facilities (EPA 2005g). Mercury is a neurotoxin, fetal exposure to which has been shown to cause IQ loss and other developmental afflictions at relatively low levels. The program was to begin in 2010 and be implemented in two phases. From 2010 to 2017, the annual total allocation of emissions allowances was to be 38 tons. Thereafter, the annual allocation was to fall to 15 tons. For reasons described below, in *New Jersey v. EPA* (2008) the United States Court of Appeals for the DC Circuit vacated the CAMR in its entirety.

Executive Order (EO) 12866 requires all federal agencies to prepare an economic impact analysis of a regulation if, among other considerations, the expected annual effect on the economy is $100 million or more (Office of Management and Budget [OMB] 2007). The term "the economy" has been interpreted broadly and includes any benefit or cost of the rule (OMB 2003). The economic impact assessment is to be described in what is known as the regulatory impact analysis (RIA). The CAMR RIA, which was published two months prior to the adoption of the rule, is 566 pages long with 12 chapters and 4 appendices (EPA 2005a).

The core of the RIA is Chapter 10, which is 148 pages long and lays out the steps for estimating the improvements in health status resulting from reductions in mercury emissions. This chapter summarizes the major components of the RIA and its findings. The RIA also describes analyses conducted for compliance with other executive orders and legislation, such as an analysis of the impacts of the regulation on minority and low-income communities (environmental justice), and on state and local governments. The RIA summary also draws on the CAMR final rule *Federal Register* notice, which compares estimated benefits and costs, and explains why certain regulatory design choices were considered and subsequently adopted in the final rule.

In the next section, I summarize the legislative and regulatory history that led EPA to adopt the CAMR. This background provides context for understanding why certain regulatory approaches were analyzed for the RIA and others were not. Understanding the charged debate surrounding the CAMR also illuminates why certain methodological choices made for the RIA were subject to considerable scrutiny.[2]

Legislative and Regulatory Background

The 1990 Clean Air Act (CAA) Amendments required a substantial study of electricity generating units (EGUS) to determine whether they emit hazardous air pollutants (HAPS) that pose a risk to human health. EPA was then required to determine whether it was appropriate and necessary to regulate any HAP identified by the study, which was released in 1998. Mercury from coal-fired EGUS was determined to be the HAP of greatest concern (EPA 1997). Although much of the emitted mercury does not deposit in the United States, in some areas coal-fired EGUS contribute upward of 70 percent of deposition.

In the waning hours of the Clinton administration, EPA found or "listed" mercury from coal-fired boilers under CAA Section 112 (EPA 2001b). The existing regulations are prescriptive in nature and typically require emissions controls at all affected sources. In March 2005, EPA reversed course by delisting mercury from Section 112, choosing instead to regulate it under Section 111 (the catch-all New Source Performance Standards section). Under Section 111, EPA chose to use a cap-and-trade approach to control mercury from existing coal-fired boilers and, for new boilers, to additionally impose restrictions on their emissions per megawatt-hour generated.

This decision is controversial for at least two reasons. The first is simply that mercury is a toxic pollutant that may have local effects and therefore may be a poor candidate for unconstrained emissions trading on a national scale (Bellas and Lange 2005; Chapter 6 of this report). The second is based on a legal interpretation that mercury should be regulated as a HAP under Section 112, which arguably would require EPA to use prescriptive source-level regulations to control mercury emissions.

EPA addressed the possibility of regulating mercury under Section 112 by proposing a rule that would have required compliance with a performance standard (EPA 2004c). This proposal and the proposed rule under Section 111 (i.e., the cap-and-trade approach) were released at the same time.[3,4] Given EPA's interpretation of how to set the performance standard under Section 112, allowable emissions from the proposed 112 rule are less than those from the proposed cap-and-trade rule in the early years, but are greater in the later years. In part, it was this difference in the emissions level between the two proposed rules that EPA used to justify its delisting of mercury and the transfer of mercury to Section 111, which the agency interprets as more clearly allowing emissions trading as an emissions control option.

Key Elements of the Cap-and-Trade Rule

The key regulatory feature of the CAMR is that it is a cap-and-trade program. The federal government sets the cap, and the states must ensure that compliance with the cap is achieved. The states are provided a budget of allowable mercury emissions that represents an apportionment of tradable allowances that they may distribute to affected entities. Therefore, the sum of the states' budgets equals the cap. Each allowance provides a one-time right to emit an ounce of mercury. Each state's budget is based on a historic measure of the type and quantity of coal combusted in the state.

The states have discretion as to whether they will participate in the cap-and-trade program and, if they do, how and whether to fully allocate their budgets. If a state does not participate in the cap-and-trade program, it must adopt an alternative regulation that limits total mercury emissions so that they do not exceed the state's budget. Many states planned to take this alternative approach. Some were already planning to adopt stringent performance-based technology standards for coal-fired boilers when the CAMR was adopted.

Sources in states that participate in the cap-and-trade program under the CAMR may buy and sell rights to emit mercury (i.e., allowances) from one another. A source is in compliance with the CAMR if it holds rights to emit mercury in excess of its emissions. Trades are not spatially restricted, so it is possible and permissible for annual emissions within a state to exceed that state's budget for that year. Furthermore, plants may retain unused emissions rights for compliance in future periods. With the substantial decline in the annual allocation (cap) in 2018, sources were expected to bank their allowances before then and slowly use them up afterward.

Another key feature of the rule is that it does not distinguish among the forms of mercury emitted. Most mercury is emitted in one of three forms. *Elemental* mercury is relatively inert and has a wide deposition pattern, whereas *reactive* (also known as speciated, ionic, or oxidized) mercury deposits relatively close to its source (EPA 2005a). *Particulate* mercury is bound to fine particulate matter and also deposits close to its emissions source. Most of the mercury emitted by coal-burning facilities (greater than 97 percent) is in the elemental and reactive forms. Reactive mercury is easily converted to methylmercury in aquatic systems; methylmercury is the most biologically available form and causes the greatest damage. Although some elemental and particulate mercury converts to reactive mercury in the atmosphere or after deposition into an aquatic environment, the share of these forms converted to methylmercury is much lower than that for reactive mercury.

Common Features of Benefits and Costs Assessments

Regulatory Baseline

The regulatory baseline includes the existing state and federal air pollution control regulations adopted prior to April 2004. Most importantly, the regulatory baseline used for the benefit and cost analysis in support of the adopted CAMR also assumes the adoption of the Clean Air Interstate Rule (CAIR) (EPA 2005f).[5] The CAIR requires reductions in nitrogen oxide (NO_x) and sulfur dioxide (SO_2) emissions from coal-fired power plants in the eastern United States (EPA 2005b). These reductions are implemented through cap-and-trade programs.

The close timing of the adoption of the CAIR and the CAMR was not coincidental. EPA argued that the regulated sources could better plan their compliance strategies if the adoption of these rules was coordinated. SO_2 emissions controls reduce mercury emissions directly, and certain NO_x controls facilitate the collection of mercury by sulfur controls. For example, the combination of a wet scrubber designed to control sulfur emissions and a selective catalytic reduction system, which reduces NO_x emissions, can reduce mercury emissions from bituminous coal by up to 95 percent.

Regulatory Options

Three alternative regulatory scenarios were analyzed in the final mercury RIA.[6] The key difference is in the level of the annual allocation of mercury allowances from 2015 to 2017. Under each option, the annual allocation from 2010 to 2014 is 38 tons and, in 2018 and thereafter, 15 tons. Under Option 1, the annual allocation from 2015 to 2017 is 38 tons. Under Options 2 and 3, the 2015 to 2017 allocations are 15 tons and 24 tons per year, respectively. The costs of all three regulatory options are analyzed, whereas the benefits are estimated only for Options 1 and 2.

The RIA does not explain why these particular regulatory design options were analyzed; however, some explanation is provided in the CAMR final rule *Federal Register* notice (EPA 2005f). For example, 2015 serves as a benchmark year because this is when SO_2 and NO_x allowance allocations under the CAIR decline. The 38-ton allocation from 2010 to 2014 was chosen because this is the expected level of mercury emissions in 2010 from coal-fired boilers with the adoption of the CAIR. EPA was concerned that controls specifically designed to reduce mercury emissions from these plants, namely activated carbon injection, would not be commercially available until 2018, and thus retained the 38-ton allocation from 2015 to 2017.[7] However, this concern did not prevent EPA from including activated carbon injection as a control option in its modeling of the expected regulatory compliance with the first phase of the CAMR.[8]

The RIA also does not describe why only a 15-ton cap was considered for 2018 and beyond. However, President Bush's Clear Skies Initiative—which would have required reductions in SO_2, NO_x, and mercury from the electricity sector—also specified an annual allocation of 15 tons for this period.[9] The RIA and *Federal Register* notice also do not explain why total allocations of 24 and 15 tons were analyzed for the 2015 to 2017 period. Furthermore, these documents do not explain why the same key design features were imposed on all three regulatory options, such as banking, spatially unrestricted trading, or even the cap-and-trade program itself (as Section 111 also allows for the adoption of technology-based performance standards). That said, the cost-effectiveness advantages of a relatively unrestricted cap-and-trade program with banking are touted in the *Federal Register* notice, and the absence of an explanation for why particular regulatory designs were considered in the RIA is not unique to the CAMR RIA.

Period of Analysis and Discounting

The RIA focuses on the estimated benefits and costs attributable to the reduction in mercury emissions in 2020. However, the benefits of the rule depend on the lag between the change in mercury emissions and the uptake of mercury by fish. Social discount rates of 3 percent and 7 percent are used to compare costs and benefits that occur in different years per EPA and OMB guidelines (EPA 2000; OMB 2003). The private discount rate assumed in the electricity-sector modeling is based on a representative market price of capital for the industry and exceeds the two social discount rates. The private discount rate dictates the price path of mercury allowances over time, and thus the amount of mercury allowances banked in each period.

Rule Implementation

Although the states have discretion for how the reductions required under the CAMR are achieved, EPA assumed in the RIA that all of the mercury emissions allowances available under the rule are distributed to the regulated entities and, generally speaking, that all of the states would adopt the cap-and-trade program.

Summary of Analysis

Table 5.1 shows the major analytical elements of the analyses reported in the CAMR RIA, which are all discussed briefly below. The CAMR RIA required new and novel analyses to forecast how the rule would eventually influence health outcomes. For example, it represented the first attempt to link changes in mercury deposition to changes in fish uptake and then to changes in the ingestion of mercury by pregnant women, using data on recreational fishing patterns.

The structure of the RIA differs somewhat from the structure implied by Table 5.1 and the discussion below. The RIA begins with a general discussion of what is known about the health and ecological damages associated with mercury and some of the pathways of mercury that eventually lead to these effects. This is followed by a description of the analysis forecasting the response of regulated entities to the three design proposals for the CAMR. This analysis is followed by a description of the air quality modeling, exposure modeling, consequent health changes, and economic valuation of those changes. The discussion below collapses the review of the health and ecological effects with the description of the estimate of the avoided damages as a result of the CAMR.

Key Components of the Cost Assessment

Modeling Compliance Behavior

EPA used a large-scale model of the U.S. electricity sector, the Integrated Planning Model (IPM), to analyze the response of coal-fired power plants to the CAMR. In addition to providing an estimate of compliance behavior and subsequently the temporal and spatial pattern of mercury emissions, this exercise also provides an estimate of the cost of the program. The major components of the cost estimation procedure are outlined under the first step of analysis in Table 5.1.

Representative facilities (model plants) in IPM are assumed to supply electricity in least-cost order to produce electricity to meet regional load demand projected by the Energy Information Administration (EIA). In addition to regional markets for electricity, IPM endogenously models markets for key inputs to the sector, such as regional markets for coal. Coal-fired boilers in IPM may install common post-combustion controls to reduce SO_2, NO_x, and mercury emissions, and the qualities of the coal used by these plants affect their mercury emissions (in addition to their SO_2 emissions). Each regulated source is assumed to comply with the CAMR in a least-cost manner, including by buying or selling mercury emissions allowances.

Table 5.2 duplicates Table 7.3 in the RIA and shows the IPM forecast of how national mercury emissions will be affected as a result of the CAMR and the two alternative regulatory scenarios. Under Option 1, an 18 percent reduction in total mercury emissions occurs in 2010 relative to the total annual allocation of mercury allowances, whereas emissions are 9.3 tons greater than the annual allocation in 2020, implying that it is worthwhile for regulated entities to bank a significant number of mercury allowances for future use.

The manner in which these reductions in 2010 are achieved is not transparent from the tables provided in the RIA. An increased use of postcombustion controls does not explain this reduction because the capacity that installs these technologies relative to the CAIR is less than 1 percent of U.S. coal-fired generating capacity (as shown in Table 7.9 of the RIA). Furthermore, the total gen-

Table 5.1
Major Steps in Conducting Cost–Benefit Analysis

1. **RESPONSE OF REGULATED ENTITIES TO RULE AND MONETIZATION OF SOCIAL COST**
   - Forecast of private compliance cost with IPM, along with sensitivity analyses
   - Forecast of monitoring cost (outside of RIA)
   - Adjustment to derive social cost estimate (distortionary taxes, and so on)
   - Analysis of cost of rule to small businesses and governments
2. **CHANGE IN FISH TISSUE CONCENTRATION FROM CHANGE IN EMISSIONS**
   - Prediction of change in mercury deposition (CMAQ)
   - Estimation of distribution and change in fish mercury concentrations (Mercury Maps)
   - Estimation of lag between deposition and fish concentration change (watershed case studies)
3. **CHANGE IN EXPOSURE FROM CHANGE IN FISH TISSUE CONCENTRATION**
   - Estimation of source of caught and consumed freshwater fish
     - Via population centroid approach
     - Via angler destination approach
   - Assumptions linking women of childbearing age, and thus fetal exposure, to population of freshwater fishermen.
   - Estimation of fish consumption by women of childbearing age.
   - Estimation of fertility rate of women who consume freshwater fish (for estimation of the number of exposed infants).
4. **CHANGE IN IQ LEVELS FROM CHANGE IN EXPOSURE**
   - Estimation of change in mercury ingestion and hair concentrations
   - Estimation of relationship between maternal hair concentrations and IQ
   - Analysis of benefits of rule to subsistence, low-income, and minority populations
5. **MONETIZATION OF SOCIAL BENEFITS OF HIGHER IQ**

Table 5.2
Projected Emissions of Mercury with the Pre-CAIR Base Case, CAIR Base Case, and CAMR Options

Cases (cap in 2010/cap in 2018 in tons)	**2010**	**2015**	**2020**
Old base case: without CAIR	46.6	45.0	46.2
New base case: with CAIR	38.0	34.4	34.0
Option 1 (38/15)	31.3	27.9	24.3
Option 2 (15/15)	30.9	25.7	20.1
Option 3 (24/15)	31.1	27.4	21.1

Source: Table 7.3 in U.S. EPA 2005a.
Notes: Some of these estimates are inconsistent with values reported elsewhere in the RIA

Table 5.3
Annualized National Private Compliance Cost and Present Value Cost
(in Billions of 1999$)

	Cost (billions$)			
CAMR option	**2010**	**2015**	**2020**	**Present value (2007-2025)**
Option 1 (38/15)	$ 0.16	$ 0.10	$ 0.75	$ 3.9
Option 2 (15/15)	0.16	0.36	1.04	6.0
Option 3 (24/15)	0.16	0.18	1.04	5.2

Source: Table 7.7 in U.S. EPA 2005a.
Notes: Costs are relative to the "with CAIR" base case.

eration of electricity from coal is forecast to fall by only 0.2 percent in 2010 and 0.8 percent in 2020. Rather, the source of these additional reductions is attributed to a mix of coal-switching and changes in the timing and location of generating units.[10] The RIA does not explore the extent to which each of these control options contributes to the reduction in mercury emissions.

The private cost of complying with the CAMR is measured as changes in the production cost of owners of electricity generators. Investment costs are annualized using a capital recovery factor (the specifics of which are not described in the RIA). The RIA does not describe how the monitoring and recordkeeping costs of the CAMR were estimated, but it reports a $76 million estimate of these costs (dollar years not provided). These costs are not included in the values reported below. The social cost is equal to the private cost adjusted for increases in local taxes (a transfer payment).[11] Therefore, the analysis does not measure the reduction in consumer surplus as a result of the contraction in the supply of electricity (the quantity of electricity consumed is not responsive to price). Furthermore, the analysis estimates neither the change in social welfare in response to changes in distortionary tax levels to hold government revenues unaffected nor the exacerbation of such welfare effects from changes in the overall price level (i.e., the tax interaction effect; OMB 1992; Goulder et al. 1997).

EPA's estimate of the private costs of the CAMR and the two alternatives relative to the CAIR baseline are provided in Table 5.3. In EPA's *Guidelines for Preparing Economic Analyses,* the "annualized cost" is the "amount one would have to pay at the end of each period [in the time frame of analysis] to add up to the same cost in present value terms as the stream of costs being annualized" (EPA 2000, 35). The annualized value is thus constant for each year over the time frame of the analysis. If this is the definition used in the RIA, then for each year for which an annualized value is provided (2010, 2015, and 2020), this is the year to which all costs over the time frame of analysis are discounted (i.e., a future present value) and then annualized over the time frame of interest from the perspective of that year. However, it is unclear why this value would be lower for 2015 if this is the case.

Furthermore, the RIA does not report the discount rate used to calculate the present value of the private cost of these three options (the right-most column of Table 5.3), nor does it include a discussion of how costs are estimated for the years between the model simulation years.[12] It does appear that these really are changes in the production costs in the different simulation years from the adoption of the CAIR.

Another notable adjustment to the IPM cost estimates is the treatment of the use of banked allowances when estimating costs. The RIA estimates the costs and benefits of the rule in a particular year. However, if allowances are banked, costs may occur in one year to build the bank, allowing the banked allowances to then be used for compliance in a later period. To address this issue, "EPA estimated the cost of using banked allowances by taking the average cost of mercury control in the first phase of the program discounted to 2020, multiplied by the number of banked allowances used" (EPA 2005a, 7-33).[13] If the cost of building the bank was credited to later years, and then deducted from the costs in the earlier years when the bank was being built, perhaps this accounting led to an estimate of declining costs from 2010 to 2015 (although it still is not clear why this would be the case). Also, note that this treatment of the bank for the purposes of estimating costs is not also applied to the treatment of the benefits. This point is discussed in greater depth below.

The RIA does not present a table of comparable social costs, but reports annualized costs assuming either a 3 percent or 7 percent discount rate. Specifically, the RIA states: "the social costs of CAMR incremental to CAIR are $151 million in 2010 and $848 million in 2020 using a discount rate of 3%, and are $157 million in 2010 and $896 million in 2020 using a discount rate of 7%" (EPA 2005a, 7-13). These values are greater than those reported for annualized private costs in 2020 but lower than those reported for private costs in 2010. Presumably these values are in 1999 dollars, but the RIA is not clear on this point; the CAIR final rule *Federal Register* notice suggests that they are. The relationship between the social annualized cost and the private annualized cost is not transparent. For example, the change in the share of total costs that are tax payments, which is one adjustment to the private cost, is not provided. Furthermore, it may even be the case that the annualized factor for the social discounting exercise uses a 30-year time horizon, rather than the 18-year time horizon used to calculate the present discounted value for the private costs.[14]

Sensitivity Analyses for Cost Assessment

A considerable number of modeling assumptions affect projections of the private and social costs of the CAMR. Two sensitivity analyses were conducted. One assumed that a less costly mercury-specific reduction technology will become available in 2013. The other used alternative natural gas and coal prices, as well as electricity demand growth, as forecast by EIA. The private cost of the CAMR is lower with the lower-cost mercury reduction technology. Despite the fact that demand growth and the price differential between coal and natural gas are greater under the alternative EIA assumptions, the present discounted value of the private cost of the CAMR (from 2007 to 2025) is also lower with the EIA assumptions relative to the native EPA assumptions. This is attributable to higher baseline emissions, with coal being relatively less expensive than natural gas in the EIA sensitivity analysis. The RIA does not report social cost estimates for these sensitivity analyses.

Distributional Analyses: Costs

In compliance with the Regulatory Flexibility Act and subsequent legislative adjustments to it, EPA is required to analyze the effects of any proposed regulation on small businesses and jurisdictions that may be subjected to it. A small jurisdiction is one that serves fewer than 50,000 people, and a small electricity generating business produces up to four billion-kilowatt hours of electricity. EPA identified 42 small entities that are forecast to incur a 1 percent increase in their compliance cost as a percentage of revenue and 11 that are forecast to see an increase in their compliance cost of greater than 3 percent of their revenue. Changes in compliance cost include changes in production costs, fuel prices, allowance sales, pollution control retrofits, and so on, as well as changes in revenues from electricity sales. The RIA notes that the states may use their discretion in how they allocate mercury allowances to limit the rule's impact on small entities.

A similar analysis was conducted for compliance with the Unfunded Mandates Reform Act, which focuses on the effect of regulation on state and local government expenditures. The analysis found that the CAMR would increase the compliance cost of 44 government-owned entities by over 1 percent, and of 14, by more than 3 percent, of revenues from electricity generation in 2020.

Limitations of Cost Analysis

The RIA acknowledges a few limitations of the cost analysis. For example, IPM fixes the quantity of electricity demanded by consumers. Modeling the effect of consumer response to higher electricity prices would yield a lower estimate of the cost of the CAMR. The model also does not capture many low-cost options for abating mercury, such as upgrading existing particulate fabric filters to capture additional mercury. To the extent that these control options are available, they would suggest a lower cost of complying with the CAMR than forecast by the model. Considering all of the sources of uncertainty and their effects on costs, the RIA concludes that the predicted cost likely overestimates the actual cost of the program.

Key Components of the Benefits Assessment

The RIA describes a number of possible harms caused by low-level chronic methylmercury exposure. These include neurological effects from prenatal, postnatal, and adult exposure, as well as cardiovascular, immunological, and genetic damages. The primary pathway for exposure in the general population is through fish consumption. To say that methylmercury bioaccumulates essentially means that once an animal (a fish, bird, human, and so on) consumes methylmercury, it remains in the body. This leads to conditions in which the methylmercury concentration in fish is up to a million times that found in the surrounding water. Potential ecological damages are briefly described in an appendix to the RIA and include potentially significant harms to plant, mammalian, avian, and invertebrate species. However, none of these harms is quantified for the purposes of the RIA given the difficulty in linking changes in exposure as a result of the rule to the damages mercury may cause.

The RIA finds that neurological damages from prenatal mercury exposure are those for which there is sufficient epidemiological evidence to claim consequential effects at levels found in the environment. These damages include reductions in the ability to concentrate, fine motor skills, and verbal memory. The RIA finds that mean methylmercury exposure levels typical of women of childbearing age in the United States are 1/100 of the levels believed to begin causing these harms.

For reasons described below, the RIA uses changes in IQ as the metric for measuring neurological damages from mercury. There are number of analyses that need to be undertaken to link changes in mercury emissions to changes in IQ through fetal exposure. These include the second through fifth steps described in Table 5.1.[15] Although the rule is expected to lead to mercury reductions beginning in 2015, only those benefits from mercury emissions reductions in 2020 are estimated.

Tracing the Pathway from Emissions to Fish Concentrations

As shown in step 2 in Table 5.1, there are three main steps that require analysis to understand how the response of regulated entities to the CAMR (mercury emissions) influences the concentrations of mercury in fish. These include the pattern of mercury deposition, changes in methylmercury concentrations in aquatic systems, and the uptake of methylmercury by freshwater fish.

The deposition analysis was conducted using EPA's Community Multiscale Air Quality (CMAQ) modeling system. This model was used, in part, because it accounts for differences in the transport and deposition behavior of different mercury species and the influence of other chemical constituents in the atmosphere on the shares of those species. Inputs to the model include emissions inventories for domestic sources of mercury and other pollutants as well as forecasts of facility-level emissions of the different mercury species from the IPM analysis. A global transport and deposition model was used to determine background mercury emissions, including mercury from upwind international transport, for input into the CMAQ model.

The CMAQ model has been used for numerous regulatory analyses of sulfur, nitrogen, and other conventional pollutants. Although there is considerable scientific support for many of the modeling decisions present in CMAQ, the model had only recently been modified to track mercury emissions, and a number of assumptions had to be made regarding the behavior of mercury in the atmosphere because of a lack of data. For example, the absence of a network of dry deposition monitors for mercury prevents the calibration of the model to historic conditions. It is unclear how consequential these assumptions may be.

The model used to capture how the change in mercury deposition leads to changes in fish methylmercury concentrations, Mercury Maps (MMaps), is a blunt tool. It essentially assumes a proportional relationship between current deposition and subsequent methylmercury concentrations in fish, so that the percentage reduction in deposition is equal to the percentage reduction in fish concentrations. As such, it does not account for associated or expected changes in aquatic chemistry as a result of changes in the deposition of other pollutants or nonatmospheric sources of mercury and how these changes would affect current methylmercury concentrations in fish. The model also does not account for the lag time between changes in deposition and changes in methylmercury concentrations in fish, or variations in that lag time resulting from other chemical and biological conditions in the aquatic ecosystem.

The presumed baseline level of mercury concentrations in fish for the purposes of the RIA analysis is unclear. According to Chapter 3 of the RIA, the fish methylmercury concentration data used to set a baseline in the model is from the early 1990s. This is consistent with the discussion in Chapter 5. However, Chapter 10 suggests that only fish tissue concentration samples from 1999 to 2003 are used to establish the baseline, which seems more justifiable given that it is known that mercury deposition declined during the 1990s. Whatever baseline year was used, the concentrations for that year were assumed to represent an equilibrium state for the purposes of modeling the effect of changes in deposition on fish mercury concentrations.

To estimate the lag time between deposition changes and changes in methylmercury concentrations in fish, EPA conducted case studies of five freshwater aquatic ecosystem types in the United States.[16] Methylmercury directly enters aquatic ecosystems through deposition or effluent and can also be created from inorganic mercury by methylating bacteria in the aquatic environment. The ecosystems analyzed in these case studies are representative of the distribution of different geological, biological, and chemical characteristics known to influence methylmercury concentrations. However, they are not intended to be representative of ecosystems that have characteristics that are at the extreme range of those qualities known to influence methylmercury concentrations in fish. The findings from these case studies show that the lag between deposition changes and steady-state changes in methylmercury concentrations in fish is between 5 and 50 years. The typical response times are between 5 and 30 years. However, the lag times may even be shorter, as one study referenced by the RIA suggests that the residence time prior to conversion to methylmercury of recently deposited mercury is briefer than for mercury already in the system.

Despite this effort, considerable modeling uncertainty remains in the forecast changes in methylmercury concentrations and changes in mercury deposition. The RIA considers this source of uncertainty greater than the uncertainty associated with the change in deposition that would result from this rule.[17] To account for this uncertainty, the RIA estimates the reduced damages from this rule using 5-, 10-, 20-, and 50-year lag times between the changes in deposition and the changes in fish mercury levels. These lags are assumed to be spatially uniform for the different benefit estimates. The 10- and 20-year lags are the preferred estimates in the RIA and are those used to generate the benefit estimates provided below.

Tracing the Pathway from Fish Concentrations to Consumption

The pathway for mercury damages modeled in the RIA is through fish catch and subsequent consumption by women of childbearing age. As shown in step 3 of Table 5.1, modeling the flow of mercury through this pathway requires an understanding of the number of freshwater fish that women of childbearing age consume and where those fish are caught.[18]

The analysis in the RIA is limited to the effect of changes in mercury deposition on the concentrations in freshwater fish. This is because changes in domestic mercury emissions were only anticipated to affect mercury concentrations in freshwater fish and because there is no commercial market for them. Mercury loading in saltwater species is believed to result primarily from international sources of mercury emissions, and thus the effect of the CAMR on this source of mercury exposure was deemed too small to consider as part of the RIA.

The analysis was further limited to freshwater fish catch and consumption in the eastern half of the United States, which includes the states from North Dakota down to Texas and eastward. This restriction is due to the dearth of data on fish tissue concentrations in the western half of the United States. This limitation was not expected to significantly affect the estimate of the reduced damages attributable to the rule, as the air quality modeling shows that the greatest reduction in deposition is in the eastern United States.

No representative data are available on the quantity and source of recreationally caught freshwater fish consumed by different subpopulations, including women of childbearing age. Therefore, these relationships must be estimated and doing so requires a number of different data sources and assumptions, beginning with an estimate, of the number of exposed infants. Two approaches to estimating changes in prenatal exposure are taken in the RIA. In the population centroid approach, the number of infants born to women aged 15 to 44 is estimated by census block, using state-level average annual fertility and projected population growth rates. For each census block, the number of expectant mothers who live with a recreational fisher is then assumed to be proportional to the number of adult recreational fishers in the state. Already this approach embodies two assumptions: women who live with a recreational fisher are no more likely than others to have a child, and the share of households with a recreational angler is equal to the share of adults who are recreational anglers.

Next, the average mercury concentration in freshwater fish consumed by households with anglers is estimated. This is where the analysis at the census block level is meaningful. First, the average mercury concentration of regularly consumed freshwater fish is estimated for five distance bands from the centroid of each census block for both stream- and lake-caught fish. The average for each of these bands is then weighted by the number of fishing trips to lakes and streams by each distance category and by two income classes and two residential classes (urban and rural). The lake, stream, and distance shares are by state, whereas the residential and income classes are by census block.

The second approach to estimating the prenatal exposure via freshwater fish consumption focuses on the destination of freshwater fish consumption and is called the angler destination approach. Average fish mercury concentrations are estimated by watershed in the study area. The United States Geological Survey watershed designation used in the RIA yields watersheds with an average size of 1,600 square miles. The assumed proportion of lake and stream catch days in each watershed is equal to estimated state-level proportions, whereas the share each watershed contributes to the total number of fishing days in the state is estimated using household-level data on angler fishing location choice. The number of anglers per watershed is then estimated using the national average number of fishing days that a typical angler takes. This estimate of the number of anglers per watershed is further disaggregated using state-level data on the percentage of female anglers of childbearing age and the percentage of married men between the ages of 18 and 44 who are anglers. The watershed-specific sum of these two subpopulations of anglers is multiplied by the state-level fertility rate to arrive at an estimate of the number of infants exposed per watershed. The estimates of exposed infants per watershed are then aggregated to the state level.

These two approaches provide both an estimate of the number of infants exposed prenatally to mercury via freshwater fish consumption and the average mercury concentration in the fish their mothers consume. What is still needed is an estimate of the quantity of their mothers' consumption of freshwater fish. All women of childbearing age who are anglers or live with an angler are assumed to consume the same quantity of freshwater fish regardless of circumstance (location, presence of mercury consumption advisory, and so on).[19] The bluntness of this assumption is due to a lack of data. As described below, however, assuming equal consumption is inconsequential to the central benefit estimates given other aspects of the analysis (e.g., the linear form of the mercury–IQ dose–response relationship).

Tracing the Pathway from Fish Consumption to Health Status

The next major step in the benefits estimation, the fourth step in Table 5.1, requires estimating how a change in mercury ingestion by pregnant mothers leads to changes in childhood IQ. The analysis focuses on changes in IQ because of the availability of critically evaluated epidemiological studies of the relationship between fetal methylmercury exposures and IQ (EPA 2005a, 9-1). Elsewhere, the RIA states that IQ serves as a surrogate for the neurobehavioral endpoints affected by mercury exposure (EPA 2005a, 11-2). The RIA further notes that valuation studies exist that have established a relationship between IQ and earnings (EPA 2005a, 9-1). However, it seems that the purpose of noting this is to justify the choice of IQ as a metric for neurological function, rather than to claim that the only quantified changes in health outcomes were those for which monetized benefits exist. The RIA mentions that there is a stronger and more noticeable effect of maternal mercury levels on other neurological functions, such as the ability to acquire and retain information provided verbally, that do not influence IQ (EPA 2005a, 9-9). However, no valuation studies have measured willingness to pay to avoid these effects.

Hair concentrations are used as a proxy for fetal exposure because epidemiological studies have established a relationship between maternal hair concentrations and IQ loss. The procedure for estimating the relationship between hair concentrations and IQ loss is summarized in Chapter 9 of the RIA. It is estimated via a meta-analysis using three well-known studies on the relationship between maternal exposure and reductions in neurological function. A linear functional form for the estimated relationship was chosen, in part, because it avoided the complication of having to estimate mercury exposures from other sources, including exposures from coal-fired plants in previous years.[20] The relationship between changes in mercury ingestion and changes in mercury concentrations in hair used for the RIA is assumed to be linear and is taken from the epidemiology literature.

Monetizing the Benefit of Improved IQ

The cost per IQ decrement is based on a single study that estimates the average percentage decrease in future earnings from a lower IQ and the percentage reduction in schooling (Salkever 1995). The average lifetime wage was estimated using the 1992 Current Population Survey and apparently was unadjusted for labor productivity growth based on time of birth (see the discussion below regarding the lag between mercury reductions and changes in mercury concentrations in fish). For the estimate of lost wages and reduced IQ, a national cohort was used rather than one representative of children of mothers who are likely to consume freshwater fish. Although the RIA notes that the loss in future earnings should be viewed as a lower bound on the estimate of an individual's willingness to pay for avoiding a lower IQ (EPA, 2005a, 10-47), it does not provide a sense of the difference between the wage loss measure and a true willingness-to-pay measure. The expected reduction in schooling resulting from lower IQ suggests that as IQ rises, expenditures on schooling fall. Therefore, the cost of a year of educational services was also estimated. The private loss in earnings is adjusted for the expected reduced cost of education services.

Emissions Scenarios for Benefit Estimation

Having connected all of the steps required to estimate the effect of changes in emissions to changes in IQ, the damages caused by mercury emissions from the power sector were then estimated for six different year and emissions control scenario combinations. One comparison provides an estimate of the increase in IQ from reducing mercury emissions in 2020 attributable to the CAIR. However, this benefit estimate is based on the difference between the effect of mercury emissions in 2001 and that of mercury emissions in 2020 with the CAIR. The typical approach in a cost–benefit analysis would be to compare benefits and costs with and without the regulation over the same period, as this is a requisite for an appropriate baseline (EPA 2000; OMB 2003). That said, the purpose of this comparison is to provide an estimate of the mercury benefits of the CAIR, and the purpose of the RIA is to estimate the benefits of the CAMR.[21] The estimate of the mercury benefits of the CAIR is used as a benchmark for evaluating the benefits of reduced IQ damages in 2020 resulting from the adoption of the CAMR (both Options 1 and 2) relative to a regulatory baseline in 2020 with the CAIR. The two additional damage estimates assume that there are no emissions from coal-fired plants in 2001 and in 2020, presumably to provide a measure of the total IQ reductions attributable to the sector.[22] Finally, note that the benefits of the CAMR were never estimated in a baseline that did not include the CAIR, despite the fact that the CAIR had only very recently been adopted.[23]

Accounting for Thresholds in Mercury Health Effects

The initial drafts of the RIA subject to interagency review presumed that any maternal mercury exposure would lead to reduced IQ in children (Griffiths 2008). However, some viewed this as inconsistent with existing EPA regulations that embraced the concept of a reference dose below which damages from ingesting mercury were unlikely or not appreciable. The reference dose was last updated by EPA in 2001 in collaboration with the National Academy of Sciences and is based on the same epidemiological studies used to estimate the mercury–IQ decrement dose–response curve that is used in the RIA (EPA 2001a).[24]

To address this inconsistency, EPA added a chapter to the RIA (Chapter 11) that adjusted the existing benefits assessment to account for the presence of a threshold (Griffiths 2008). The threshold was assumed to be equal to the reference dose even though the reference dose was not set at a level at which damages from mercury exposure are first expected to appear, but rather at a level below which it is known that there are no appreciable risks.

The sole adjustment to the approach taken in Chapter 10 was to estimate the number of women of childbearing age whose current consumption of mercury was already below EPA's reference dose, and thus provide a basis for estimating the population of women consuming mercury above this dose. To this point, all of the analytical assumptions created conditions where one only needed to know the change in the average mercury ingestion of the average woman of childbearing age to estimate the rule's benefits because the functions linking fish consumption to IQ were all linear. To account for the threshold, one must remove from consideration those women who already consume below it.

The National Health and Nutrition Examination Survey (NHANES) was used to estimate the distribution of mercury concentrations among women. The NHANES data provide insufficient information to determine which individuals in the sample may be consumers of freshwater fish; therefore, a very simple approach was taken. The distribution of mercury hair concentrations in women who consume freshwater fish, as estimated in Chapter 10 of the RIA, was assumed to be the same as the distribution in the NHANES sample, with the caveat that only that range of the NHANES distribution where hair mercury levels were above the estimated minimum concentrations in women who consume freshwater fish was used. No other information was used from the NHANES data, including information on the spatial variation in hair mercury concentrations in women of childbearing age.

Only 23 percent of women in the NHANES sample had hair mercury concentrations above the minimum estimated level in women who consume freshwater fish. Generally speaking, the women estimated to be exposed to mercury from freshwater fish were assumed to be distributed evenly across this range of the NHANES distribution. One modification to this assumption was that women with mercury exposures from freshwater fish consumption above the observation at the 77th percentile in the NHANES data were distributed over a higher NHANES range so that they were not treated as having hair mercury levels lower than the levels predicted solely from their freshwater fish consumption. Furthermore, for those freshwater fish-consuming women with mercury hair concentrations exceeding the maximum concentration in the NHANES sample, mercury from freshwater fish consumption was treated as their sole source of mercury exposure.

Once this distribution of mercury concentrations in the hair of women who consume freshwater fish was estimated, the next step was to estimate how the changes in mercury exposures estimated in Chapter 10 affected this distribution.[25] Two approaches were taken, each of which relied on the changes in exposure estimated under the population centroid approach. The first assumed that the reduction in exposure as a result of the CAMR was randomly distributed across the range of women with exposures above the reference dose, whereas the other assumed that the reduction in exposure was perfectly correlated with the current level of exposure.[26] The threshold damage model employed for this analysis uses the same dose–response function estimated for Chapter 10, except that it is truncated below the threshold (i.e., the dose–response function "jumps up" at the level of the threshold). The benefits under the random assignment approach are 21 percent of the benefits estimated in Chapter 10, whereas under the assumption of perfect correlation, they are 34 percent of the benefits estimated in Chapter 10. The RIA refers to the 21 percent and 34 percent values as "scaling factors."

Summary of National Benefits

Table 5.4 reports the estimated benefits resulting from the emissions reduction in 2020 for CAMR Options 1 and 2, assuming no threshold. The estimates are sensitive to the approach for estimating the concentration of mercury in consumed fish, the social discount rate, and the time lag to changes in fish mercury concentrations. These values are much lower than the cost of the rule in 2020 (compare to Table 5.3). Furthermore, according to the RIA, the total damage from 2020 mercury emissions from coal-fired power plants given the CAIR is at most $3 million (in 1999$), assuming a 3 percent discount rate, the population centroid approach, a lag between deposition and fish uptake of 10 years, and no threshold. If accurate, this value provides an upper bound on the benefit of reducing 2020 mercury emissions from coal-fired power plants. However, as discussed above, the estimation method did not consider potential ecological damages or other health effects that may be caused by mercury.[27]

Table 5.5 provides estimates of the benefits under CAMR Options 1 and 2 that account for the possibility of a threshold at the reference dose. These values are reported in Tables 11.7 and 11.8 of the RIA .[28] The range of estimates in each cell reflects different assumptions about the time lag between changes in deposition and changes in fish mercury concentrations. The lower end of the range represents a 20-year lag; it is unclear what the upper end is, although it falls between the 5- and 10-year lags.

Distributional Analyses: Benefits

In addition to requiring a cost–benefit analysis of major regulations, EO 12866 requires federal agencies to consider the impact of regulations on low-income and minority populations. One of the distributional analyses looked at the effect of the rule on subsistence consumers. Mothers in census blocks whose estimated average daily consumption was in the top 5 percent of freshwater fish consumers were treated as a subsistence population (greater than 25 grams of fish consumption per day). The analysis first required an assumption of the distribution of individual consumption rates for fish. Each census block from the population centroid approach was then randomly assigned a fish consumption rate drawn from this distribution. One drawback of this random-assignment approach is that it ignores the possibility of any correlation between areas

Table 5.4
Summary of Benefits from the CAMR in 2020 under Alternative Approaches to Estimating the Affected Population
(in 1999$)

Approach	CAMR option	Discount rate	Estimate of fish tissue response times: 10-year lag	20-year lag
Population centroid approach	Option 1	3%	$ 2,086,359	$ 1,687,988
		7	1,425,357	787,840
	Option 2	3	3,112,816	2,527,403
		7	2,126,610	1,179,624
Angler destination approach	Option 1	3	2,995,451	2,457,145
		7	2,046,429	1,146,832
	Option 2	3	4,586,570	3,762,327
		7	3,133,448	1,756,004

Source: Table 10-30 in EPA 2005a

Table 5.5
Benefits for the CAMR Accounting for U.S. EPA Mercury Reference Dose
(in Millions of 1999$)

CAMR option	Scaling factor	Discount rate	Benefits
Option 1	21%	3%	$0.36 – $0.63
		7	0.17 – 0.42
	34%	3	0.58 – 1.00
		7	0.27 – 0.68
Option 2	21%	3	0.53 – 0.97
		7	0.25 – 0.65
	34%	3	0.85 – 1.56
		7	0.41 – 1.05

Source: Tables 11-6 and 11-7 in EPA 2005a.

with high fish consumption and areas where there may be significant changes in fish mercury concentrations. The results of this analysis show that about 90 percent of infants realize less than a 0.002-point increase in IQ resulting from 2020 CAMR emissions reductions. However, about 1 percent of infants affected by the reduction in emissions in 2020 realize an increase of more than 0.008 IQ points.

The second analysis focused on low-income households. The population centroid approach was adjusted to account for the fertility, population distribution, and fishing patterns of households with incomes less than $10,000 in 2000. These households were assumed to have a daily freshwater fish consumption level in the top 5 percent of all freshwater fish consumers.

The benefits of the CAMR were also estimated for two ethnic populations that have a strong cultural identification with freshwater fish consumption. The Chippewa are a Native American community (tribe) residing primarily in the upper Midwest around Lakes Superior and Huron. The Hmong are immigrants from Southeast Asia that settled primarily in Minnesota and Wisconsin. Although freshwater fish is a large part of the diets of other ethnic groups, the RIA focuses on these two because their fishing behavior has been formally studied. The population centroid approach is modified to estimate the effects of the CAMR on these populations, accounting for their spatial distribution, fertility rates, the distance they travel to fish, whether they consume any freshwater fish, and if so, their average daily intake.

Table 5.6 reports the benefits under CAMR Option 1 for these four subpopulations. Note that the benefits estimates are quite low for the Hmong and Chippewa populations. Although members of these communities consume more than the average freshwater fish consumer, the effect of the CAMR is low in part because the rule does not reduce mercury emissions where they live.[29] Three additional sensitivity analyses were conducted for the estimates for these subpopulations, but the RIA only provided the average IQ increase from the CAMR under these alternative assumptions and not monetized benefit estimates.

Particulate Reduction from Fabric Filters to Control Mercury

One component of the technology specifically designed to reduce mercury emissions that is modeled in IPM, activated carbon injection (ACI), is a fabric filter. Activated carbon is injected into the flue gas stream, where it bonds with the mercury, and the fabric filter collects the activated carbon. The fabric filter also collects other fine particulate matter in the flue gas stream. Fine particulate matter causes a number of respiratory illnesses and is associated with premature mortality. Although reduced mortality is not the sole benefit of reducing particulate matter, it forms the lion's share of the benefit estimates for rules to control direct and secondary particulate emissions. For this reason, the RIA also estimates the benefit of reduced mortality attributable to the collection of additional particulate matter by the fabric filter.

The estimated benefits from reduced particulate matter emissions in 2020 as a result of the CAMR range from $40 million to $44 million, depending on the discount rate used.[30] This value is much larger than the estimated benefits associated with reducing the target pollutant, mercury, and it is due to a very small percentage of the total capacity of coal-fired generation installing ACI

Table 5.6
Estimate of Benefits of the CAMR (Option 1) on High-Risk Populations
(in 1999$)

		Estimate of fish tissue response times	
Community	**Discount rate**	**10-year lag**	**20-year lag**
Subsistence population	3%	$ 573,373	$ 463,559
	7	391,716	216,358
Low-income subsistence population	3	572,354	454,554
	7	391,020	212,155
Hmong	3	3,477	3,511
	7	2,375	1,546
Chippewa	3	6,698	6,331
	7	4,576	2,955

Source: Tables 10-34, 10-37, 10-42, and 10-43 in EPA 2005a.

(only about 3 to 4 percent in 2020). Recall that most of the reductions in mercury up until that time are expected to derive from conventional pollutant controls along with changes in the types of coal used.

Limitations of the Benefits Analysis

The end of Chapter 10 of the RIA discusses the effect of likely uncertainties and biases in the benefits assessment. On net, the RIA concludes that these biases imply that the monetized estimates of the CAMR are probably understated. These uncertainties and biases include the following:

- Strong assumptions had to be made about the levels and spatial distribution of mercury concentrations in fish given the lack of a sampling frame for fish mercury concentration observations.
- The population centroid approach imposes critical assumptions on the percentage of women who are or who live with freshwater anglers. Similarly, the angler destination approach makes strong conjectures about how the population of anglers is associated with the population of women of childbearing age.
- The possibility that women may currently avoid eating fish and, therefore, may consume more in the future as mercury concentrations decline is ignored.
- The daily fish consumption rate is assumed to be constant across the population of interest (although, given other proportionality assumptions, this matters only for the estimates that assume a threshold).
- The assumptions of proportionality between mercury consumption and mercury concentration in hair, between hair concentrations and IQ loss, and between IQ loss and wage loss each introduce modeling uncertainty.
- Other sources of neurological benefits—such as improved language development—from reduced mercury emissions are not quantified. These endpoints have been shown to be more sensitive to mercury exposure.
- Benefits from potential cardiovascular, genotoxic, immunotoxic, and ecological damages are not quantified. In particular, as the RIA notes, accounting for the willingness to pay to avoid cardiovascular damages (reduced life expectancy) would raise the benefit estimate considerably.
- Mercury exposure from other consumption pathways that may be affected by the rule (commercial fish and recreationally caught shellfish) are not accounted for.
- There are concerns regarding the applicability of the wage estimate used to value reduced IQ, including its assumed constancy over time and across the population. No mention is made in this particular section of the RIA about using wage data as a proxy for willingness to pay for IQ improvements. However, as noted above, this limitation is mentioned in the section of the RIA where the wage-derived benefit estimate is introduced.

One caveat to the analysis that is notably absent in the discussion at the end of Chapter 10 is the effect of the potential asymmetric treatment of the drawdown in banked allowances on the estimates of the benefits and costs. The estimate of the cost of the CAMR in 2020 was supposedly adjusted to reflect the cost of building the bank of allowances that are used in 2020. Of the fore-

Table 5.7 Summary of Annual Benefits, Costs, and Net Benefits of the CAMR (in Millions of 1999$)

Description			2020
Social cost	3% discount rate		$ 848.0
	7% discount rate		$ 896.0
Social benefits	3% discount rate		
		EPA reference dose	$0.4–$1.0
		No threshold	$1.7–$3.0
	7% discount rate		
		EPA reference dose	$0.2 –$0.7
		No threshold	$0.8 –$2.0
Unquantified benefits and costs			U
Annual net benefits (benefits–costs)			
	3% discount rate		
		EPA reference dose	$ –848 + U
		No threshold	$ –846 + U
	7% discount rate		
		EPA reference dose	$ –896 + U
		No threshold	$ –895 + U

cast mercury emissions in 2020, about 40 percent come from drawing down the bank. However, the benefit estimate of reduced emissions in 2020 does not account for the fact that emissions were lower prior to 2020 to build up the bank. So in 2020, the cost of building the bank may be accounted for, but the benefit of delaying emissions until later is not.

Had this additional benefit been accounted for, the difference between the estimated benefits and costs in 2020 would be lower. Given the magnitude of the benefit of increasing IQ and the restriction of the benefit estimate to IQ effects, an appropriate adjustment would not meaningfully affect the ratio between costs and benefits. However, on a more fundamental point, this accounting discrepancy highlights the problem with focusing on the costs and benefits of a rule in a single year.

Had the stream of benefits from the rule been accounted for over the period until the bank had essentially been drawn down (or until discounting made the inclusion of an additional year essentially moot), this problem would not have arisen. Furthermore, if one restricts oneself to a particular year of analysis, it is unclear how one should account for the costs and benefits of shifting emissions over time via allowance banking. Moreover, it is possible that the impression provided by the cost–benefit analysis may be conditional on the year and accounting approach chosen regardless of how the bank is treated.

Comparing the Costs to the Benefits

Nowhere does the RIA summarize the benefits and costs of the CAMR and compare them to one another. This summary is available in the *Federal Register* notice for the final rule for the CAMR and is reproduced in Table 5.7 (EPA 2005g). Despite the concern about the consistency between EPA's initial analysis of the benefits of the CAMR and EPA's assumption of a threshold (reference dose) for mercury consumption, the CAMR final rule *Federal Register* notice reported the benefits of the rule under both characterizations of the damages from mercury. The range of benefit estimates captures the range in estimates of the lag between deposition and changes in concentrations in fish. The range of benefit estimates also reflects the range of the two different methods for estimating the affected population (compare to Table 5.4). The ranges of values under the reference dose assumption fold in the range of uncertainty in the distribution of reduced mercury exposure captured by the scaling variable (compare to Table 5.5). Despite preamble text suggesting otherwise, the benefits of reduced mortality from reduced particulate matter are *not* accounted for in this summary table. On the cost side, it is unclear how the costs estimated in the RIA were manipulated to arrive at the estimates found in the final rule *Federal Register* notice. Finally, note the explicit recognition in the table that certain benefits and costs of the rule may not have been estimated.

Conclusion

The RIA represents a snapshot of what is known about the benefits and costs of controlling the pollutant being regulated. Clearly, what is known about the pollutant is evolving, and there may be significant disagreements in the interpretation of the relevant scientific literature. Furthermore, some may advocate a more cautious approach to regulating the pollutant based on its possible effects. In that case, estimating the benefits of reducing potential damages would be appropriate and informative as to whether the possible effects are substantial enough to warrant tighter regulation.

These observations are particularly salient in the regulation of mercury. Public comments on the final CAMR and delisting regulatory notices criticized the limited scope and methods of EPA's damages assessment, in part based on the legal argument that uncertain benefits are relevant to determining the stringency of mercury regulation. The comments also argued that EPA should consider recent studies on mercury's health effects in its analyses. EPA evaluated these studies and their applicability to the agency's decision to delist mercury (in EPA 2006a, 2005h). EPA also accepted petitions from environmental groups and states to reconsider its decision to delist mercury and the CAMR itself (EPA 2005i, 2005j).

The delisting reconsideration was granted in part because of concerns regarding EPA's "methodology and conclusions concerning why utility mercury emissions...are not reasonably anticipated to result in hazards to public health" (EPA 2006b, 33390). Technically speaking, these concerns about whether all relevant sources of damages from mercury were accounted for relates to U.S. EPA (2005c), which supported the delisting decision, and not the RIA. However, in response to the petitions and other comments received regarding the CAMR and the delisting, EPA significantly updated the benefits assessment presented in the RIA (EPA 2006a, 2005h). For example, EPA conducted a bounding exercise—essentially additional sensitivity analyses—to determine the maximum potential IQ benefits that would come from reduced exposure through both freshwater and saltwater fish consumption.

Postscript

In *New Jersey v. EPA* (2008) the United States Court of Appeals for the DC Circuit invalidated the CAMR. The basis for the court's decision was that EPA's method for delisting mercury as a HAP did not follow the procedures described in the CAA.[31] It is notable that the decision did not address the legality of cap-and-trade per se, the possibility of regulating mercury using cap-and-trade under

Section 112, or the acceptable level of the standard that would be adopted under Section 112. Rather, the ruling addressed the procedure through which EPA reversed its earlier finding that had classified mercury as a HAP, which made moot any questions regarding subsequent decisions about how to regulate mercury under Section 111.

■ ■ ■

Notes

1. The views expressed in this paper are those of the author and do not necessarily represent those of EPA. In addition, although the research described in this paper may have been funded entirely or in part by EPA, it has not been subjected to the agency's required peer and policy review. No official agency endorsement should be inferred.

The author began employment at EPA after the publication of the Clean Air Mercury Rule Regulatory Impact Analysis and, at the time of this writing, has never worked on the development of mercury regulations at EPA.

2. The following discussion is not intended to suggest that certain methodological choices were made so that the findings of the analysis would support one regulatory approach over another.

3. A combined RIA was prepared for the proposed rules under Sections 111 and 112 (EPA 2003, 2004a, 2004b), although only the EPA webpage refers to this collection of documents as an RIA (EPA 2008). The analysis focused on the Section 112 proposal. The regulatory baseline for the 112 analysis did not include the Clean Air Interstate Rule (CAIR) or a regulation like the CAIR. The cost–benefit analysis under Section 111 was brief, did not quantify any benefits of reduced mercury, and included a CAIR-like rule in its regulatory baseline. The Section 111 analysis adapted those conducted for the president's concurrent legislative proposal for controlling the three pollutants of interest.

Note that the CAMR rule docket contains an analysis of the expected compliance behavior under the Section 111 approach that assumes caps equivalent to those that were eventually adopted (EPA 2004d). This analysis was added to the docket on February 24, 2004, a few months after the rule was signed (on December 15, 2003). However, it is not described in any of the RIAs or *Federal Register* notices regarding the rule.

4. EPA also proposed a cap-and-trade program under Section 112 that had a very similar structure to the one proposed under Section 111 (i.e., the same allowable emissions, affected sources, and so on), with the important exception that the federal government would allocate the emissions rights to the affected sources whereas, as described below, it is the states that make the allocations under 111 (EPA 2004e).

5. None of the analyses of the CAMR considers a setting in which the CAIR is not also adopted.

6. OMB guidelines require that at least three regulatory options be analyzed: the preferred, a more stringent option, and a less stringent option (OMB 2003). For the CAMR, the least stringent was the preferred option.

7. The ability of SO_2, NO_x, and particulate matter controls to reduce mercury notwithstanding.

8. In announcing the CAMR final rule, EPA stated that "such technologies are adequately demonstrated for use in the 2010 to 2018 time-frame to allow for compliance with the CAMR Phase II cap" (EPA 2005g, 28618).

9. The *Federal Register* notice proposing the rule simply states that the 15-ton cap was chosen to reduce mercury emissions from coal-fired power plants by 70 percent from current levels by 2018 (EPA 2004c). However, the notice does not explain why the 70 percent value was chosen.

10. Another possible source of reduction, increasing the operation of units with sulfur and nitrogen controls relative to those without, is not mentioned as a way in which mercury emissions are reduced under the CAMR.

11. The RIA states that the social cost includes "the costs of added insurance" (EPA 2005a, 7–13). However, it is not clear what this insurance is for or whether it is already included in the private cost.

12. IPM is run for a select number of years within the modeling forecast horizon. In the case of the RIA, these years are 2007, 2010, 2015, 2020, and 2025.

13. It is not clear that this adjustment was actually made. If so, it was not done as part of a post-processing of IPM output. For this statement about adjusting costs to account for the bank to be true, this adjustment must be made so regularly to IPM estimates of production costs that it is part of the standard reporting of production cost changes from the model.

14. The RIA states that one needs to know the "life of the capital" for calculating an annualized cost and reports that it is assumed to be 30 years. However, this seems to be a non sequitur for the social welfare discounting exercise at hand.

15. If this were a rule to propose changes in the level of National Ambient Air Quality Standards, much of the analysis in the RIA described in the next three subsections would be found in an integrated science assessment and a risk assessment. The science assessment would explore the question of what damages are caused by the pollutant, whereas the risk assessment would explore how large these damages are given expected and potential levels of the pollutant.

16. The forecast changes in mercury concentrations from the ecological models are also used to evaluate the changes forecast by the MMaps model.

17. Although it is unclear whether what is meant here is uncertainty in the change in deposition given an expected response of the regulated entities, or the change in deposition given uncertainty about the response of regulated entities; presumably it is the former.

18. The description in this section relies exclusively on Chapter 10 of the RIA and EPA (2005c). (The analyses within EPA 2005c were conducted for the purposes of the delisting decision, although the related *Federal Register* notices are not particularly clear about this fact.) Although Chapter 4 of the RIA also describes angler and fish consumption patterns, it relies on older data and methods than the data described in Chapter 10. Furthermore, Chapter 4 states, "If fish consumption rates differ significantly across regions, then this may suggest that exposure through fish consumption may also differ regionally (of course this will also depend on the regional variability of mercury concentrations in fish consumed).... However...the patchiness of data characterizing regional variability in fish consumption rates tends to prevent a comprehensive treatment of this issue in... a national-scale benefits analysis" (EPA 2005a, 4-45). Nevertheless, this is exactly the type of analysis that is described in Chapter 10.

19. One might expect fishers to respond to reductions in fish mercury concentrations by consuming more than they had in the past. However, this approach does not address the possibility of such (un)averting behavior in response to reduced mercury concentrations in fish. Alone, ignoring this effect yields an overestimate of the benefits of reduced mercury emissions (i.e., some rebound in maternal concentration levels would occur as consumption would increase when average mercury concentrations are lowered).

20. Given the assumptions made in the RIA, damages from mercury are due solely to emissions in a particular year. For example, the benefits of the CAMR in 2020 are solely attributable to the emissions reductions in 2020 beyond the CAIR regulatory baseline. If a nonlinear functional form were used to estimate the change in IQ from a change in mercury deposition, one would have to estimate the reductions in IQ effects as a result of the CAMR prior to 2020 to estimate the change in IQ resulting from emissions reductions in 2020 from the CAMR.

21. The RIA states that the 2001 mercury emissions, and thus fish concentrations, were "assumed [to] remain constant at their observed levels," implying that 2001 levels can be treated as representative of 2020 levels without the CAIR. However, the reductions in mercury from the CAIR may actually be lower than this approach suggests because mercury emissions had already fallen between 2001 and 2007 and would likely continue declining as a result of programs like the Title IV SO_2 cap-and-trade program. As we see in Table 5.2, the IPM forecasts total mercury emissions of 46.6 and 46.2 tons in the baseline in 2010 and 2020, respectively. These estimates are both about 5 percent lower than the estimate of 2001 emissions (see Table 8-2 of the RIA). Furthermore, as we see in Table 5.2, the CAIR is expected to reduce mercury emissions by 12.2 tons in 2020. This reduction roughly compares to the 14.2-ton difference between the estimates of mercury emissions from EGUS in 2001 and 2020 assuming that the CAIR has been adopted. However, given its propensity to deposit locally and convert to methylmercury, reactive mercury is the critical form for this comparison. It appears that EPA did not provide a forecast of the share of mercury emissions that is reactive in 2020 absent the CAIR.

The share of reactive mercury meaningfully influences the dangers that mercury emissions cause. Note that the RIA shows that the difference in mercury emissions between the CAIR in 2020 and 2001 is 14 tons, whereas the reduction resulting from the CAMR relative to the CAIR in 2020 is 9 tons. However, the difference in the damages from mercury in 2020 with the CAIR relative to damages from 2001 emissions are estimated to be 5 to 15 times higher than the difference in damages from the CAMR relative to the CAIR in 2020. In part, this difference is explained by the fact that reactive mercury is about 10 tons higher in 2001 relative to the forecast reactive mercury emissions with the CAIR in 2020, whereas the CAMR is forecast to reduce reactive mercury emissions by only 1.3 tons relative to the CAIR in 2020.

22. In a curious comparison, the RIA shows that the difference in the damages from emissions in 2001 and emissions under the CAIR in 2020 is greater than the total damages caused by emissions in 2001 and thus greater than the total benefit of eliminating 2001 emissions (see Tables 10-6 and 10-7 of the RIA). However, it appears that the difference between 2001 emissions damages and 2020 damages with the CAIR treats 2001 emissions as if they occur in 2020. Therefore the two estimates are measured in different years. A proper comparison of these two values requires discounting the 2020 damage estimate to 2001. Furthermore, by assuming that 2001 emissions occur in 2020, the damage estimate is inflated given that the population is expected to increase in 2020.

23. Part of EPA's justification for delisting mercury is that it could consider the effect of other requirements of the CAA when determining whether it was required to regulate mercury under Section 112 of the act. EPA claimed that it had erred in not considering the effects of Title I regulations (i.e., revised ambient air quality standards, which provides the authority for the CAIR) on mercury emissions when it originally decided to list mercury as a HAP (U.S. EPA 2005b, 16003). However, it is unclear if this legal issue influenced the choice of the baseline in the RIA.

24. Perhaps reflecting the collaborative nature of RIA authorship, the introduction to Chapter 9 of the RIA defends the approach to estimating IQ reductions without a threshold (Chapter 9 describes the approach used to estimate the relationship between maternal mercury concentrations and IQ). It further notes that the reference dose is a level at which damages are non-appreciable, and not one where the damages are zero.

25. It appears that the relationship between the distribution of total mercury exposure resulting from freshwater fish consumption and the changes in those exposures because of the CAMR estimated in Chapter 10 were not available to the authors of Chapter 11. However, they had access to ranges of changes (across eight bands), and assumed that the distribution of changes within those bands was uniform (see Table 11.3 and surrounding discussion of the RIA).

26. Presumably the reductions in exposure could not be linked to the estimates of total baseline exposure from freshwater fish (i.e., that those with large exposure reductions are probably those with large exposures from freshwater fish) given the data limitations described in note 25.

27. The RIA also reports two sensitivity analyses that varied the relationship between IQ and maternal mercury exposure (EPA 2005a, 10-97). One of the sensitivities assumed a smaller effect of maternal mercury exposure on IQ, whereas the other assumed a larger effect. It is unclear from the RIA how the two alternative estimates of this relationship were derived, but it appears that they are based on alternative treatments of the data used in the meta-analysis estimating the relationship between maternal hair mercury concentrations and IQ (see Table 9 in Ryan 2005).

28. Table 11.7 of the RIA also includes benefits estimates that assume a threshold at the World Health Organization's and the Canadian government's reference doses for mercury consumption. These two alternative reference doses are about twice as high as EPA's reference dose, and therefore the benefits estimates using these reference doses are lower.

29. The benefits to these communities under Option 1 are about 5 percent of the benefits that would result from eliminating 2020 mercury emissions from coal-fired boilers, whereas, for the general population, the benefits of the CAMR are about 20 percent of what they would be if emissions from coal-fired boilers in 2020 were eliminated.

30. The RIA also considers the case in which significant improvements in ACI occur, such that fewer fabric filters are needed to achieve a certain percentage reduction in mercury emissions. With this sensitivity, the benefit from reduced particulate matter emissions falls by 96 percent. One might also expect that the entire cost of the rule would fall if ACI was more effective. As shown in Section 7 of the RIA, the cost of the rule in 2020 is 25 percent lower in this case.

31. The court found that EPA unlawfully delisted the pollutant, failing to implement a formal process to reverse the previous finding, and therefore the pollutant must continue to be regulated under Section 112. To reverse the ruling, EPA's only recourse at this point is to petition the DC Circuit for a rehearing en banc or through an appeal to the U.S. Supreme Court.

References

Bellas, A., and I. Lange. 2005. Are Tradable Permits for Mercury Worthwhile? *The Electricity Journal* 18(2): 85–90.

Goulder, Lawrence H., Ian W.H. Parry, and Dallas Burtraw. 1997. Revenue-Raising Versus Other Approaches to Environmental Protection: The Critical Significance of Preexisting Tax Distortions. *Rand Journal of Economics* 28(4): 708–731.

Griffiths, C. 2008. Personal communication with the author, May 28.

New Jersey v. EPA, 05-1097 (DC Cir., Feb. 8, 2008).

Office of Management and Budget (OMB). 1992. OMB Circular A-94, Guidelines and Discount Rates for Benefit-Cost Analysis of Federal Programs. www.whitehouse.gov/omb/circulars/a094/a094.html (accessed December 6, 2008).

———. 2003. OMB Circular A-4, Regulatory Analysis. www.whitehouse.gov/omb/circulars/a004/a-4.pdf (accessed October 30, 2008).

———. 2007. Executive Order 12866. Regulatory Planning and Review. www.whitehouse.gov/omb/inforeg/eo12866/index_eo12866.html (accessed October 30, 2008).

Ryan, L.M. 2005. *Effects of Prenatal Mercury Exposure on Childhood IQ: A Synthesis of Three Studies.* Report to the U.S. Environmental Protection Agency. EPA-HQ-OAR-2002-0056-6048 and EPA-HQ-OAR-2002-0056-6049. www.regulations.gov (accessed November 24, 2008).

Salkever, D. 1995. Updated Estimates of Earnings Benefits from Reduced Exposure of Children to Environmental Lead. *Environmental Research* 70: 1–6.

U.S. Environmental Protection Agency (EPA). 1997. *Mercury Study Report to Congress.* Washington, DC: EPA.

———. 2000. *Guidelines for Preparing Economic Analyses.* Washington, DC: EPA.

———. 2001a. *Integrated Risk Information System: Methylmercury (MeHg)* (CASRN 22967-92-6). www.epa.gov/iris/subst/0073.htm (accessed November 1, 2008).

———. 2001b. Regulatory Finding on the Emissions of Hazardous Air Pollutants from Electric Utility Steam Generating Units. *Federal Register* 65(245): 79825–79831.

———. 2003. *Regulatory Flexibility Act Analysis for Utility MACT Proposed Rulemaking.* Office of Atmospheric Programs, Clean Air Markets Division. Washington, DC: EPA. www.epa.gov/ttn/atw/utility/utilityflexact_121503.pdf (accessed November 5, 2008).

———. 2004a. *Benefit Analysis for the Section 112 Utility Rule.* Office of Air Quality Planning and Standards, Air Quality Strategies and Standards Division. Research Triangle Park, NC: EPA. www.epa.gov/ttn/atw/utility/proposal utilitymactbenefitsanalysisfinal.pdf (accessed November 5, 2008).

———. 2004b. *Economic and Energy Impact Analysis for the Proposed Utility MACT Rulemaking.* Office of Atmospheric Programs, Clean Air Markets Division. Washington, DC: EPA. www.epa.gov/ttn/atw/utility/utilitymact econ_ 012804.pdf (accessed November 5, 2008).

———. 2004c. Proposed National Emission Standards for Hazardous Air Pollutants; and, in the Alternative, Proposed Standards of Performance for New and Existing Stationary Sources: Electric Utility Steam Generating Units; Proposed Rule. *Federal Register* 69(20): 4651–4752.

———. 2004d. *Proposed Hg Trading Rule + IAQR\IPM Run Output EPA216_PM54—Proposed Hg Trading Rule + IAQR—Regional Summary Report.* EPA-HQ-OAR-2002-0056-0338. Washington, DC: EPA. www.regulations.gov (accessed May 15, 2008).

———. 2004e. Supplemental Notice for the Proposed National Emission Standards for Hazardous Air Pollutants; and, in the Alternative, Proposed Standards of Performance for New and Existing Stationary Sources: Electric Utility Steam Generating Units; Proposed Rule. *Federal Register* 69(51): 12397–12472.

———. 2005a. *Regulatory Impact Analysis of the Clean Air Mercury Rule.* Office of Air Quality Planning and Standards. Research Triangle Park, NC: EPA.

———. 2005b. Revision of December 2000 Regulatory Finding on the Emissions of Hazardous Air Pollutants from Electric Utility Steam Generating Units and the Removal of Coal- and Oil-Fired Electric Utility Steam Generating Units from the Section 112(c) List; Final Rule. *Federal Register* 70(59): 15994–16035.

———. 2005c. *Technical Support Document: Methodology Used to Generate Deposition, Fish Tissue Methylmercury Concentrations, and Exposure for Determining Effectiveness of Utility Emission Controls.* Washington, DC: EPA.

www.epa.gov/ttn/atw/utility/deposition_fish_tissue_mehg_exposure_oar-2002-0056-6186.pdf (accessed November 5, 2008).

———. 2005d. *CAMR 2005 Option 1 IPM Output—Regional Summary Report.* EPA-HQ-OAR-2002-0056-6144. Washington, DC: EPA. www.regulations.gov (accessed May 15, 2008).

———. 2005e. *IPM Output CAIR 2004 Final—Regional Summary Report.* EPA-HQ-OAR-2002-0056-6147. Washington, DC: EPA. www.regulations.gov (accessed May 15, 2008).

———. 2005f. Rule to Reduce Interstate Transport of Fine Particulate Matter and Ozone (Clean Air Interstate Rule); Revisions to Acid Rain Program; Revisions to the NO_x SIP Call; Final Rule, *Federal Register* 70(91): 25161–25405.

———. 2005g. Standards of Performance from New and Existing Stationary Sources: Electric Utility Steam Generating Units: Final Rule, *Federal Register* 70(95): 28605–28700.

———. 2005h. *Technical Support Document: Revision of December 2000 Regulatory Finding on the Emissions of Hazardous Air Pollutants From Electric Utility Steam Generating Units and the Removal of Coal and Oil-Fired Electric Utility Steam Generating Units from the Section 112(c) List: Reconsideration.* Washington, DC: EPA. www.epa.gov/ttn/atw/utility/TSD-112-final.pdf (accessed October 10, 2006).

———. 2005i. Revision of December 2000 Regulatory Finding on the Emissions of Hazardous Air Pollutants From Electric Utility Steam Generating Units and the Removal of Coal- and Oil-Fired Electric Utility Steam Generating Units From the Section 112(c) List: Reconsideration. *Federal Register* 70(208): 62200–62213.

———. 2005j. Standards of Performance for New and Existing Stationary Sources: Electric Utility Steam Generating Units: Reconsideration. *Federal Register* 70(208): 62213–62221.

———. 2006a. Response to Significant Public Comments Received in Response to: Revision of December 2000 Regulatory Finding on the Emissions of Hazardous Air Pollutants From Electric Utility Steam Generating Units and the Removal of Coal- and Oil-Fired Electric Utility Steam Generating Units from the Section 112(c) List: Reconsideration (70 FR 62200; October 28, 2005) and Standards of Performance for New and Existing Stationary Sources: Electric Utility Steam Generating Units: Reconsideration (70 FR 62213; October 28, 2005). Washington, DC: EPA. www.epa.gov/ttn/atw/utility/final_com_resp_053106.pdf (accessed October 10, 2008).

———. 2006b. Revision of December 2000 Clean Air Act Section 112(n) Finding Regarding Electric Utility Steam Generating Units; and Standards of Performance for New and Existing Electric Utility Steam Generating Units: Reconsideration. *Federal Register* 71(111): 33388–33402.

———. 2008. Clean Air Mercury Rule. Washington, DC: EPA. www.epa.gov/ttn/atw/utility/utiltoxpg. html (accessed November 5, 2008).

■ ■ ■

CHAPTER 6

The Mathematics of Mercury

CATHERINE A. O'NEILL[1]

The title for this chapter owes a debt to Cass Sunstein, who, in an article entitled "The Arithmetic of Arsenic," set out to consider the strengths and limitations of cost–benefit analysis (CBA) in the context of a concrete case study, the U.S. Environmental Protection Agency's (EPA) regulation of arsenic in drinking water.[2] Here I similarly aim to wade into the "muck and mire" of EPA's recent effort to regulate mercury emissions from coal-fired utilities to glean what lessons I can for regulatory analysis.[3]

In the first part, I provide a brief background on the nature of mercury contamination and the history of mercury regulation. In the second, I critique EPA's regulatory impact analysis (RIA) for its rule regulating mercury emissions from coal-fired utilities. Seven issues for regulatory analysis that the mercury rule brings to the fore are also identified. Finally, I close with a few observations for improving regulatory analysis assuming, in accordance with the premise for this report, that the existing executive orders—or something very close to them—continue to direct this analysis.

Mercury Contamination and Regulation

The Problem[4]

Mercury has long been known to be highly toxic to humans. Exposure to even small amounts of methylmercury can lead to irreversible neurological damage, placing the developing fetus and children at particular risk. Methylmercury exposure has also been associated with adverse cardiovascular effects in adults and is toxic to other species as well. It has been associated with an array of adverse effects in loons, kingfishers, ospreys, bald eagles, river otters, and mink.

Once released into the environment, mercury's behavior is complex and includes local, regional, and global components. Anthropogenic sources of mercury increasingly account for these releases, although natural processes contribute as well. Anthropogenic emissions in the United States are currently dominated by coal-fired utilities; they are deposited to surrounding land and water at varying distances from these sources. Mercury that enters water bodies becomes methylated by microorganisms present in these aquatic environments. Methylmercury is an extremely bioavailable form of mercury, readily taken up by fish in these waters. Methylmercury bioaccumulates in fish tissue, which in turn is a source of exposure to those species that consume fish. Fish consumption is the primary route by which humans are exposed to methylmercury.

Many fish species that humans rely on for food are highly contaminated with methylmercury. However, humans vary considerably with respect to fish consumption practices, and fish species vary considerably with respect to methylmercury concentration. As a consequence, exposure can differ considerably among people. Some Native Americans, Asian Americans, and low-income subsistence fishers are highly exposed. Members of fishing tribes consume fish in greater amounts, at higher frequencies, and in accordance with different seasonal or cultural constraints than do members of the general population. Members of fishing tribes in the Great Lakes region and elsewhere also rely on fish species—including walleye, muskellunge, lake trout, and northern pike—that are relatively highly contaminated.[5]

Based on studies of methylmercury's adverse human health effects, EPA has derived a reference dose (RfD) for methylmercury of 0.1 microgram per kilogram of body weight per day.[6] This RfD represents a threshold for exposure—in other words, the amount that EPA believes can be ingested each day over the course of a lifetime without adverse health effects.[7] According to a recent study, some 15.7 percent of women of childbearing age in the United States had blood mercury levels above EPA's RfD, thus posing a risk to a developing fetus.[8] Importantly, this study also found marked differences among women in groups characterized by race or ethnicity. Whereas 15.3 percent of self-identified "white" women of childbearing age had blood mercury levels above the RfD, this number more than doubles, to 31.5 percent, for women who identified themselves as "other," a category composed primarily of Native Americans, Pacific Islanders, those of "Asian origin," and those of "mixed race."[9]

As a consequence of mercury contamination, health and environmental agencies have had to issue fish consumption advisories recommending that children and women of childbearing age reduce or eliminate entirely their consumption of some fish species. In the 1990s, advisories about mercury were increasingly issued throughout the United States, with some states placing all of their lakes, rivers, and coastal waters under advisory. In 2001, widespread methylmercury contamination prompted the Food and Drug Administration and EPA to issue the first-ever national fish consumption advisory.

The Law

Federal Indian law

Many tribes in the Great Lakes region and elsewhere are party to treaties with the United States that recognize tribes' fishing rights. By means of these treaties, the tribes reserved their aboriginal rights to take fish throughout their customary fishing areas, while ceding vast portions of the land that now composes the United States.[10] Although the language differs from treaty to treaty, the guarantee each secures is similar. For example, the Treaty of 1837 between the Lake Superior Chippewa and the United States provides: "The privilege of hunting, fishing, and gathering the wild rice, upon the lands, the rivers and the lakes included in the territory ceded, is guaranteed to the Indians ...."[11]

Courts interpreting the treaties as a matter of U.S. law have upheld and elaborated the treaty promises. In *Lac Courte Oreilles Band of Lake Superior Chippewa Indians v. Wisconsin*, the court explained that, by dint of the 1837 and 1842 treaties, the Chippewa were:

guaranteed the right to make a moderate living off the land and from the waters in and abutting the ceded territory and throughout that territory by engaging in hunting, fishing, and gathering as they had in the past and by consuming the fruits of that hunting, fishing, and gathering, or by trading the fruits of that activity for goods they could use and consume in realizing that moderate living.[12]

As the court here recognized, the treaty protections include not only tribal members' right to fish in the ceded area, but also their right to consume the fish they catch, or to sell it to others for others' consumption. Logically, if the fish to which tribes have rights are permitted to become so contaminated as to be unfit for human consumption, these treaty-guaranteed rights are greatly compromised.[13]

When it entered into the treaties with the fishing tribes, the United States bound itself and its successors to protect the tribes' right to take fish.[14] Indeed, as courts have observed, "the Indians viewed a guarantee of permanent fishing rights as an absolute predicate to entering into a treaty."[15] Notably, courts have affirmed that these treaties are the "supreme law of the land."[16] Federal agencies, including EPA, are required to consider and comply with the treaties when they make decisions affecting the rights secured by the treaties.[17] Federal agencies are bound, as well, by the trust responsibility and other legal obligations uniquely owed to tribes and their members.

Federal environmental law

Section 112 of the Clean Air Act comprises the comprehensive scheme for reducing hazardous air pollutants (HAPS), including "mercury compounds."[18] Although HAPS had been addressed by the Clean Air Act since 1970, they remained largely unregulated as the 1990 amendments were taking shape. Frustrated at this widely heralded failure, Congress enacted sweeping reforms to this section designed to address the inaction and delay that had plagued earlier versions of the act.

Congress set up a two-step process for regulating HAPS. First, EPA was directed to issue technology-based standards (known as maximum achievable control technology (MACT) standards) for those source categories listed under Section 112.[19] Congress established a 10-year schedule by which EPA was to list the source categories primarily responsible for emitting HAPS and to promulgate a MACT standard for each source category. Congress stipulated that sources were to be given a tight, three-year timeline to comply with the resulting emissions limits (with the possibility of, at most, a one-year extension). Second, EPA was directed to issue additional standards within eight years if this MACT standard left unaddressed any residual risk to human or environmental health. That is, under a Section 112 MACT-based approach, EPA is required in this second step to issue further regulations if necessary "to provide an ample margin of safety to protect public health... or to prevent... an adverse environmental effect."[20]

Several provisions of the 1990 Amendments evidenced particular concern for pollution problems involving mercury. among these, Section 112(n) tackled HAP emissions from utilities. EPA was directed to conduct and transmit to Congress two studies, one focusing on HAPS more generally and one focusing on mercury from these sources. Again, Congress specified tight deadlines for these tasks. Congress directed EPA to consider these studies and list utilities among the source categories to be regulated under Section 112 if it found such regulation to be "appropriate and necessary."[21]

EPA's Mercury Regulation

During this period, EPA took steps to regulate the major sources of anthropogenic mercury. In the 1990s, it issued standards for two of the top three categories of emitters—medical waste incinerators and municipal waster combustors—requiring that these sources reduce their mercury emissions on the order of 90 percent. In 2000, EPA listed the third of these major contributors, coal-fired utilities, among the source categories to be regulated under Section 112 of the Clean Air Act, having made the requisite finding under Section 112(n) that it was "appropriate and necessary" to do so. As a consequence of this listing, it was widely expected that EPA would require similarly significant reductions in utilities' mercury emissions. Crucially, it was also widely expected that these reductions would be realized quickly, given a deadline for promulgation of the MACT standard to which EPA had agreed to settle a lawsuit, and given the tight timeline for sources to comply with the standard specified by the act. Thus, up until the time EPA announced its proposed rule for coal-fired utilities in December 2003, observers looked forward to a MACT standard that would require coal-fired utilities to achieve roughly 90 percent reductions in their mercury emissions, and to do so by 2007.

Instead, EPA set out two alternative proposals to address mercury from coal-fired utilities: a cap-and-trade program (to be issued either under Section 112 or under Section 111), and a watered-down version of a MACT standard (one that would require only approximately a 55 percent reduction in emissions) under Section 112.[22] EPA's proposed rule was highly controversial. It fomented a record number of public comments, congressional hearings and requests for oversight, and considerable criticism from almost every quarter.

In its final rule, which it dubbed the Clean Air Mercury Rule (CAMR), EPA abandoned any pretense of providing a MACT standard. Rather, it opted for a cap-and-trade program, promulgated under Section 111. The CAMR instates a cap on mercury emissions from utilities in two phases. The Phase I cap is set for 2010 to require no additional reductions beyond those achieved as "co-benefits" of a companion rule, the Clean Air Interstate Rule (CAIR), governing criteria pollutants in the eastern portion of the country. Thus, the CAMR's first-phase cap is set to allow utilities to emit 38 tons of mercury per year—down from roughly 48 tons per year emitted by these sources at the outset of the program.

The Phase II cap is set for 2018 to allow utilities to emit 15 tons of mercury per year. However, given structural features of the cap-and-trade program, the 70 percent reduction in emissions that this second-phase cap represents will not actually be realized until well after the year 2020,[23] and perhaps even as late as the 2030s.[24] Note, too, that the cap-and-trade program, issued as it was under the auspices of Section 111, makes no provision for addressing any residual risk to human health or the environment, as would have been required under Section 112.

The rulemaking process was marked by procedural irregularities and reversals-of-course on EPA's part.[25] For example, in the wake of EPA's 2000 finding that the regulation of utilities was "appropriate and necessary," a high-level multistakeholder working group labored diligently to determine an appropriate MACT standard. But sometime in the spring of 2003, EPA senior political appointee Jeffrey Holmstead ordered staff to develop a cap-and-trade program instead and the working group was disbanded without producing any further information on the feasibility, costs, or benefits of the MACT-based approach. In addition, in 2003, the agency had predicted that mercury-specific control technology would be available by 2007 and could achieve up to 90–95 percent reductions in emissions.

But by the time of its final rule in 2005, EPA changed its mind and claimed that such technology would "not be commercially available until 2010 or later." One consequence of this checkered history is that the RIA, which ordinarily would have accompanied the proposed rule, was missing. Instead, at this point, EPA offered a rough assessment of costs and benefits for the proposed rule that did not account for the benefits of reducing mercury itself (it focused mainly on the co-benefits of reducing particulate emissions). In fact, as Professor Rena Steinzor explains, Holmstead's abrupt decision to eschew MACT and embrace cap-and-trade "caught the Agency's economists off guard," and left them to scramble to produce the supporting economic analysis.[26] Thus, the RIA was not published until March 2005, alongside the final CAMR.[27]

The final CAMR met with a flurry of criticism. Congress issued a rare request for reconsideration. State after state declined to participate in EPA's cap-and-trade program, calling instead for more meaningful and immediate emissions reductions within their borders. Several states, tribes, and environmental groups sued EPA, and industry groups joined the fray on the other side. Ultimately, the DC Circuit in *New Jersey v. EPA* vacated the CAMR in February 2008, and rehearing en banc was denied shortly thereafter.

EPA's Regulatory Impact Analysis

EPA concluded that the total annualized cost of the CAMR in 2020 will be approximately $848 million, whereas the total annual monetized benefits will be $0.4 million to 3 million.[28]

EPA elsewhere offered alternative figures for both the costs and the benefits of the rule.[29] A figure of $50 million in benefits versus $750 million in costs was attributed to EPA officials in the press at the time that the final CAMR was announced.[30]

EPA calculated the costs of the CAMR in terms of coal-fired utilities' capital investments and operating expenditures for pollution controls together with costs stemming from additional fuel expenditures. The benefits of the CAMR were calculated in terms of the change in IQ decrements suffered by humans exposed in utero to mercury in recreationally caught freshwater fish from U.S. waters that EPA deemed attributable solely to utility emissions, after accounting for the implementation of the CAIR.[31] EPA concluded that "a typical child of freshwater fishers lost approximately 0.06–0.07 IQ points because of mercury exposure in 2001."[32]

EPA tallied these benefits by estimating the present value of the lifetime loss in earnings attributable to each point decrease in IQ, less the amount saved in educational costs avoided for each point decrease in IQ. EPA assumed that these benefits would not accrue until 10 to 20 years after the year 2020, given the lag in time that it estimated would occur between the mercury emissions reductions required by the CAMR and the expected environmental response, namely the reduction in fish tissue methylmercury.

EPA's Regulatory Impact Analysis: A Critique

Sunstein's examination of the regulatory analysis for the arsenic rule led him to conclude that, although CBA ought not determine regulatory outcomes, CBA is nonetheless "indispensable" to the decisionmaking process given the need to compile and organize the relevant data, to assess the effects of regulation in a way that is transparent rather than opaque, and to reveal "exactly why the decision to regulate … is genuinely difficult—and why, and where reasonable people might differ."[33] My examination of the regulatory analysis for the mercury rule has left me less sanguine about CBA.

At the outset, it must be stated that EPA faced a daunting task, given, among other things, the complexities and uncertainties of the problem at hand. Against this backdrop, any attempt to assess the impacts of mercury contamination and regulation would be susceptible to criticism. Nonetheless, I identify seven concerns raised by EPA's analysis of the CAMR that focus on principle and practice. The critique that follows is not exhaustive, but is meant to highlight a selection of issues: those that the CAMR uniquely or emphatically brings to the fore; those that are especially contentious, as between proponents and skeptics of CBA as a decisional tool; and those on which progress might be made through critical attention.

Slimmed Pickings

From the outset, the CAMR RIA served to obscure the range and contours of the alternatives on the table. Although proponents offer CBA as a means of enabling decisionmakers and the public to comprehend the various possible courses of action and to select thoughtfully among them, the CAMR RIA provides a cautionary tale: an RIA's usefulness in this regard depends mightily on how the questions are structured and how the alternatives are fashioned. Rather than informing deliberation, the RIA for the mercury rule was structured in a manner that thwarted comparison among the relevant options.

EPA framed its inquiry by asking, What are the incremental costs and benefits of the CAMR in 2020, assuming implementation of the CAIR? In so framing the question, EPA subtly crafted a new baseline—the world in 2020—by which time the benefits of the CAIR, the companion rule that addressed criteria pollutants in the eastern United States, would have been realized. This move, in turn, determined important aspects of both the alternatives to be analyzed and the outcomes of that analysis. Notably, it permitted EPA to exclude from consideration the chief alternative to EPA's preferred approach, which would have imposed its requirements prior to the RIA's 2020 baseline. Further, it permitted EPA to reassign to the CAIR a sizeable category of benefits otherwise attributable to mercury regulation.

In its RIA, EPA purported to consider various alternative scenarios, including its preferred option—a cap-and-trade approach with caps of 38 and 15 tons per year in 2010 and 2018, respectively. In addition to its preferred option, EPA considered an option assuming an identical cap-and-trade approach but with slightly different caps and an option assuming that utility-attributable mercury emissions were to be eliminated entirely in 2020. However, EPA only ran the numbers for these three alternatives relative to its new 2020 baseline, in which the benefits of the CAIR had already been realized.[34] Importantly, EPA did not include an alternative reflecting the primary competing regulatory approach, a Section 112 MACT-based approach. EPA's choices shaped the resulting analysis in important ways.

Costs of Delay

EPA prevented consideration of a crucial difference between the alternative approaches to mercury regulation that were at issue, namely time. EPA's CAMR significantly delays the reductions in mercury emissions relative to a Section 112 MACT-based approach. Indeed, EPA's CAMR delays meaningful emissions reductions for well over a decade—and perhaps as many as two or more decades—relative to the expected Section 112 MACT-based approach. Recall that, under the CAMR, the 70 percent reduction in emissions promised by the Phase II cap will likely not actually materialize until well after 2020, and perhaps into the 2030s. Under Section 112, by contrast, the roughly 90 percent reduction in emissions expected under a MACT standard would have been required by the end of 2007. In fact, this reprieve to sources was one of the most controversial aspects of EPA's rule. But EPA's RIA simply defined away this matter of delay.

Here, as elsewhere, the costs of delay are potentially large in dollar terms and unconscionable in human terms.[35] A sense of these costs is afforded if one considers methylmercury's neurodevelopmental effects. In view of this impact alone, the failure to control mercury emissions from coal-fired utilities can have irreversible consequences, affecting the intelligence and life prospects of the children in each new birth cohort who are exposed in utero to harmful levels of mercury.

Assuming, generously, that the CAMR will result in substantial reductions in mercury emissions by 2023, this represents a delay of 15 years relative to the compliance date for the 90 percent reductions expected under a MACT-based approach in 2007. This 15-year delay will visit permanent harm on millions of children. That is, based on calculations by Drs. Leonardo Trasande, Philip J. Landrigan, and Clyde Schechter, between 4,748,820 and 9,558,495 children will be born with cord blood mercury at levels associated with a loss of IQ in the 15-year period during which utilities enjoy a reprieve from regulation.[36]

This 15-year delay translates into $19.5 billion in the form of losses in future earnings for these children.[37] In each case, these estimates represent the harms attributable solely to mercury emissions and exposure from U.S. utilities.[38] Although these comparisons represent a rough cut,[39] they nonetheless provide a glimpse of the considerable costs—in terms of life prospects for our children, and in terms of social utility—of delay. In fact, the more recent work of Trasande and his colleagues adds to this estimate. By calculating the additional societal costs resulting from the increase in cases of mental retardation (MR; defined clinically as an IQ less than 70) suffered by those children exposed in utero to utility-attributable mercury emissions during the years 2005–2020, they found that more immediate and stringent emissions reductions could prevent an additional 4,450 cases of MR and save an additional $13.1 billion.[40]

Note that these comparisons reflect losses based on data from the general population; data more specific to particular, highly exposed populations provide another window on the costs of delay. Whereas Trasande et al. considered a general population, and concluded that the most highly exposed 5 percent of children in each birth cohort would suffer losses in IQ ranging from 1.60 to 3.21 points, John Persell of the Leech Lake Band of Chippewa considered Great Lakes tribal populations, and concluded that the average child in each birth cohort would suffer losses in IQ ranging from 6.2 to 7.1 points.[41] Persell employed a similar method to Trasande et al. but considered exposure consistent with fish consumption practices appropriate to these fishing peoples (e.g., tribal fish consumption rates; tribal exposure frequencies, including bolus doses, given extraordinary intake during certain seasons or in accordance with certain ceremonial practices;

locally important species, such as lake trout, whitefish, and walleye; and tribal data on local fish tissue methylmercury concentrations).[42]

Additionally, Native peoples in the Great Lakes and elsewhere have recounted in qualitative terms the numerous other costs of a delay in mercury regulation, including impacts to tribal health along interrelated physical, social, cultural, and spiritual dimensions.[43] The Aroostook Band of Micmacs, for example, described these additional costs of delay in comments to EPA, emphasizing the permanent, intergenerational nature of the loss: "Although many of our Tribal members continue to fish and consume fish despite [Maine's statewide] fish consumption advisory, there are many Tribal families that no longer engage in cultural practices associated with fishing, and are thus not passing these traditions to new generations of Tribal members. The loss of our cultural ceremonies, language, and songs associated with fishing represents a significant impact on our Tribe, and results in permanent loss of the culture which defines our Tribe."[44]

In the context of mercury regulation, the temporal aspects of EPA's choice were serious and central. Because a child exposed to mercury can suffer lifelong, irreversible harms, and because each year of inaction meant that a new birth cohort of children would be exposed, the public debate about mercury regulation should have (and did, in public fora) centered around not only the magnitude of the emissions reductions to be required, but also the timing of those reductions. Rather than using its RIA to reflect and inform this public debate on the temporal dimensions of the regulatory alternatives, however, EPA used its RIA to obscure and preempt this debate.

The agency never provided a direct comparison between a Section 112 MACT-based approach and the Section 111 cap-and-trade approach that composes the final CAMR. And by shifting baselines, it presented obstacles to anyone trying to gauge this comparison. When pressed, moreover, as to why it had not estimated the costs and benefits of a Section 112 MACT-based approach, EPA responded that it did not do so because it had already decided to put forth a Section 111 cap-and-trade approach.[45]

Benefits Shell Game

By crafting its new baseline, EPA could also reassign an entire category of co-benefits, permitting these to be attributed not to a mercury rule, but to the CAIR. At some point between its CBA for the proposed rule and its CBA for the final rule, EPA decided to reallocate the co-benefits of controlling emissions from utilities, moving them from the mercury rule to the CAIR. Recall that, had the agency proceeded with a Section 112 MACT-based approach, sources would have been required to control for mercury within three years, that is, as early as 2007.

Controls designed to reduce mercury emissions would have garnered co-benefits in the form of reduced particulate emissions, beginning in 2007 when sources came into compliance with the MACT standard. EPA estimated these co-benefits to amount to roughly $15 billion. EPA had initially assigned these co-benefits to the mercury rule, an assignment that contributed significantly to EPA's finding a favorable benefit-to-cost ratio of 16 to 1 for its proposed MACT standard for coal-fired utilities. In its final estimate for the CAMR, however, EPA found the costs of mercury regulation to far outstrip the benefits.

As observed by James E. McCarthy of the Congressional Research Service, "[t]he primary change appears to be a reassignment of the $15 billion in particulate matter co-benefits to the CAIR rule. By making implementation of mercury controls simultaneous with CAIR, the co-benefits are attributed to CAIR, instead of to the mercury rule.... Some of this change is simply a paper exercise: the co-benefits are taken from one rule and given to another."[46]

In fact, by shifting the baseline such that the CAIR is incorporated as a given, a portion of the benefits of mercury emissions reductions themselves come to be seen as co-benefits of the CAIR. EPA's baseline in effect siphoned off from a mercury rule all but the incremental benefits of reductions in mercury after 2020 in a post-CAIR world. But an earlier baseline would have attributed much of this same roster of benefits and co-benefits to the regulation of mercury. This is not to suggest that these benefits ought to have been double-counted—which would clearly be inappropriate—but to highlight the considerable impact of EPA's choices on the apparent bottom line for the regulation of mercury.

As a consequence, the CAMR RIA seems less a tool to facilitate thoughtful comparison among the benefits offered by the various options and more a device to belittle the benefits afforded by regulating mercury from coal-fired utilities at all.

Cost or Benefit?

The CAMR RIA demonstrates that CBA is not a means merely of tallying up what are obviously costs and obviously benefits. Rather, impacts must be assigned to the cost or the benefit side of the ledger, an assignment that will often require a judgment of value. In fact, there may be real disagreement over whether a given impact should be understood as a negative or a positive consequence.

In the RIA, EPA recognized that one consequence of mercury contamination is neurological damage to humans exposed in utero, manifested in part by a decrease in IQ. EPA counted as a benefit of regulation, then, that this adverse impact would be alleviated. It measured this benefit in terms of the loss in future earnings that would be expected to accompany a decrease in IQ. But EPA understood neurological damage to have a silver lining: children with lower IQs will seek fewer years of education, and so save society the costs of educating these individuals (measured as the direct costs of educational services together with the opportunity costs of work forgone).[47] A cost of regulation, by EPA's lights, was that it would eliminate this positive effect of mercury contamination. As Steinzor puts it, from EPA's perspective, "the good news is that stupider children need less school and earn just a little more money because they are working rather than sitting in a classroom."[48]

But members of the public saw things differently; they understood mercury's neurodevelopmental impacts to be an unmitigated harm. In comments to EPA, the Children's Health Protection Advisory Committee, for example, lamented the fact that children exposed prenatally "will likely have to struggle to keep up in school and might require remedial classes or special education."[49] The Bad River Band of Lake Superior Tribe of Chippewa Indians cited mercury's links to "learning problems" and other effects, and concluded that "[i]t is unacceptable to continue to let our children be exposed to such a dangerous toxin."[50] These and other commenters used value-laden terms to describe mercury's harms in the real world and to decry the fact that much of the damage is visited on children, who are particularly vulnerable members of society.

Mercury contamination affects humans and the ecosystems of which they are a part in numerous and diverse ways, some of which are poorly understood and some of which are differently appreciated. There may be wide agreement among economists and the public about whether many of the relevant effects ought to be viewed as negative or positive consequences: an increase in consumers' electricity bills is a *cost* of regulation; a decrease in neurological damage to children is a *benefit* of regulation. But, as the RIA shows, there may be profound disagreements even here, at this most basic step in the method of CBA.

Economists seek to ensure that "the widest practicable range of benefits and costs" has been included in each CBA.[51] Whether a given impact constitutes a cost or a benefit, however, tends to be treated as if it were obvious—a brute fact about the world.[52] But nothing in economists' methods provides an objective basis for making the call. Is it a good or a bad thing when children with diminished IQs opt to enter the workforce directly rather than pursue further education? The assignment of such an impact to one side of the ledger or the other turns out to be more a matter of art than science.[53]

To economists, the assignment that EPA made in its RIA may well be unobjectionable. Society does save an amount of money when the children exposed to mercury grow up to demand fewer years of schooling. And this amount cuts in the opposite direction of the loss society incurs when these children are left with a diminished earning capacity. If one is going to count the latter, economists might argue, one ought, for the sake of comprehensiveness, to weigh this against the former.[54] But, although economists might find EPA's call defensible in terms of method, the implications of EPA's assignment are clearly disturbing to many: it makes the case for more, rather than less, of a contaminant that leaves us with neurologically damaged children.

In the end, this aspect of the CAMR RIA highlights an important criticism of CBA: in the context of environmental policy decisions, economists' work has not been (and cannot be) confined to the value-free realm of "questions about the correct measure of benefits and costs."[55] Although offered in the positivist tradition, as an objective social scientific tool,[56] CBA's practitioners cannot avoid making judgments of value as well as findings of fact.

A Partial Accounting

The CAMR RIA provides an accounting of the costs and benefits of mercury regulation that is partial—in both senses of the term. The RIA shows CBA to be a tool that is highly malleable, given the context in which it is employed for environmental policy analysis. The RIA also shows CBA to produce an incomplete assessment of the benefits of environmental regulation, given the current state of the method.

CBA is highly malleable

The RIA's benefits analysis illustrates CBA's extraordinary malleability. EPA seems here to have taken every opportunity to choose inputs and make assumptions that minimize the apparent value of the benefits to be gained from reducing mercury. Examples litter the RIA:

- EPA narrowly circumscribed the exposed population: it counted only prenatally exposed individuals whose mothers eat freshwater fish caught by recreational anglers on inland U.S. lakes. Missing are all those exposed during childhood,[57] all those exposed via ingestion of freshwater fish caught commercially on inland U.S. lakes, and all those exposed via ingestion of nonfreshwater fish caught recreationally or commercially in coastal or other waters.[58] By EPA's own estimate, the exposed population it modeled for its primary benefits analysis "represents only 13% of total fish consumption in the U.S."[59]

- EPA chose a fish consumption rate, eight grams per day, that is less than half that of the general population according to its own more recent guidance (let alone the much greater rate for those who rely on fish for subsistence or who look to fish for cultural reasons).[60]
- EPA opted for a dose–response curve to relate maternal mercury levels to IQ decrements in children exposed in utero that is roughly one-third of that employed by Dr. Trasande and his colleagues—a team of specialists in pediatric medicine.[61]
- EPA based its calculation of the loss that would accompany an IQ decrement on dated figures for total lifetime earnings that produced a value roughly half of that employed by Dr. Trasande and his colleagues. If EPA's 1992 earnings data were to be presented in 2000 dollars for purposes of comparison, this value would be $472,465.[62] Transande et al. used data from 2004, which estimated total lifetime earnings at $1,032,002 for men and $763,468 for women.[63]
- EPA deemed too speculative the cardiovascular impacts of methylmercury exposure, whereas other analysts felt compelled to account for this consequence. The alternative benefits assessment undertaken by Glenn Rice and James K. Hammitt, of the Harvard Center for Risk Analysis, shows the significance of this single exclusion.[64] Whereas they estimated the benefits of mercury regulation to be $119 million, if one considers only the averted IQ decrements for those exposed in utero, as EPA did, this number soared to $4.9 billion, if one considers averted cardiovascular impacts in adults.[65]
- EPA undercounted those in "high-risk" populations. EPA constructed an estimate of the number of Chippewa children who will be exposed in utero, in an effort to account for high-risk populations, but used a census-based approach that, by its own estimate, likely undercounted the exposed population by some 50 percent.[66]

Even this short list makes two points. First, given the uncertainty and variability that characterize many of the necessary informational inputs, the occasions for choice were many. Second, in the George W. Bush EPA, the judgment calls all went one way. That is, although any given input to the CAMR CBA might have fallen somewhere along a plausible range, EPA seemed always to have selected the low end of the range when it came to assessing benefits. As a consequence, EPA's final benefits tally is so low that it anchors the various estimates produced at the time. The next lowest estimate, that by Ted Gayer and Robert Hahn of the American Enterprise Institute (AEI)–Brookings Joint Center for Regulatory Studies, is an order of magnitude greater than EPA's.[67]

Of course, EPA also offered sensitivity analyses, in which it purported to consider bounding assumptions for many of the relevant parameters. But the bottom line for EPA's primary benefits analysis was undeniably affected by judgment calls of the sort canvassed here.

CBA incompletely accounts for benefits

The RIA also illustrates the inability of CBA to produce a complete account of the benefits. Any benefits of mercury regulation that had not been—or cannot be—monetized simply went unaccounted for.

The CAMR RIA assessed the benefits of mercury regulation solely in terms of one human health endpoint, IQ decrements, "because it [had been] monetized."[68] This criterion served to winnow the benefits analysis. EPA counted only benefits to human health, and so excluded all benefits in terms of ecological health.[69] EPA considered only human physiological health, narrowly understood, and so excluded benefits in terms of economic, social, political, cultural, and spiritual well-being for the fishing tribes and, indeed, for other commercial and recreational fishers. Of these human physiological health benefits, EPA counted only neurodevelopmental effects and so excluded cardiovascular and other health effects.

Because the method calls for an accounting in dollars, the RIA's quantitative tally simply ignored any benefit of reducing mercury contamination that had not been monetized. If an impact—say, the fraying of the social fabric of a fishing tribe when fish, fishing, and the associated practices are no longer a part of members' daily lives and no longer a source of the intergenerational transfer of traditional ecological knowledge—had not been (or could not be) monetized, it was entered in the ledger as a "0" value. To be sure, EPA acknowledged that reducing mercury would bring about additional benefits that had not been quantified. But several concerns remain, including the point that such qualitative descriptions and caveats may tend to get left behind, whereas the quantitative account comes to dominate the public debate.

Even if one believed that, theoretically, every benefit can be monetized, in practical terms, every benefit has not been monetized. So, for the moment at least, we do not have a true cost–benefit analysis, but only what Professors Frank Ackerman and Lisa Heinzerling have termed a complete cost–incomplete benefit analysis.[70] Indeed, the CAMR RIA appears to bolster the claim that CBA operates in practice as a one-way ratchet, systematically understating the benefits of environmental and other regulations.[71] Whereas the RIA's estimate of the costs of mercury regulation is likely at least to be close (although, in the case of the CAMR, as elsewhere, it has already become clear that EPA's initial estimate of the costs is too high[72]), its estimate of the benefits is sure to be off. Given the current state of the method, much that is at stake is simply missing from the CBA calculus. And what is missing belongs overwhelmingly on the benefits side of the ledger.

The CAMR RIA raises the concern that, given the current state of the method, CBA produces a much less complete accounting of regulatory benefits than it does of regulatory costs. This asymmetry, moreover, can be exacerbated when those wielding the calculator are hostile to environmental regulation.[73]

You Are What You Earn

The CAMR RIA illustrates some of the difficulties with CBA's dollar metric. Many of the benefits of mercury regulation resist monetization. These benefits are realized in the form of children's life prospects undiminished by neurological damage; in the form of political and cultural self-determination on the part of the fishing tribes; in the form of treaty obligations honored by the federal government; and in the form of intact and functioning aquatic ecosystems. The problem of incommensurability—in this context, the point that society arguably values what is at stake in efforts to address mercury contamination in ways that cannot be captured in monetary terms—presents particularly challenging issues for proponents of CBA.[74]

The CAMR RIA assessed the benefits of mercury regulation solely in terms of the loss in future income that is estimated to accompany a decrease in IQ of those children exposed in utero. EPA derived its estimate by determining the present value of lifetime earnings for a person born in the

United States, which it calculated to be $366,021 (in 1992 dollars).[75] It then estimated the monetary value of a loss of an IQ point, assuming a 2.379 percent decrease in future earnings per one-point decrease in IQ, leavened by a 0.1007 percent decrease in future years of schooling, and its attendant costs.[76] The end result was an estimate that the average present value of net earnings losses per IQ point decrease is $8,807 (in 1999 dollars).[77]

The agency conceded that the loss-in-earnings method fails to account for many facets of the harms to humans as a result of methylmercury contamination.[78] For example, this method does not account for any increased medical costs that go along with neurological damage. Nor does it account for the anguish and suffering occasioned by this damage. EPA thus allowed that there might be a measurement problem, but suggested that it was one that could be corrected, in theory, if one were to use a better method of valuation, such as willingness to pay (WTP).[79]

But the problem is not simply a matter of getting an imprecise answer to the question, as proponents of CBA suggest. Rather, for many, it is a matter of asking the wrong question. A loss-in-earnings approach does not comport with many beliefs and ideals to which our society is deeply committed. For example, this approach is reductionist and nonegalitarian: it rests on a view that a person's worth is determined by his or her earning power. As such, it effectively values more highly those who are young, male, white, and rich.

As Ackerman and Heinzerling have argued, the implications for public policy are highly unpalatable in a society that holds dear the "ideals of democracy and equal treatment under the law, let alone the sacredness of every human being."[80] Moreover, as Ackerman and Heinzerling have pointed out, a particularly egregious consequence of the loss-in-earnings approach "is that it implies that the lives of retired people are worth nothing—or perhaps less than nothing, since they consume scarce goods and services without earning or producing any marketed goods themselves."[81] Taken to its logical conclusion, they observe, this perspective would suggest a net social benefit to a policy that kills off a lot of older people.[82]

As repugnant as this conclusion might sound to many people in the United States, it is probably even more profoundly at odds with the perspectives of the groups most affected by mercury contamination, namely, various Native peoples. For these peoples, elders are not the least valued, but among the most prized members of the community.[83] Their contributions—as holders of traditional knowledge, custodians of cultural practices, keepers of historical records, and guardians of the youngest tribal members—are recognized as irreplaceable, an important asset comprising the intergenerational legacy of the tribe.[84] Importantly, their value to the tribal community comes not chiefly from market-based employment, but from other contributions.[85] In fact, if elders must participate as earners in the market economy, their ability to perform traditional duties can be compromised.[86]

Proponents of CBA have proffered some responses to versions of this criticism. They have pointed out that EPA in practice, as in the CAMR RIA, employs an average figure for lifetime earnings, which does not distinguish among beneficiaries of mercury regulation on the basis of their earning potential. So, in effect, impacts to elders or to those born to tribes with astronomical unemployment rates (and so whose lifetime earning prospects are bleak) are valued as if they enjoyed the earning potential of the "average American"—that is, in the same dollar amount. Thus, they might argue, CBA, in practice, values each individual equally.[87]

Although this response may allay some of the relevant concerns, it does not address the more fundamental problem that, for many, what is at stake in addressing mercury contamination is understood in ways that are not commensurable with money. That is, even if analysts were to gauge the value of lifetime earnings by the highest earner in the United States, and so increase EPA's $8,807 figure several-fold, this problem would not be resolved. The problem is not that $8,807 is an incorrect answer to the question because it gives too small a dollar amount; it is that the question seeks an answer in dollars at all. CBA's requisite of monetization continues to pose serious hurdles for those who believe that one cannot price every facet of human and ecological health as if it were traded on markets—and that the attempt to do so is not only absurd, but an affront to things held sacred.

The Minnesota Chippewa Tribe provided comments to EPA that arguably suggest precisely this, that is, that the tribe values an environment uncontaminated by mercury differently than it values money:

> *Over the last several decades this toxic substance, mercury, has caused many human and ecological problems for Indian people. The potential impacts to Tribes who traditionally consume fish as a large part of their diet is alarming. . . . And, the human health impacts of mercury and other contaminants bear hardest on those who cannot speak for themselves, our children. Mercury is [also] known to seriously impact fish eating wildlife such as loons and mink. These animals are a value to the ecosystem they inhabit and they are clan symbols for Tribal members. If these animals are threatened, Tribal culture is threatened.*
>
> *For our Tribe, the stakes are high in this fight to limit mercury emissions. The science is clear, mercury is toxic and negatively impacting many facets of the health, well being, and social fabric we all value. With this in mind, it is unclear to me why there is a controversy surrounding efforts to limit mercury emissions to the best of our technical capacity, and in the most expedient fashion. If it is a cost and benefit question then I must ask what profits are worth the health of our children and grandchildren?*

Other tribal commenters spoke more directly to this point, stating that "the cost–benefit analysis performed by the EPA is wholly deficient with respect to tribes" because many impacts to tribes were "unquantifiable" by the method of CBA.[88]

Economists have attempted to respond to the unease with efforts to "price the priceless." They explain that the concept of *economic value* refers to a theoretical construct in which analysts infer monetary values from choices made by individuals reflecting "how important aspects of the environment are to them."[89] Thus, economists point out, they are not actually putting a price tag on, say, the Great Lakes. Rather, they are inferring the value—in monetary terms—of the Great Lakes to some person by looking at what she gives up (or says she would give up) to see the Great Lakes, such as the cost of travel to get to a viewing point on the shore, or to ensure that the Great Lakes are not contaminated by mercury, such as the additional cost of electricity supplied by a source that does not emit mercury. "To economists, the importance of things (tangible or intangible) is revealed by what a person will give up to obtain them. The lower bound on the value of the item obtained is equated to what is given up. If the thing given up was money, the value can be expressed in monetary units; otherwise, it is expressed in the natural units of the thing given up."[90] Economists, therefore, are confident that they can overcome the objections of those like the Minnesota Chippewa Tribe and render, in dollars, every facet of human understanding and experience—the importance of everything whether tangible or intangible.

Although more might be said about economists' efforts in this regard, two points might usefully be considered in view of the current context. First, economists describe a process of translating values to dollars that may do more than merely translate. As Professor Mark Sagoff has observed, economists take individuals' *preferences* to be their primary data, but preferences themselves are not observable facts about the world.[91] Rather, economists must *infer, discover,* and *elicit* preferences from people's behavior or statements. To do this in the context of environmental regulatory policy analysis, they must construct hypothetical projects or questions about which people are supposed to have a measurable WTP (because the point in policy analysis is to gather information on questions for which real markets do not exist). Economists have given considerable attention to the context in which individuals are placed to elicit preferences, seeking to conduct experiments that generate numbers as if there were a real market.[92] In fact, they have devoted a fair amount of research to designing surveys that produce numbers representing what people are *actually* willing and able to pay, given the hypothetical role in which they have been put. But notice that this virtual market, as Professor Louis Wolcher has explained, "becomes the framework that [economists] impose on the concrete flow of historical time."[93] This imposition may, in fact, be difficult to square with the actual position that people occupy within the concrete flow of historical time. Consider, for example, an economist's question to an Ojibwe parent about his WTP for his child's mercury chelation therapy, in order to infer the ways in which the existence of fish, uncontaminated with mercury, are important to him and to his people.[94] How does this question speak to the real and relevant history in which the fishing tribes already gave up vast tracts of land—not to mention other sacrifices—to secure their continued right to fish and consume fish as they had?[95]

Second, economists work to infer preferences from individuals' behavior, but, as Sagoff has demonstrated, people act, choose, vote, and even buy for reasons that are often complex, and not always self-evident.[96] In a multicultural society, moreover, these reasons are surely plural and diverse. Although there may be some advantages to be gained from the pursuit of a unitary metric along which comparisons can be made, there are also surely some losses. In fact, as the CAMR demonstrates, a need to reduce every relevant consideration to dollars may work as an obstacle to reasoned analysis, inasmuch as it flattens important qualitative dimensions of the effects of contamination and regulation that, as Sunstein once said, "are important in both life and law."[97] Thus, even if one assumes that an economist can assign a dollar value to the importance of fish to the Minnesota Chippewa Tribe, it seems that vital information has been sacrificed in the process. A dollar figure simply doesn't tell us as much as we might learn when we are told that mercury contamination threatens mink and loons, which are clan symbols for tribal members.

Proponents of CBA have grappled with some of the issues raised by this discussion, but they have yet to adequately allay all of the concerns raised by the requisite of monetization, for the quite good reason that this is not easy—and perhaps not possible—to do. Some proponents have usefully begun to explore analytical techniques that abandon a quest to monetize every impact and look instead to structure deliberation among the options in terms of natural units, concrete time and place, and real people. The CAMR RIA illustrates the real work that will need to be done if regulatory analysis is to surmount the limitations of its current dollar metric.

Justice Denied

The CAMR RIA illustrates that questions of justice present terrain that is not adequately comprehended by a CBA-dominated analysis. Given the route of exposure involved, those who consume relatively large quantities of fish will be among those most exposed to mercury in the environment. Various Native peoples, Asian Americans, and low-income subsistence fishers are disproportionately among the most highly exposed; as such, the burdens of mercury contamination are not equally distributed in the United States. CBA, however, is insensitive to questions of distributive justice—a point proponents concede.[98] That is, CBA is a tool that is meant to get at the costs and benefits of a decision in aggregate terms, at the societal level. It is not designed to inquire into who will bear the costs and who will reap the benefits of any particular decision, nor whether the decision ameliorates or exacerbates current inequities. Yet various executive orders instruct EPA to attend to matters of equity and justice. Executive Order 12866 itself directs each agency to seek the regulatory approaches that "maximize net benefits" and includes among these benefits "distributive impacts" and "equity."[99] Executive Order 12898 requires each agency to "make achieving environmental justice a part of its mission" and directs each agency to identify and address the "disproportionately high and adverse human health or environmental effects" of its actions. EPA did, in the context of the CAMR RIA, attempt an analysis of equity and disproportionate impacts.[100]

In the preamble to the final CAMR, EPA recognized that, in the absence of regulation, certain groups, including "low-income and minority populations," will disproportionately suffer adverse health effects, given their fish consumption practices.[101] EPA further acknowledged that these practices may have "economic, cultural, and religious" dimensions.[102] EPA explained that Executive Order 12898 requires it to "assess whether minority or low-income populations face risks or a rate of exposure to hazards that are significant and that 'appreciably exceed or is likely to appreciably exceed the risk or rate to the general population.'"[103] EPA's environmental justice inquiry consisted of two parts.

First, EPA satisfied itself that the relevant groups would be no worse off and, in fact, somewhat better off with the CAMR than with the status quo. On the positive side, EPA expected the rule "to lead to beneficial reductions in air pollution and exposures generally."[104] CAMR was also expected to have "a small negative impact through increased utility bills," which would be "shared among all members of society equally."[105] So those highly exposed would be better off with the CAMR than in the absence of the CAMR.

Second, EPA considered what it posed as a further question of distributive justice: whether the CAMR makes these groups *too much* better off. "To further examine whether high fish-consuming (subsistence) populations might be disproportionately benefited by the final rule (i.e., whether distributional equity is a consideration)... EPA conducted a sensitivity analysis [using fish consumption rates for Ojibwe in the Great Lakes region] focusing on the distributional equity issue."[106] EPA found the benefits to this group to be modest in absolute terms. Assessing the question through the lens constructed in its RIA, EPA found that "this group would accrue total benefits... of $6,300 to $6,700 in 2020 when using a 3 percent discount rate."[107] Thus, EPA concluded, "although Native American subsistence populations (and other high fish-consuming populations) might experience relatively larger health benefits from the final rule compared with general recreational anglers, the absolute degree of health benefits are relatively low (i.e., less than a 1.0 IQ point change per fisher for any of the locations modeled)."[108]

The first part of EPA's inquiry is laudable, so far as it goes. That is, attention to distributive justice seems at least to require that an agency assess whether a rule actually makes things worse. EPA was thus correct to ask whether its rule increased or decreased exposure to those whose exposure "appreciably exceeds" that of the general population.[109] EPA was also correct to consider who would pay for a given regulation, in other words, to ask whether the costs of its rule would be shouldered primarily by the poor or whether, as it found here, they would be "shared among all members of society equally."

But the agency stopped too short. EPA declined to ask whether "somewhat better off" meant "adequately protected." EPA's assurance that some degree of beneficial reductions in exposures would occur did not speak to its own calculation that as many as 45 percent of Native Americans would be left exposed above EPA's RfD for mercury, considering utility-attributable mercury emissions alone[110]—a rate of exposure that is surely "significant," and thus ought to have been a matter of concern under EPA's environmental justice analysis. EPA also declined to ask whether more significant and timely emissions reductions would go further toward ameliorating the fact that Native people "face risks or a rate of exposure" to methylmercury that "appreciably exceed(s) the risk or rate to the general population."

The second part of EPA's inquiry is troubling. EPA's take on the environmental justice inquiry, that is, its concern that high fish-consuming populations not be disproportionately benefited by the final rule, however, is not out of step with that urged by proponents of CBA. Proponents are fond of the claim that low-income communities and communities of color are the "net gainers" from environmental regulations.[111] Professor Sunstein, for example, cites a study of the effects of air pollution regulation in California, which found that the largest emissions reductions occurred in the poorest neighborhoods, but that much of the cost of these reductions was borne by those wealthy enough to purchase new cars, which were required to be outfitted with $1,000 to $2,000 worth of pollution control equipment.[112] These relatively wealthy individuals, according to Sunstein, had to pay "emissions penalties that many of the poor are avoiding."[113] The view that the poor in this example are "net gainers" and the rich are "net losers" is worth examining.

The poor might be viewed as net gainers if one considered only a snapshot in time, devoid of historical and social context. Considering only this snapshot, one might find that a quantum of benefits, for example, an amount of emissions reductions, or a decrease in neurological damage, or "total benefits of $6,300 to $6,700," would accrue to those who are poor, whereas only a lesser quantum of benefits would accrue to those who are rich. With no more context than this, a rule with this result appears inequitable on its face—a boon to the poor. But as soon as one contextualizes the inquiry, one learns that the poor communities and communities of color in the California study enjoyed the greatest emissions reductions relative to the "especially high pollution levels" to which they had previously been subjected—levels that meant 25 percent greater exposure to nitrogen dioxide (NO_2) in poor communities compared with wealthy ones for years prior to the air quality regulations studied.[114]

Thus, the notion of gain cannot reasonably or ethically be understood apart from an examination of the status quo. If one is concerned, as environmental justice advocates have suggested we ought to be, that the benefits and burdens of economic life have been systematically maldistributed, with the poor and people of color disproportionately among those suffering the harms of contamination, then one should question the characterization of regulation that remedies inequities in NO_2 exposure as a "net gain" to the poor. A problem with this view, then, is that, if pursued seriously, it could always be invoked to disqualify efforts to ameliorate a current maldistribution—or at least to support a claim that low-income communities and communities of color are "disproportionately benefited" by such efforts. More fundamentally, it presumes an allocation of entitlements, with the right to pollute at current levels comprising the relevant baseline.

To be fair, EPA has had relatively less time to develop its analytical techniques for the relevant environmental justice questions. The precise contours of an environmental justice or "equity" analysis are not completely specified on the face of the relevant executive orders, so EPA has worked to elaborate the requirements of this inquiry.[115] The understandings suggested by the environmental justice literature, however, have been countered by proponents of welfare economics-based approaches.[116] Professor W. Kip Viscusi, for example, has challenged the concern "that hypothetical individual risks not be too great" and urged that "a more meaningful and compelling risk equity concept is to have equity in terms of the cost per life saved rather than equity in terms of risk outcomes."[117]

In its analysis of the CAMR, EPA embraced such economists' understanding of the equity issues at play, substituting it for the conception developed in the environmental justice guidance and literature. This embrace led EPA to worry that the Ojibwe and other fishing peoples might be "disproportionately benefited" by the CAMR, a worry that ignores the current maldistribution of the burdens of mercury contamination; denies a long history of efforts to colonize and assimilate Native peoples; and displays a callousness to the impacts on real people—impacts on human well-being with aspects both practical and profound, given the "economic, cultural, and religious" significance of fish that EPA acknowledges.[118] In so doing, the agency presumed a contaminated baseline in which fish consumption advisories and large methylmercury body burdens are the starting points from which departures must be justified. This presumption, it should be noted, deviates considerably from the baselines embedded in the relevant statutory and legal directives, including those recognizing tribes' reservation of their fishing rights. From the tribes' perspective, this reassignment of entitlements is unsupportable legally or morally.[119]

Enhanced Oversight?

There is reason to doubt that the CAMR RIA served as a transparent vehicle to inform agency decisionmaking and permit oversight. Proponents of CBA hold out hope that, by increasing transparency, CBA will lead ultimately to better regulatory policy. Professors Matthew Adler and Eric Posner, for example, make this case: "[o]ne overlooked virtue of CBA is that it, more than other decision procedures, increases the transparency of agency decisions, thus facilitating oversight by elected officials and the public."[120] Although those outside EPA were perhaps unusually engaged in the debate surrounding mercury regulation, the CAMR RIA arguably did little or nothing to inform this debate. Part of the problem in this instance surely stems from the fact that the RIA came only late in the day: it was only made available when EPA published the final rule (with several rounds of revisions to EPA's estimates of both the cost and benefit estimates following months later). In addition, the CBA for the final rule bore almost no resemblance to the CBA for the proposed rule, given EPA's decision to abandon a Section 112 MACT-based approach in favor of its cap-

and-trade approach in the final rule. But there is reason to question whether even a more timely RIA would have enhanced the regulatory process by informing debate.

The CAMR RIA illustrates some of the issues and trade-offs in terms of complexity and accessibility and, with accessibility, meaningful oversight. As a preliminary observation, EPA's imposing RIA presents obvious barriers to access in terms of sheer heft. It is 566 pages long and includes a host of technical charts, graphs, and tables. It draws on (although, maddeningly, is not always consistent with) an additional layer of lengthy technical support documents, which are in turn supported by elaborate computer models. As Professor Steinzor observes, even if one wanted to understand only how EPA arrived at its dollar value for IQ points, one would be up against it: "[n]o one but an experienced team of economists with weeks of free time on their hands could possibly hope to evaluate these or any of the assumptions made in the [RIA]."[121]

To be sure, one must try to understand a considerable amount of information when one contemplates the problem of mercury contamination. Mercury is a complex pollutant, and its regulation no simple matter. Any reasonable effort to grapple with the issues will necessarily itself be complex, demanding sustained attention by anyone who hopes to comprehend what is at stake and for whom. But, although the CAMR RIA is touted as being highly sophisticated and complex, some of this complexity arguably was manufactured. For example, EPA constructed two highly elaborate scenarios—the "angler destination" and "population centroid" approaches—to determine how many people in the United States are exposed to methylmercury by consuming fish.

These circuitous approaches had EPA piecing together data on everything from the number of fishing licenses issued to the number of miles people are presumed willing to travel from their homes to go fishing. After pages of analysis, EPA derived two alternative estimates of what it viewed as the relevant exposed population (prenatally exposed infants born in 2001 whose mothers consume recreationally caught fish): 434,000 and 587,000 individuals (respectively).[122] EPA then ran these alternative numbers through most (but not all) of its scenarios for its benefits estimate.

But how much have we learned from this sophisticated presentation of these two scenarios? EPA here dazzled with detail, but never addressed the question begged by its approach, namely, is it appropriate to consider exposures only from the narrowly circumscribed universe of "recreationally caught freshwater fish" from inland waters when there are clearly other sources of exposure (fish caught in coastal waters, for example) to mercury emitted by U.S. utilities?[123] In fact, if EPA had not needed to shore up its choice to limit its benefits analysis to just this fraction of total fish consumption, EPA could readily have used the National Health and Nutrition Examination Survey (NHANES) results,[124] which provide empirical data on just how many women in the United States have elevated blood methylmercury levels associated with intake of contaminated fish, and from which an EPA scientist had already calculated that some 630,000 children were born each year in the period 1999–2000 with umbilical cord blood mercury levels above EPA's RfD.[125] The NHANES results had already been quoted in the media, cited in congressional hearings, and relied on to inform public debate; as such, they were probably more familiar and accessible to the public than either the "angler destination" or the "population centroid" approach constructed for the RIA.

Whatever the gains from such complexity, losses in accessibility and a consequent impairment of oversight are likely. This may be true even for high-level agency administrators and members of Congress. And it is certainly true for many members of the public, some of whom will be the ones left to bear the burden of methylmercury contamination left unaddressed. Here, those most affected by EPA's decision included, as the agency recognized, Native Americans, Southeast Asian

Americans, and low-income people who rely on fish for food. Yet, as Eileen Gauna has observed, members of such groups are generally less likely to have the technical expertise to pore over the agency's RIA or the financial means to hire "an experienced team of economists."[126]

The more sophisticated and voluminous the materials supporting regulatory decisions become, the larger these obstacles to public participation will loom. How can a low-income woman who fishes for food be expected to have the time to locate, digest, and comment on hundreds of pages of documents or to have the money to hire someone to do it for her? In view of this reality, proponents' claim that CBA will ensure transparent decisions and facilitate informed public oversight seems somewhat fanciful. Yet those affected often possess unique expertise: they may be the only ones able to alert an agency to relevant exposure data (for example, a survey of Ojibwe fish consumption practices) or to educate it about pertinent impacts (for example, the interrelated impacts to human and ecological health, from the perspective of the Minnesota Chippewa Tribe, when mink and loons, their clan symbols, are harmed by methylmercury contamination). With diminished oversight by elected officials and the public comes a loss in accuracy. Regulatory decisionmaking is not enhanced, but compromised.

In fact, it is not only CBA's sophistication that may thwart public participation and oversight; a potentially more problematic hurdle stems from the formal demands of the method. As noted above, the public was in fact highly engaged in the mercury rulemaking. EPA received a record number of public comments on its proposed rule, and additional comments on its subsequent rulemaking activities. Yet, because many of these points were not lodged in the form of, say, a quibble with the dollar value that EPA placed on an IQ decrement, they were taken not to speak to the CBA.[127] That is, because of the formal demands of the method, many comments appeared irrelevant. EPA made no attempt to translate such comments—for example, the Minnesota Chippewa Tribe's concern for methylmercury's threat to tribal culture—into a form that could be entered in the CBA ledger (if, indeed, translation were possible).

Nor did EPA recognize that such comments were sometimes protests to the use of the CBA method at all. These statements by the public were simply not registered by the CBA-centered RIA. Without any real conversation in this regard, it is hard to imagine that the CBA here actually helped decisionmakers and the public understand why the issues involved in regulating mercury were "genuinely difficult" and "why, and where, reasonable people might differ," as Sunstein hopes.[128]

Unconnected to Legal and Moral Obligations

The CAMR RIA addressed itself to questions unrelated to EPA's legal obligations to the tribes and untethered to its legal mandate under the Clean Air Act. Although proponents may see a role for CBA even if (or perhaps precisely because) the relevant statutes or other legal directives eschew a cost–benefit test, a tally of costs and benefits in these instances stands wholly apart from the appropriate bases for an agency's decision. This point raises questions about the appropriate role of CBA in regulatory analysis.

Consider, for example, the matter of tribal fishing rights, which are secured in many cases by treaty and protected in all cases as a matter of the federal trust responsibility. EPA recognized early on, in its preamble to the proposed rule, that "Native Americans . . . may rely on fish as a primary source of nutrition and/or for cultural practices."[129] EPA should have immediately been aware of the unique constellation of legal obligations and, arguably, normative considerations that governed its work. EPA was also reminded during the public comment period that tribes' treaty-protected fishing rights were impacted by the mercury rule and was alerted to the precise ways in which mercury contamination threatens the tribes' treaty fisheries.[130]

These threats might be thought of in three categories.[131] First, methylmercury contaminates fish tissue, harming directly the health of those tribal members who consume (or whose mothers consume) fish, in the form of neurological and cardiovascular damage. Second, methylmercury contaminates fish tissue and renders it less saleable to others, thereby impairing the tribes' treaty-protected rights to earn "a moderate living" by fishing. Third, methlymercury impairs various physiological functions in the fish and inhibits their ability to reproduce, ultimately causing depletion of the fisheries resource on which tribes are entitled to depend.

Although the RIA offered estimates of the impact of mercury contamination on tribal members' health, it said nothing—and the final rule said nothing—of the other dimensions of the treaty-protected rights that are threatened by mercury contamination. Indeed, the word *treaty* appears nowhere in the RIA.[132] In the end, it is unclear how or even whether EPA viewed its analysis as engaging the tribes' legally protected rights to fish.

EPA's inattention to tribal rights in the CAMR RIA may be attributable in part to the Bush administration's steadfast commitment to a predetermined set of objectives for regulating utilities' mercury emissions. Scholars who followed the rulemaking process have observed that it revealed an agency intent on providing a reprieve from regulation to coal-fired utilities; enamored of a cap-and-trade approach to regulating mercury; and determined to salvage as much as possible of the president's Clear Skies Initiative, which had failed repeatedly to persuade Congress.[133] The portrait of an agency so wedded to this agenda that it felt itself unfettered by the relevant statutory directives is arguably supported by the DC Circuit's stern rebuke to EPA in *New Jersey v. EPA*. On this view, the RIA may well have been pressed into service to justify the administration's predetermined ends.[134] That the RIA arguably did not serve here to cabin the agency's discretion is probably a source of disappointment for those proponents, such as Professor Sunstein, who see this role for CBA.

As I suggest in this chapter, however, questions remain as to whether and how CBA ought to figure in agencies' decisions, particularly those structured by laws that reject an efficiency criterion. That the CAMR RIA did not serve to remind EPA of the relevant treaties and other legal directives is perhaps unsurprising. As Sid Shapiro and Chris Schroeder have observed, a preoccupation with CBA "unhinges" regulatory analysis from the legal directives that govern agency decisions.[135] Environmental statutes, they point out, "almost never" embrace a cost–benefit criterion.[136] Instead, these statutes direct EPA to set a standard based on the best available technology or to balance several considerations and values, exclusive of cost. As such, they require EPA to ask questions that differ from those asked in a CBA, for example: What level of emissions control is "achieved by the best performing 12 percent of existing sources?"[137] How does EPA's decision bear on Ojibwe rights to "make a moderate living . . . from the waters . . . [by] fishing . . . as they had in the past?" Indeed, Shapiro and Schroeder point out, "since cost is not a consideration in setting the level of regulation in [many] statutes, CBA is irrelevant to the outcome."[138] The same, of course, could be said of the legal mandates that protect tribes' fishing rights, including the treaties and the federal trust responsibility.

But, as Douglas Kysar suggests, there may be a deeper problem. CBA may work subtly to unseat these legal mandates. Although moderate proponents have disavowed any designs on supplanting other decisional criteria with an efficiency-driven "super-procedure," there may nonetheless be reason for concern. CBA's method upends a host of determinations, including basic allocations of entitlements and rights, that have been made in democratic fora.[139] CBA proceeds as if the relevant determinations—U.S. recognition in treaty of tribal resources and rights, or the federal commitment in the Clean Air Act to require the maximum achievable reduction in HAP emissions—were up for grabs, to be renegotiated via economists' disciplinary lens, that is, on the basis of one's WTP.

How can the lessons of this case study assist in shaping regulatory analysis for the future?

Toward Improved Regulatory Analysis

There is surely wide agreement that, in the end, regulatory analysis ought to be designed to improve the quality of regulatory decisions. To improve regulatory decisions, we need to employ our best analytical tools. These tools ought to assist us, insofar as possible, in making an accurate and nuanced assessment of the problem at hand and the potential solutions to it.

Proponents seem to worry that, without CBA, there are no tools for rigorous regulatory analysis. Sunstein argues that CBA is "indispensable" to regulatory decisionmaking and states that "[w]ithout some effort to ascertain the effects of regulation, agencies are making a mere stab in the dark," intimating that it is CBA or nothing.[140] Richard Revesz and Michael Livermore similarly portray the options as being "gut-level decisionmaking" on the one hand or "economic analysis" on the other.[141]

But we need not "abandon reasoned analysis"[142] if we draw on multiple analytical tools from differing disciplinary perspectives. In fact, we could expect to enhance the quality of our analysis. Environmental problems are complex, as the case study of EPA's mercury regulation shows, and the expertise of multiple disciplines will need to be brought to bear to begin to solve them. Economics is one discipline that can make contributions, but it is not the only one.

Shapiro and Schroeder have recently outlined a pragmatic, problem-oriented approach to regulatory analysis that embodies this understanding.[143] This approach would be interdisciplinary, with the analytical tools of each discipline offered as an aid to deliberation, but with no single analytical approach purporting to incorporate every relevant consideration. Decisionmakers and the public would be expected to defer to each discipline—including economics—on matters within its sphere of competence, but to look elsewhere when the nature of the question dictated.

This approach would be problem-oriented, in that it would recognize that the regulatory questions are, in many instances, structured in advance by the governing statutes and laws. The regulatory analysis would, therefore, be framed so as to produce answers that are usable within the relevant legal structure. Finally, this approach would be sensitive to issues of justice, including intergenerational justice. These issues would not be defined by the normative commitments of welfare economics, nor would they be considered only as an afterthought to a decision evaluated on the basis of an efficiency criterion.

And we need not forgo rational analysis if we look to multiple individuals from differing cultural traditions to understand the impacts of contamination. Such an approach to regulatory analysis would enhance rationality because, as Sagoff urges, it would promote decisions that are reasoned, intelligent, and the product of open-minded deliberation that, importantly, countenances qualitative evidence, including evidence about purposes, values, and beliefs.[144] Instead of accepting only those inputs that can be quantified, such an alternative analytical approach would accept useful information and arguments from various disciplines, traditions, and sources. Instead of impoverishing the debate by excluding all effects that cannot readily be monetized, it would facilitate and enrich deliberation. And instead of limiting its knowledge base to the expertise of a single group or intellectual tradition, it would enhance accuracy by considering the often unique contributions of those affected.

Many economists, in fact, have propounded a view that is not at odds with this interdisciplinary approach to regulatory analysis. They have evidenced an understanding that their disciplinary contributions are important, but not outcome-determinative—"a tool, not a rule" for regulatory decisionmaking. This understanding, in fact, has supported research into, for example, the interplay among quantitative and qualitative inputs to decisions; the possibility of assessing impacts in terms of their natural units; and the practice of the art and science of economics. Further work in this vein seems useful, so that the tools of economic analysis inform, but do not take over decisions.

Ultimately, the task for regulatory analysis will be to harness the insights of economics, while avoiding the losses that attend a strict adherence to CBA as currently practiced. This task is necessary, for example, to contemplate the effect of mercury contamination on the generation of girls in the Leech Lake Chippewa tribe who, in the absence of meaningful regulation, will be advised to reduce or eliminate fish from their diets for more than half of their lives—throughout their childhood to age 20 (when they are vulnerable to neurodevelopmental toxins) and then throughout their childbearing years to age 44 (when they might expose a developing fetus to irreversible neurological damage).

If the losses that this would entail are understood in terms of loss in earnings, decisionmakers learn only that these girls will suffer a setback that is worth $5,372, in 1999 dollars. If, on the other hand, the losses that this would entail are understood in the ordinary, qualitative terms of public discourse, decisionmakers might come to appreciate the multiple and interrelated dimensions of the harms to these girls and to their people, with all their physiological, social, economic, cultural, spiritual, and political facets. To ensure that decisionmakers are not deprived of a rich and nuanced understanding, we ought to arm them with the information that economics can provide, but also with the information that economics can't provide.

■ ■ ■

Notes

1. I am indebted to David Driesen, Lisa Heinzerling, Amy Sinden, and my colleagues who participated in a work-in-progress colloquium at Seattle University, for helpful comments. I am also grateful to Mike Morita, for his excellent research assistance. Portions of this chapter draw on material published in an earlier article, Environmental Justice in the Tribal Context: A Madness to EPA's Method, 38 *Environmental Law,* 495 (2008).

2. Cass R. Sunstein, The Arithmetic of Arsenic, 90 *Georgetown Law Journal,* 2255 (2002) [hereinafter Sunstein, Arithmetic].

3. Thomas O. McGarity, Professor Sunstein's Fuzzy Math, 90 *Georgetown Law Journal,* 2341, 2376 (2002) (emphasizing the importance of wading into the "muck and mire" to examine cost–benefit analysis in context).

4. Unless otherwise noted, the account of mercury contamination and exposure that follows is taken from two prominent studies, which were to guide EPA during its rulemaking process: Committee on the Toxicological Effects of Methylmercury, National Research Council (NRC), *Toxicological Effects of Methylmercury* 175–81 (2000) [hereinafter NRC, *Methylmercury*]; Office of Air Quality Planning and Standards and Office of Research and Development, U.S. Environmental Protection Agency, *Mercury Study Report to Congress* (1997), available at www.epa.gov/ttn/ caaa/t3/reports/volume1.pdf [hereinafter *Mercury Study Report to Congress*].

5. *See* Catherine A. O'Neill, Mercury, Risk, and Justice, 34 *Environmental Law Reporter* (Environmental Law Institute) 11,070, 11,078–11,079 (Dec. 2004) [hereinafter O'Neill, Mercury, Risk, and Justice] (comparing levels of methylmercury in fish species consumed by various groups).

6. *Mercury Study Report to Congress,* supra note 4, at O-2.

7. NRC, *Methylmercury,* supra note 4, at 2 n.2.

8. Kathryn R. Mahaffey et al., Blood Organic Mercury and Dietary Mercury Intake: National Health and Nutrition Examination Survey, 1999 and 2000, 112 *Environmental Health Perspectives.* 562, 565 (2004).

9. Id. at 565. Among the categories of "race/ethnicity" employed by the National Health and Nutritional Examination Survey analyzed by Mahaffey et al. are the categories "[n]on-Hispanic white" and "[o]ther." With respect to the latter, the authors explain, "[p]articipants who designated themselves as Other include Native American Tribal people, individuals of Pacific Island origin, persons of Asian origin, and persons of mixed race who did not designate another category." Id.

10. Some tribes' rights to fish are not secured by treaty, but instead are protected by executive orders and other federal laws. *See,* e.g., *Parravano v. Babbit,* 70 F.3d 539, 546–47 (9th Cir. 1995).

11. Treaty with the Chippewas, art. V, July 29, 1837, 7 Stat. 536. *See also* Treaty with the Chippewas, art. II, Oct. 4, 1842, 7 Stat. 592.

12. 653 F. Supp. 1420, 1426 (W.D. Wis. 1987).

13. A similar logic supported the district court's finding in the second phase of *United States v. Washington,* which interpreted the treaties guaranteeing to the fishing tribes in the Pacific Northwest the right "to take fish." *United States v. Washington,* 506 F. Supp. 187 (W.D. Wash. 1980) (Phase II), vacated, 759 F.2d 1353 (9th Cir. 1985). There, the court reasoned that "implicitly incorporated in the treaties' fishing clause is the right to have the fishery habitat protected from man-made despoliation.... The most fundamental prerequisite to exercising the right to take fish is the existence of fish to be taken." Id. at 203. Although this opinion was vacated on what were essentially procedural grounds, its unassailable logic remained available to EPA in its deliberations. Since EPA's issuance of the final CAMR, note that the district court has reiterated this understanding in the particular context of the state's duty to refrain from diminishing fish runs by constructing or maintaining culverts that block fish passage. *U.S. v. Washington,* No. 9213RSM, slip op. at 11 (W.D. Wash. Aug. 27, 2007) (Subproceeding 01-01) (finding that the treaty negotiators "specifically assured the Indians that they would have access to their normal food supplies now and in the future" and that "[t]hese assurances would only be meaningful if they carried the implied promise that neither the negotiators nor their successors would take actions that would significantly degrade the resource.").

14. *See,* e.g. *United States v. Washington,* 520 F.2d 676, 685 (9th Cir. 1975) ("[N]either the treaty Indians nor the state... may permit the subject matter of these treaties to be destroyed.").

15. *United States v. Washington,* 873 F. Supp. 1422, 1437 (W.D. Wash. 1994).

16. *Worcester v. Georgia,* 31 U.S. (6 Pet.) 515, 531 (1832).

17. *See*, e.g., *Nw. Sea Farms v. U.S. Army Corps of Engineers*, 931 F. Supp. 1515, 1520 (W.D. Wash. 1996); but cf. *George E. Warren Corp. v. EPA*, 159 F.3d 616, 624 (D.C. Cir. 1998).

18. Clean Air Act, 42 U.S.C. § 7412(b) (2000).

19. Clean Air Act, 42 U.S.C. § 7412(d)(2) (2000).

20. Clean Air Act, 42 U.S.C. § 7412(f)(2)(A) (2000).

21. Clean Air Act, 42 U.S.C. § 7412(n)(1) (2000).

22. Proposed National Emissions Standards for Hazardous Air Pollutants; and, in the Alternative, Proposed Standards for Performance for New and Existing Stationary Sources: Electric Utility Steam-Generating Units; Proposed Rule, 69 *Federal Register* 4652 (Jan. 30, 2004) [hereinafter EPA, Proposed Mercury Rule]. For a discussion of the proposed rule, see, generally, Lisa Heinzerling and Rena I. Steinzor, A Perfect Storm: Mercury and the Bush Administration, 34 *Environmental Law Reporter* (Environmental Law Institute) 10,297 (April 2004); Lisa Heinzerling and Rena I. Steinzor, A Perfect Storm: Mercury and the Bush Administration, Part II, 34 *Environmental Law Reporter* (Environmental Law Institute) 10,485 (June 2004); and O'Neill, Mercury, Risk, and Justice, supra note 5. EPA's Office of the Inspector General criticized the proposed MACT standard as anemic. Office of the Inspector General, U.S. Environmental Protection Agency, Additional Analyses of Mercury Emissions Needed before EPA Finalizes Rules for Coal-Fired Utilities, at "At a Glance," and 11–16 (Feb. 3, 2005), available at www.epa.gov/oig/reports/2005/20050203-2005-P-00003.pdf ("Evidence indicates that EPA senior management instructed EPA staff to develop a Maximum Achievable Control Technology [MACT] standard for mercury that would result in national emissions of 34 tons annually, instead of basing the standard on an unbiased determination of what the top performing units were achieving in practice. . . . [T]he standard likely underestimates the average amount of mercury emissions reductions achieved by the top performing 12 percent of utilities, the minimum level for a MACT standard required by the Clean Air Act.").

23. U.S. Environmental Protection Agency, Methodology to Generate Deposition, Fish Tissue Methylmercury Concentrations, and Exposures for Determining Effectiveness of Utility Emission Controls 3, Tables 1.1 and 1.2 (2005), available at www.epa.gov/ttn/atw/utility/eff_fnl_tsd-031705_corr_oar-2002-0056-6301.pdf [hereinafter EPA, CAMR Effectiveness TSD] (figures for emissions reductions presented in kg/yr; author's conversions). According to EPA's models, under the CAMR in 2020, total national mercury emissions will be approximately 25 tons. This amounts to a 48 percent reduction from 1999 baseline emissions of approximately 48 tons.

24. *See* James E. McCarthy, *Mercury Emissions from Electric Power Plants: An Analysis of EPA's Cap-and-Trade Regulations.* Congressional Research Service Report 7–8 (Updated Jan. 13, 2006). "It appears that full compliance with the 70% reduction might be delayed until 2030," and noting that "EPA has not provided an estimate of the year in which the 70% reduction will be attained. The Integrated Planning Model [IPM], which the agency uses to calculate regulatory impacts, runs to the year 2030 and assumes that all allowances will be used by the end date. Discussions we held with EPA staff indicate that some think the allowances will be used more quickly (perhaps as early as 2025), while others think use of allowances will be stretched into the 2030s." Id. at n.24.

25. Rena I. Steinzor, *Mother Earth and Uncle Sam: How Pollution and Hollow Government Hurt Our Kids.* University of Texas Press. 103–125 (2008) [hereinafter Steinzor, *Mother Earth*] (recounting numerous irregularities and abrupt changes in course of mercury rulemaking at EPA).

26. Steinzor, *Mother Earth,* at 120.

27. Office of Air Quality Planning and Standards, U.S. Environmental Protection Agency, *Regulatory Impact Analysis of the Clean Air Mercury Rule Final Report* (2005), available at www.epa.gov/ttn/atw/utility/ria_final.pdf [hereinafter EPA CAMR RIA].

28. Standards of Performance for New and Existing Stationary Sources: Electric Utility Steam Generating Units; Final Rule, *Federal Register* 70 28,606, 28,642 (May 18, 2005) [hereinafter EPA, Final CAMR]. These figures assume a 3 percent discount rate. EPA also calculated costs and benefits assuming a 7 percent discount rate, arriving at $848 million in costs and $0.2 million to $2 million in benefits on this assumption. Id.

29. Upon reconsideration, EPA concluded that "the upper bound estimate of aggregate economic benefits of reduced IQ decrements from eliminating utility-attributable mercury exposure in 2020 after CAIR are approximately $50 million." U.S. Environmental Protection Agency, *Technical Support Document: Revision of December 2000 Regulatory Finding on the Emissions of Hazardous Air Pollutants from Electric Utility Steam Generating Units and the Removal of Coal- and Oil-Fired Electric Utility Steam Generating Units from the Section 112((c) List: Reconsideration* 35–37 (2005) available at www.epa.gov/ttn/atw/utility/tsd_oar-2002-0056-6303.pdf. [hereinafter EPA, *Reconsideration TSD*]. EPA presented benefits as "$50 million plus some additional amount from the consumption of commercial freshwater, estuarine, and aquaculture fish by the general public." Id. at 35. In addition, EPA had estimated costs of compliance to be $560 million in the technical documents accompanying the final rule; but later, upon reconsideration, suggested that $750 million "reflects our best estimate." Id. at 37, n.20; note that these figures differ from the $848 million figure cited above in that the $560 million and the $750 million figures reflect the costs to industry of compliance, whereas the $848 million figure reflects the cost to society at large. See McCarthy, supra note 24, at 11. EPA also cited a range of benefits from $0 to $190 million in its *Reconsideration TSD,* but later revised the "upper bound" of its benefits estimate to $210 million. EPA, *Reconsideration TSD* at 32–33; McCarthy, supra note 24, at 11.

30. Shankar Vedantam, New EPA Mercury Rule Omits Conflicting Data, *The Washington Post,* March 22, 2005, at A1 ("[EPA] officials said the health benefits were worth no more than $50 million a year while the cost to industry would be $750 million a year.").

31. EPA allowed, additionally, that the controls installed to reduce mercury under the CAMR could be expected to result in a slight reduction in emissions of fine particulate matter, saving up to seven lives annually, for monetized benefits of $1.4 million to $40 million per year. EPA, Final CAMR, supra note 28, at 28,642; EPA, CAMR RIA, supra note 27, at Section 12.

32. EPA, CAMR RIA, supra note 27, at 10-3.

33. Sunstein, Arithmetic, supra note 2, at 2259.

34. EPA also considered two other alternatives, each of which referenced a 2001 baseline: an estimate of the benefits that would result if utility-attributable mercury emissions were to be eliminated entirely in 2001 (assuming "base case" conditions), and an estimate of the benefits that would result from the co-benefits to be realized in 2020 through the implementation of the CAIR alone relative to this 2001 baseline. But EPA did not present any estimates of the benefits that would flow from its three main alternatives relative to this 2001 baseline. Instead, EPA reset the baseline (with a new "base case" as of 2020, assuming that the CAIR had already been implemented) and presented the benefits of the three alternative scenarios only vis-à-vis this new 2020 baseline.

35. *See,* generally, David M. Driesen, Is Cost-Benefit Analysis Neutral?, 77 *University of Colorado Law Review* 335, n.94 (2006) (citing sources discussing the costs of delay in issuing environmental health and safety regulation); see also, William J. Nicholson and Philip J. Landrigan, Quantitative Assessment of Lives Lost Due to Delay in the Reg-

ulation of Occupational Exposure to Benzene, 82 *Environmental Health Perspectives* 185 (1989) (estimating that an 11-year delay in regulating occupational exposure to benzene resulted in some 30 to 490 excess deaths to those exposed between 1978 and 1987).

36. Leonardo Trasande et al., Public Health and Economic Consequences of Methyl Mercury Toxicity to the Developing Brain, 113 *Environmental Health Perspectives* 590 (May 2005) [hereinafter Trasande et al., Economic Consequences of Mercury]; see also, Trasande et al., Applying Cost Analyses to Drive Policy That Protects Children: Mercury as a Case Study, 1076 *Annals of the New York Academy of Sciences* 911, 919 (2006) [hereinafter Trasande et al., Cost Analyses and Mercury Policy]. Trasande et al. enlisted recent data from the Centers for Disease Control and Prevention, finding that between 316,588 and 637,233 children are born each year with cord blood mercury levels greater than 5.8 µg/L, a level associated with loss of IQ; they assumed reductions on the order of 70–90 percent, in line with legislative proposals on the table at the time of the CAMR; and they concluded that the failure to control U.S. coal-fired utilities would cost $1.3 billion each year, tallied mainly in terms of lost future earnings. Id.

37. Id.

38. Note that no discount rate has been applied for those born in 2023 compared with 2007.

39. First, these figures calculate the benefits that would result from the complete elimination of utility-attributable mercury emissions, which overstates both the effect of a MACT-based approach (which would result in 90 percent reductions) and the effect of the CAMR (which would result in 70 percent reductions). In this respect, these figures probably underestimate the number of children harmed by the delay. (This assumption comports, however, with EPA's assumption for its IPM runs and for its upper-bound benefits analysis, so provides a useful basis for comparison.) Second, these figures compare benefits that would result when significant reductions are assumed to be achieved, respectively, in 2007 and in 2023, but this simplifying assumption does not account for the more modest reductions under the CAMR that are predicted to occur earlier, resulting from the operation of the Phase I cap in 2010 and, in some models, from structural features of the cap-and-trade program—namely, its banking mechanism. EPA estimates mercury emissions to be reduced by 21 percent in 2010 (from 48 to 38 tons). EPA, *CAMR Effectiveness TSD,* supra note 23. The Congressional Research Service puts emissions reductions at 35 percent in 2010 (from 48 to 31.3 tons). McCarthy, supra note 24, at 7, Table 2. In this respect, these figures probably overestimate the number of children harmed by the delay. On the other hand, it must be kept in mind that, under the second step of a Section 112 MACT-based approach, further emissions reductions might be required as early as 2015, which would mean that the figures above may come closer to an accurate estimate of the effect of a MACT-based approach after this point in time. Clean Air Act, 42 U.S.C. § 7412(f)(2)(A) (2000). Third, these figures obviously represent only a partial account of the harms wrought by delay, accounting as they do only for IQ decrements to prenatally exposed children. In this respect, these figures underestimate the costs of delay.

40. Trasande et al., Mental Retardation and Prenatal Methylmercury Toxicity, 49 *American Journal of Industrial Medicine,* 153, 156 (2006) [hereinafter Trasande et al., MR]. Note that these estimates include only the "direct" costs to society, such as the increased medical and other costs of caring for those damaged in utero by mercury from U.S. utilities, but exclude the "indirect" costs, "such as lost economic productivity due to morbidity." Id. at 156; Trasande et al., Cost Analyses and Mercury Policy, supra note 36 at 919.

41. Honorable George Goggleye, Jr., Chairman, Leech Lake Tribal Council, Speech at the International Conference on Mercury as a Global Pollutant, Madison, WI (Aug. 6, 2006); telephone interview with John Persell, Leech Lake Band Department of Natural Resources (Jan. 15, 2008).

42. Id.

43. *See,* e.g., Great Lakes Indian Fish & Wildlife Commision Staff, Tribal Perspective Shared at International Mercury Conference, *Mazina'igan: A Chronicle of the Lake Superior Ojibwe,* Winter 2006–07 at 1, available at www.glifwc.org/Publications/mazinaigan/Winter2006.pdf (noting disproportionate impact on tribes).

44. Letter from William W. Phillips, Tribal Chief, Aroostook Band of Micmacs, to U.S. Environmental Agency (Apr. 30, 2004), available at www.regulations.gov/fdmspublic/ContentViewer?objectId=09000064800ae485&disposition=attachment&contentType=pdf (providing comments on Proposed National Emissions Standards for Hazardous Air Pollutants; and, in the Alternative, Proposed Standards for Performance for New and Existing Stationary Sources: Electric Steam-Generating Units, Docket No. OAR-2002-0056-2483).

45. U.S. Environmental Protection Agency, Response to Significant Public Comments Received in Response to: Revision of December 2000 Regulatory Finding on the Emissions of Hazardous Air Pollutants From Electric Utility Steam Generating Units and the Removal of Coal- and Oil-Fired Electric Steam Generating Units from the Section 112(c) List and Standards of Performance for New and Existing Stationary Sources: Electric Utility Steam Generating Units 295–96 (2006).

46. McCarthy, supra note 24, at 11–12; note that utilities in the eastern portion of the United States would have to comply with the CAIR beginning in 2009 and 2010. U.S. Environmental Protection Agency, *Clean Air Interstate Rule: Basic Information.* www.epa.gov/interstateairquality/basic.html (last visited Jan. 19, 2008).

47. EPA, CAMR RIA, supra note 27, at 10-46 to 10-47.

48. Steinzor, *Mother Earth,* supra note 25, at 122.

49. Letter from Melanie A. Marty, Children's Health Protection Advisory Committee, to Michael Leavitt, Administrator, U.S. Environmental Protection Agency 6 (Jan. 26, 2004) [hereinafter CHPAC Comments] (providing comments to the Proposed Mercury Rule, Docket 2002-0056-5570).

50. Letter from Donald Moore Sr., Tribal Chairman, Bad River Band of Lake Superior Tribe of Chippewa Indians, to Micheal Leavitt, Administrator, U.S. Environmental Protection Agency (Apr. 19, 2004), available at www.regulations.gov/fdmspublic/ContentViewer?objectId=09000064800ac810&disposition=attachment&contentType=pdf (comments on Proposed Utility Mercury Reductions Rule, OAR 2002-0056-2118).

51. Raymond J. Kopp, Alan J. Krupnick, and Michael Toman, Cost-Benefit Analysis and Regulatory Reform: An Assessment of the Science and the Art 40 (Jan. 1997) (Resources for the Future Discussion Paper 97-19).

52. *See*, e.g., Office of Management and Budget, Circular A-4 (September 17, 2003), available at www.whitehouse.gov/omb/circulars/a004/a-4.pdf.

53. *See*, generally, Kopp et al., supra note 51; David Calonder, The Lost Art of Economics: Essays on Economics and the Economic Profession (2001); according to Chapter 9 of this report, ("In short, the nature of 'adverse ecological consequences' was incompletely specified, a not unusual occurrence for ecological impacts. Even what is 'adverse' implies some value judgment.").

54. Of course, one could always argue that additional impacts, cutting in each direction, ought to be included. For example, in the case of mercury regulation, EPA stopped short of including the societal costs when the IQ of those exposed to methylmercury dipped below 70, the clinical threshold for MR. See Trasande et al., MR, supra note 40; Trasande et al., Cost Analyses and Mercury Policy, supra note 36 (finding that the CAMR would prevent some 1,475 cases of MR and save $4.1 billion in societal costs, including lost productivity, increased special education costs, and increased health care costs).

55. Robert Stavins, as quoted by Douglas Kysar (Chapter 10 of this report).

56. Sidney A. Shapiro and Christopher H. Schroeder, Beyond Cost–Benefit Analysis: A Pragmatic Reorientation, 32 *Harvard Environmental Law Review* 433, 446–50 (2008).

57. Trasande and his colleagues cite, but do not quantify, additional adverse effects on those exposed as neonates and infants up to age two, when the blood–brain barrier remains vulnerable. Trasande et al., Economic Consequences of Mercury, supra note 36, at 594; see also Children's Health Protection Advisory Committee (CHPAC), observing "[i]n addition to exposure in utero, infants and children have ongoing dietary exposure to methylmercury. Children and infants are sensitive to mercury's effects because their nervous systems continue to develop until about age 20." CHPAC Comments, supra note 49, at 6.

58. *See,* e.g., Lisa Heinzerling et al., Mercury, Center for Progressive Reform Perspectives Series www.progres sivereform.org/perspectives/mercury.cfm (2005); Glenn Rice and James K. Hammitt, *Economic Valuation of Human Health Benefits of Controlling Mercury Emissions from U.S. Coal-Fired Power Plants* (Feb. 2005), available at www.nes caum.org/documents/rpt050315mercuryhealth.pdf (including, in assessment of costs and benefits, exposure to those consuming fish caught in coastal waters. Report for the Northeast States for Coordinated Air Use Management (NESCAUM)).

59. EPA, CAMR RIA, supra note 27, at 4-46.

60. EPA selected the mean fish consumption rate from its 1997 Exposure Factors Handbook, eschewing its more recent guidance from its Ambient Water Quality Standards Methodology, which suggested that the general population default rate should be more than twice this high, in other words, 17.5 grams per day. EPA, CAMR RIA, supra note 27, at 10-44. Among other things, EPA justified its selection of the 8-grams-per-day figure because it represents both consumers and nonconsumers of fish. But the choice to include nonconsumers has the effect of depressing the mean and, especially, high-end values, because of the inclusion of so many "0" values reflecting those who do not eat fish. Id. at 10-44. As I have argued elsewhere, this is an unsupportable choice in the context of public health regulation. Catherine A. O'Neill, Variable Justice: Environmental Standards, Contaminated Fish, and "Acceptable" Risk to Native Peoples, 19 *Stanford Environmental Law Journal* 3, 60–61, 80 (2000).

61. EPA ultimately assumes a relationship of –0.16 IQ points for each ppm of maternal hair mercury, whereas Trasande et al. calculate a relationship of –0.465 IQ points per ppm of maternal hair mercury. Note that the EPA CAMR actually gives this figure as –0.13 IQ points per ppm maternal hair mercury. EPA, CAMR RIA, supra note 27, at 9-7. However, EPA revised its estimate to –0.16 IQ points per ppm maternal hair mercury upon reconsideration. Charles Griffiths et al., A Note on Trasande et al., "Public Health and Economic Consequences of Methylmercury Toxicity to the Developing Brain," 8, n.3 (National Center for Environmental Economics, Working Paper No. 06-02, 2006), available at http://yosemite.epa.gov/ee/epa/eed.nsf/ffb05b5f4a2cf40985256d2d00740681/dd32a21a7da2bdf38525715500485642/$FILE/2006-02.pdf (explaining that this revision came in response to public comment). Transande and his colleagues present the dose–response curve in terms of ppb of mercury in cord blood; this figure can be converted into ppb of mercury in hair for purposes of comparison. Trasande et al., Economic Consequences of Mercury, supra note 36, at 591–92. The conversion here was undertaken by Griffiths et al., supra, at 8.

62. Translation by Griffiths et al., supra note 61, at 9; CAMR RIA, supra note 27, at 10-46 (citing a figure of $366,021, discounted at 3 percent).

63. Trasande et al., Economic Consequences of Mercury, supra note 36 at 592 (discounted at 3 percent).

64. Rice and Hammitt, supra note 58.

65. Rice and Hammitt, supra note 58, at xix.

66. EPA, CAMR RIA, supra note 27, at 10-113,10-120 to 10-122. Note that EPA then greatly understated the fish consumption rate for this population, a point that is taken up infra.

67. Ted Gayer and Robert W. Hahn, Designing Environmental Policy: Lessons from the Regulation of Mercury Emissions, *Regulatory Analysis* 05–01, 22, 33 (2005) (in 2004 dollars, depending on whether one assumes Model 1 or 2 for the rate of reductions, and depending on whether one employs a discount rate of 3 percent or 7 percent). Note that Gayer & Hahn assume the proposed version of the cap-and-trade program, which would have set the Phase I cap at 34 rather than 38 tons; as a consequence, they note, "[t]his may mean that our estimates slightly overstate the benefits as well as the costs of the final rule." Id. at 5–6.

68. EPA, Final CAMR, supra note 28, at 28,641 ("EPA determined that IQ decrements due to Hg exposure is one endpoint that EPA should focus on for a benefit analysis, because it can be monetized.").

69. EPA, CAMR RIA, supra note 27, at 10-1 to 10-2. To its credit, EPA acknowledged the omission of ecological benefits, which it was "unable to quantify," and the fact that the exclusion of these and other categories of benefits, taken together, means that its assessment "likely underestimate[s] the total benefits of reducing mercury emissions from power plants." Id.

70. Frank Ackerman and Lisa Heinzerling, *Priceless: On Knowing the Price of Everything and the Value of Nothing* 40 (2004).

71. David M. Driesen, Is Cost–Benefit Analysis Neutral?, 77 *University of Colorado Law Review* 335 (2006).

72. According to one analysis, EPA's estimate of the CAMR's costs "relies on estimates of mercury control costs that are 4 to 20 times higher than current projections by pollution control industry sources." McCarthy, supra note 24, at 9, 20. *See*, generally, Thomas O. McGarity and Ruth Ruttenberg, Counting the Cost of Health, Safety, and Environmental Regulation, 80 *Texas Law Review.* 1997 (2002); Winston Harrington, Richard D. Morgenstern, and Peter Nelson, On the Accuracy of Regulatory Cost Estimates, 19 *Journal of Policy Analysis & Management.* 297 (2000) (EPA tends to overestimate the costs of regulations); Ackerman & Heinzerling, supra note 70 at 37–39.

73. *See* Shapiro and Schroeder, supra note 56, at 450 (describing the call for "regulatory relief" as one of the primary motivating factors behind the adoption of CBA).

74. Ackerman and Heinzerling, supra note 70, at 39-40; Cass R. Sunstein, Incommensurability and Valuation in Law, 92 *Michigan Law Review.* 779 (1994) [hereinafter Sunstein, Incommensurability].

75. EPA, CAMR RIA, supra note 27, at 10-46. EPA uses earnings data from 1992 and employs a discount rate of 3 percent.

76. Id. (also discounted at a 3 percent rate).

77. Id. at 10-47. EPA informs, in a footnote, that the average present value of net earnings losses per IQ point decrease is $1,580, if one assumes a 7 percent discount rate instead. Id. at 10-47, n.17.

78. Id. at 10-47.

79. Id. (observing that its loss-in-earnings method is serviceable, nonetheless, because a "cost-of-illness estimate may be considered a lower bound estimate of WTP"); accord Griffiths et al., supra note 61, at 9, n.5 ("It should be noted that lost earnings from IQ loss is not the conceptually correct metric for valuing benefits of reduced mercury exposure. Ideally, we should use a measure of willingness-to-pay (WTP) to avoid neurological damage caused by mercury exposure.").

80. Ackerman and Heinzerling, supra note 70, at 72.

81. Id.

82. Lest someone think that Ackerman and Heinzerling's observation is far-fetched and would never see the light of day in a policy context, consider that economist W. Kip Viscusi undertook research that concluded that states, in fact, saved money when their citizens smoked: because smokers die early, states were saved the expense of providing elder care and other services associated with an aging population. This study was undertaken at a time when the question was very much in the public realm, as states were in litigation with the tobacco companies, seeking reimbursement for the medical costs the states incurred as a result of smoking. As Ackerman and Heinzerling note, "[a]ccording to Viscusi, the financial benefit to the states of their citizens' premature deaths was so great that, if some of his results were 'taken at face value,' then 'cigarette smoking should be subsidized rather than taxed.'" Ackerman and Heinzerling, supra note 70, at 72.

83. *See,* e.g., Swinomish Tribal Mental Health Project, *A Gathering of Wisdoms: Tribal Mental Health—A Cultural Perspective* 145–63 (1991). "Elders have a unique and honored place in Indian society.... Elders are the teachers and carriers of tradition. Their greater life experience, historical perspective, spiritual knowledge and closer ties to the old ways of tribal ancestors make them a valuable resource for younger people.... [Y]ounger people without elders may be considered 'poor.'" Id. at 154, 156.

84. Id.

85. Id. For example, "[o]lder people, especially grandparents, are often the primary teachers of children, and not infrequently are their primary care givers." Id. at 154.

86. *See,* e.g., Jamie Donatuto, environmental specialist, Swinomish Indian Tribal Community, "Risk in the Tribal Context," Presentation to EPA workgroup (Nov. 2007). This is a perspective that seems largely to be missing from the literature debating quality-adjusted life years and similar approaches that involve controversial judgments about the relative value of human life in its various stages. *See,* e.g., Robert W. Hahn and Scott Wallsten, Is Granny Worth $2.3 Million or $6.1 Million?, AEI–Brookings Joint Center Policy Matters 03-13 (undated article) available at www.aei-brookings.org/policy/page.php?id=138&.

87. In fact, they might suggest, EPA's practice can be seen as progressive—a sort of regulatory redistribution. That is, it justifies more protective mercury regulation than the actual earning potential of those most affected by the rule would warrant. *See,* e.g., Richard L. Revesz and Michael A. Livermore, *Retaking Rationality: How Cost–Benefit Analysis Can Better Protect the Environment and Our Health.* Oxford University Press 14, 82–84 (2008) (making an analogous argument in the context of EPA's use of average WTP or willingness-to-accept figures to determine the "value of a statistical life"). But this claim is circular. EPA's practice is "progressive" only if one accepts that the lives of those with little earning potential actually are worth less than those who can expect to earn more. If one believes, instead, that each individual is equally valuable, the use of average values does not appear redistributive.

88. Letter from Pearl Capoeman-Baller, Chairperson, National Tribal Environmental Council, to Michael Leavitt, Administrator, U.S. Environmental Protection Agency, Comments on the Proposed Utility Mercury Reductions Rule 5 (June 4, 2004) (providing comments to the Proposed Mercury Rule, Docket 2002-0056-2695).

89. Kopp et al., supra note 51 at 12.

90. Id.

91. Mark Sagoff, *Price, Principle, and the Environment.* Cambridge University Press 57–79 (2004) [hereinafter, Sagoff, *Price, Principle*].

92. Scott Farrow, Chapter 9 of this report.

93. Louis E. Wolcher, Senseless Kindness: The Politics of Cost–Benefit Analysis, 25 *Law and Inequality.* 147, 184 (2007).

94. *See,* e.g., Gayer and Hahn, supra note 67 at 20–21 (deriving the value of an IQ point from WTP surveys that asked precisely this question of parents in the context of lead chelation).

95. *See,* e.g., Stuart Harris et al., Presentation to the 8th International Conference on Mercury as a Global Pollutant (Aug. 8, 2006).

96. Mark Sagoff, *The Economy of the Earth: Philosophy, Law and the Environment.* Cambridge University Press (1988) (observing that people's choices as consumers are not the same as their choices as citizens); Sagoff, *Price, Principle,* supra note 91, at 64–66 (arguing that multiple reasons might explain even the choice to purchase Girl Scout cookies).

97. Sunstein, Incommensurability, supra note 74, at 797. In this earlier article, Sunstein observed that people value things, goods, relationships, and states of being in qualitatively different ways, and that these values cannot without significant loss be reduced to a single metric, such as money or utility. Not only would the metric fail adequately to describe experience, but also, crucially, it "would actually transform it, in a way that would make a great deal of difference . . . because it would elide certain qualitative differences that are important in both life and law." Id.

98. Matthew D. Adler and Eric A. Posner, *New Foundations of Cost–Benefit Analysis.* Harvard University Press, 156 (2006) ("[W]e happily concede that CBA does not track deontological, egalitarian, or non-welfare-based values."); Matthew D. Adler, Risk Equity: A New Proposal, 32 *Harvard Environmental Law Review,* 1, 2 (2008) ("[T]he net-benefits-maximization test of traditional cost–benefit analysis is insensitive to distributional considerations.")

99. Exec. Order No. 12,866 § 1(a), *Federal Register* 58 51,735 (Sept. 30, 1993).

100. *See* EPA, CAMR RIA, supra note 27, at 10-121 ("equity analysis"); EPA, Final CAMR, supra note 28, at 28,648.

101. EPA, Final CAMR, supra note 28 at 28,648.

102. Id.

103. Id.

104. Id.

105. Id.

106. Id.

107. Id.

108. Id.

109. This is not necessarily to say that EPA's conclusion was satisfactorily supported, only that it asked a correct question. See O'Neill, *Mercury, Risk, and Justice,* supra note 5 (discussing potential for hot spots under cap-and-trade actually to make things worse in some locations).

110. EPA, CAMR *Effectiveness* TSD, supra note 23. These values assume a scenario of highly contaminated fish—in other words, methylmercury contamination held at the 99th percentile—a reasonable assumption for many tribal fishers and their families, given that the species traditionally consumed are highly contaminated (e.g., walleye, pike, and others, for the Great Lakes tribes). On these assumptions, exposures will be above the RfD, considering only utilities' emissions, for all those consuming at or above the 55th percentile for this population. Id.

111. *See,* e.g., Matthew E. Kahn, The Beneficiaries of Clean Air Act Regulation, 24 *Regulation* 34 (Spring 2001) ("What we find is that better educated, wealthier populations do experience cleaner air, but that poorer, less educated populations have experienced a greater overall improvement in air quality between 1980 and 1998.").

112. Cass R. Sunstein, Willingness to Pay versus Welfare. Public Law & Legal Theory Working Paper No. 150 7 (Jan. 2007), available at www.law.uchicago.edu/academics/publiclaw/150.pdf (citing Matthew E. Kahn, The Beneficiaries of Clean Air Act Regulation, 24 *Regulation* 34 (2001)).

113. Id.

114. Id.

115. *See,* e.g., U.S. Environmental Protection Agency, *EPA Guidance for Consideration of Environmental Justice in Clean Air Act Section 309 Reviews* (1999).

116. Compare Robert Kuehn, A Taxonomy of Environmental Justice, 30 *Environmental Law Reporter* (Environmental Law Institute) 10,681 (2000); Sheila R. Foster, Meeting the Environmental Justice Challenge: Evolving Norms in Environmental Decisionmaking, 30 *Environmental Law Reporter* (Environmental Law Institute) 10,992 (2000); and Catherine A. O'Neill, Environmental Justice in the Tribal Context: A Madness to EPA's Method, 38 *Environmental Law,* 495 (2008) [hereinafter, O'Neill, Environmental Justice in the Tribal Context] with W. Kip Viscusi, Risk Equity, 29 *Journal of Legal Studies* 843 (2000); and Mathew D. Adler, Risk Equity: A New Proposal, 32 *Harvard Environmental Law Review,* 1 (2008).

117. Viscusi, supra note 116, at 843, 855.

118. *See* O'Neill, Environmental Justice in the Tribal Context, supra note 116.

119. *See,* e.g., Lynda V. Mapes, Culverts Add Obstacles to Salmon, State, Politics, *Seattle Times,* Jan. 24, 2008. Mapes reports that, although tribes' treaty rights were recently reiterated in court, the State of Washington, the defendant in the case, is citing the large costs of fixing culverts that block habitat and so deplete salmon populations. She cites Billy Frank, Jr., a Nisqually tribal elder and chairman of the Northwest Indian Fisheries Commission: "Frank, for one, likes to remind people that amid all the grumbling about the costs of fixing culverts and rebuilding salmon runs, non-Indians enjoy uncountable economic prosperity from the lands the tribes gave up in the treaties so long ago. In fighting to get the culverts fixed, tribes are simply seeking their part of the bargain, Frank said." Id.

120. Adler and Posner, supra note 98, at 101.

121. Steinzor, *Mother Earth,* supra note 25, at 122.

122. EPA, CAMR RIA, supra note 27, at 10-23, Table 10-5.

123. Compare the analysis by Rice and Hammitt, for example, which included fish caught in three coastal regions. See Rice and Hammitt, supra note 58.

124. Centers for Disease Control and Prevention, National Center for Health Statistics, National Health and Nutrition Examination Survey (NHANES), www.cdc.gov/nchs/nhanes.htm.

125. Mahaffey et al., supra note 8 at 565. Note that Mahaffey's analysis of the NHANES data was relied on by Trasande et al. *See* Trasande et al., *Economic Consequences of Mercury,* supra note 36, at 591.

126. Eileen Gauna, The Environmental Justice Misfit: Public Participation and the Paradigm Paradox, 17 *Stanford Environmental Law Journal* 3 (1998).

127. Griffiths et al., supra note 61, at fn.2 suggests that EPA revised its upper bound estimate of benefits from $168 million to $210 million in response to public comments; Griffiths cites the responsiveness summary as the document in which this upward revision occurred. Note that this document was not published until 2006, well after EPA had published its final rule.

128. Sunstein, Arithmetic, supra note 2, at 2259 ("[CBA] is indispensable to informing the inquiry and to ensuring that [agency] discretion is exercised in a way that is transparent rather than opaque.... At the very least, an understanding of the data helps show exactly why the decision about how to regulate [arsenic and similar toxic substances] is genuinely difficult—and why, and where, reasonable people might differ. This is itself a significant gain.").

129. Proposed Mercury Rule, *Federal Register* 69 4652, 4709.

130. Brief of Petitioners National Congress of American Indians, *New Jersey v. EPA* Case 05-1097 (DC Cir., Jan. 12, 2007).

131. Id. at 21–22.

132. In fact, according to the brief for the tribes, the word *treaty* appears only once in EPA's decision documents, in the Response to Significant Comments. Id. at 28.

133. *See*, e.g., Steinzor, *Mother Earth,* supra note 25; Heinzerling and Steinzor, A Perfect Storm I & II, supra note 22; O'Neill, Mercury, Risk, and Justice, supra note 5; O'Neill, Environmental Justice in the Tribal Context, supra note 116; Catherine A. O'Neill, Clear Facts about Clear Skies, *San Francisco Chronicle* at B9 (March 9, 2005).

134. Steinzor, *Mother Earth,* supra note 25, at 120; Wendy Wagner (Chapter 4 of this report).

135. Shapiro and Schroeder, supra note 56, at 37.

136. Id.

137. Clean Air Act, 42 U.S.C. 7412(d)(3)(A).

138. Shapiro and Schroeder, supra note 56, at 482.

139. Douglas Kysar (Chapter 10 of this report). ("Although portrayed as an 'objective, transparent standard for cost-effective decision-making,' welfare economics is an emphatically political program, one with foundational assumptions very much at odds with many of the premises and aspirations of environmental law.")

140. Sunstein, Arithmetic, supra note 2, at 2259.

141. Revesz and Livermore, supra note 87, at 3.

142. Id.

143. Shapiro and Schroeder, supra note 56, at 476–82.

144. Mark Sagoff, *The Economy of the Earth: Philosophy, Law and the Environment.* Cambridge University Press 12–14, 220–24 (1988).

■ ■ ■

CHAPTER 7

The CAMR: An Economist's Perspective

ALAN J. KRUPNICK[1]

The Clean Air Mercury Rule (CAMR) was issued in March 2005 (EPA 2005a) along with the supporting regulatory impact analysis (RIA; EPA 2005b). The CAMR would have put in place a cap-and-trade system for reducing mercury emissions from power plants, following up on a cap-and-trade system (mandated by the Clean Air Interstate Rule, or CAIR; EPA 2005c) that would have tightened the existing cap on sulfur dioxide (SO_2) emissions from utilities. It also would have implemented a new cap-and-trade system for utility nitrogen oxide (NO_x) emissions. The CAIR would have resulted in significant ancillary reductions in mercury emissions at these sources. The incremental mercury reductions from the CAMR, in contrast, would have been relatively small. Unfortunately, efforts of the U.S. Environmental Protection Agency (EPA) resulted in no reductions in any of these types of emissions because the District of Columbia Circuit Court vacated both the CAIR (U.S. Circuit Court of Appeals, 05-1244, *St NC v. EPA*) and the CAMR (U.S. Circuit Court of Appeals, 05-1097, *St NJ et al. v. EPA*) rules in 2008. As of this writing, regulatory activities to further control these emissions are in limbo.

My goal in this chapter is to critique the RIA that was issued along with the CAMR and used by the agency to support it. This critique takes the perspective of an economist, focusing primarily on the methods used to define regulatory options and estimate benefits—primarily those to health. Because the estimation of benefits (which here means the monetization of physical health and environmental improvements) depends on the estimation of emissions reductions, concentrations, exposure, and health effects, I devote some attention to these parts of the analyses as well. In addition, because of the localized nature of some types of mercury emissions and their likely disproportionate impact on certain vulnerable groups, such as Native Americans that rely heavily on fish in their diets, I also devote some attention to the way in which EPA handled distributional concerns.

An overarching point is that the RIA process is meant to inform the agency—as well as the public—about the efficiency (and distributional effects) of a range of steps it can take to address the particular regulatory problem at hand. That is, it is an aid to and an input into the future course of action the agency will take. When an RIA is issued contemporaneously with the rule itself, as it was in the case of the CAMR, at least on the surface this primary informational function is absent, and the RIA becomes mere justification for the agency's choices rather than a means of informing and improving the ultimate choice. Whether preliminary analyses associated with the RIA were used to inform EPA's decisions on the CAMR is unclear and beyond the scope of this chapter.

This critique assumes that the reader has at least skimmed Chapter 5 in this report, in which Evans describes the CAMR. I examine nine issues that relate to the methodology and write-up of the analyses used in the RIA. These issues relate to (a) shortcomings in EPA's analysis of noneconomic issues that nonetheless bear on the size and/or robustness of its estimates of the benefits and costs of the rule; (b) problems with EPA's application of cost–benefit analysis (CBA) techniques, as they relate to what might be thought of as best practices, taking the underlying assumptions of CBA as given; and (c) issues that are controversial and contentious between the proponents and opponents of CBA that appear in the CAMR RIA, irrespective of whether EPA handled them appropriately. I present these issues in the order in which they are presented in a standard CBA and include the scope of alternatives considered: exposure estimates; health pathways and studies considered; the valuation of health effects, including IQ loss and reductions in cardiovascular-related mortality rates; the use of discounting for benefits and costs; the treatment of uncertainties; and the treatment of equity issues.

This review is aided by reference to four alternative studies addressing at least some aspects of the benefits and costs of reducing mercury emissions. These include (a) Gayer and Hahn (2005) who, like EPA, examine the costs of mercury reductions and the benefits of reducing IQ loss associated with reduced mercury exposure for a CAMR scenario with a CAIR baseline but, unlike EPA, also examine the benefits of a maximum achievable control technology (MACT) scenario applied to electric utilities; (b) Griffiths et al. (2007), who estimate only the benefits of mercury reductions for IQ that are associated with the CAMR on top of a CAIR baseline; (c) Rice and Hammitt (2005), who, in examining the benefits of EPA's failed Clear Skies Initiative, consider cardiovascular mortality reductions associated with mercury reductions along with IQ loss reductions; and (d) Palmer et al. (2007), who examine the benefits of IQ loss reductions as well as cardiovascular mortality reductions associated with lower mercury exposures under the CAMR.

Scope of Alternatives Considered

Any regulatory CBA must begin by deciding which regulatory scenarios will be considered and the baseline against which their benefits and costs will be measured. Ideally, according to Office of Management and Budget (OMB) guidelines, the scenarios will cover a range of stringency and policy variants to help identify the most appropriate combination of stringency and design elements. Also ideally, the baseline chosen for the analysis will be one that makes reasonable assumptions about future population and economic growth and regulations already in place. The latter element is most subject to judgment. Some government modeling efforts, such as the National Energy Modeling System, take a very conservative view of the baseline, and only include policies that are actually in place at the time. Other modeling efforts, including those used by EPA, may also include proposed regulations because they are considered likely to be in place by the time the target regulation becomes finalized and is implemented.

The theoretical goal is to find a policy that is efficient—that is, one that maximizes net benefits to society. In practice, RIAS are not sophisticated or comprehensive enough to identify optimal policies. But among the alternatives considered, which are usually quite limited, the agency focuses on identifying those with the largest excess of benefits over costs, other things being equal. Many other factors are typically included in determining the agency's ultimate decision, however, which is consistent with best practices of CBA—in other words, efficiency is viewed as only one criterion among many that should underlie regulatory decisions.

Turning first to scenarios, EPA, under the Clean Air Act, believed that it had two options for reducing mercury emissions from power plants. One, under Title V of the act (EPA 1990), was to treat mercury as a toxic pollutant and regulate it using a command and control approach by setting uniform MACT standards. The other was to set a national cap on such emissions and permit utilities to trade allowances that the agency would distribute to such plants. The agency decided to take the latter approach.[2]

This choice was quite controversial, with some believing that the Clean Air Act required the MACT approach and others believing that the best way to obtain the desired emissions reductions was with a cap-and-trade approach. The reasoning behind EPA's choice is beyond the scope of this chapter because it doesn't directly bear on the conduct of CBA.

Yet, legal issues aside, the agency could have conducted an RIA that considered both approaches as options. Indeed, even taking the choice of a cap-and-trade approach as a given, the agency could have examined the consequences of a cap consistent with the greater mercury reductions that might have arisen under a MACT standard, but achieved with trading. Such an analysis would have permitted an estimate of the cost savings associated with a cap versus MACT, as well as any differences in benefits associated with levels of emissions reductions and the likely different spatial distribution of benefits under these two regulatory approaches. Nevertheless, the RIA does make a case for a cap-and-trade approach,[3] saying that relative to MACT, it is a lower-cost approach (for equivalent reductions in mercury), provides a greater incentive to the utilities and their control technology suppliers to innovate (because innovation is rewarded through selling allowances), and provides for more permanent reductions (under MACT, additional plants emit additional mercury, raising aggregate emissions, whereas with a cap, aggregate emissions are constant, but the demand for additional allowances from new plants raises the allowance price). An Electric Power Research Institute (EPRI 2004) cost-effectiveness analysis is relevant to this issue: it compares MACT and cap-and-trade approaches for the year 2020, when both would be fully implemented. The key finding is that under a cap, mercury emissions are about 2 percent lower than under a MACT approach (7 percent versus 5 percent lower, respectively), particularly in high-mercury regions.

EPA considered three options involving the initial and final size of the cap as well as its time path (Table 7.1). For all three options, EPA chose a start date of 2010 with the last adjustment to the cap in 2018. The 2010 start date is consistent with the pace of regulatory implementation, but the agency had more leeway in setting the size of the cap and the time path for the cap to be reduced. The agency could have been more or less aggressive by setting a more rapidly or more slowly declining cap, or even by starting with a lower cap[4] and ending with a lower or higher cap than shown in the table.

The agency also could have chosen to examine a different baseline or several baselines.[5] It chose the reductions expected to result from the CAIR as a baseline in part because this regulatory initiative was ahead of the CAMR initiative on the agency's calendar. But this just begs the question about why the CAMR was behind the CAIR on the calendar. The reasons for this choice are beyond the scope of this chapter and lie outside of the scope of CBA, representing more fundamental decisions by agency officials. Had the agency estimated the benefits of the CAMR without the CAIR as the baseline, both the benefits and the costs would have been much larger. On the cost side, the CAIR would have reduced mercury emissions "for free" in the process of meeting tighter caps for

Cap options (cap in 2010/cap in 2018)	Caps under three time path options 2010–2014	2015–2017	2018 and beyond
Option 1 (38/15)	38	38	15
Option 2 (15/15)	38	15	15
Option 3 (24/15)	38	24	15

Note: For all options, the start date is 2010 and the final adjustment to the cap occurs in 2018.

Source: EPA 2008. "IPM Analysis for the Clean Air Mercury Rule (CAMR)," last updated August 25th, 2008 at www.epa.gov/airmarkets/progsregs/epa-ipm/camr/index.html Assessed August 5th, 2008.

Table 7.1 CAMR Options for Annual Emissions Caps (Tons of Mercury Emissions) and Time Path

SO_2 and NO_x. Without the CAIR's free reductions (from the baseline of 50 tons to 38 tons), these costs would have been attributed to the CAMR, and the benefits of these reductions would also have been attributed to the CAMR. Overall, adherence to best practices would have involved examining a baseline with the CAIR and a baseline without the CAIR because the CAIR had not yet been implemented.

Concerning the specifics of EPA's scenarios in Table 7.1, because the CAIR brings emissions to 38 tons, the effective start date for the CAMR is 2015. Thus, the delay is longer than it appears. Indeed, the only variation among these options is in the size of the cap during the three-year period from 2015 to 2017. By any standards of best practice, these choices seem particularly narrow, both in terms of the size of the caps and the timing.

In summary, EPA's choice of regulatory approaches (MACT versus cap and trade), baseline (with versus without the CAIR), the size of the initial and final caps, and the time path appears very limited and represents lost opportunities to inform the selection of an efficient regulatory strategy. This situation was not a fault of CBA, however, but rather a consequence of decisions made prior to the conduct of such an analysis.

EPA's Exposure Estimates

EPA estimates health effects by using a highly sophisticated air quality model (the Community Multiscale Air Quality, or CMAQ, modeling system) for the eastern United States[6] to distribute reductions in air emissions from power plants and then estimate how much of the methylmercury reaches the water and fish. EPA looked only at freshwater fish because the agency believed that the literature was too uncertain about the link from emissions to ocean fish, particularly in the time period between a given emissions increase and an increase in mercury levels in tuna as well as the share of the observed increase in mercury in tuna from the United States versus other countries. Nevertheless, EPA could have included marine fish in its analysis, as there are acknowledged problems of contamination in various species of marine fish. In fact, Rice and Hammitt (2005) did include estimates of exposure from this category.

Given the focus on freshwater fish, EPA uses two approaches (population centroid and angler destination) to estimate the number of exposed pregnant women and, therefore, prenatally exposed infants. The two approaches produce results that differ by about 25 percent, so the choice between these two approaches is not central to the estimation of benefits. Then EPA assumes an amount of fish consumption by these women and assumes that fetal blood mercury equals the mercury concentration of the mother. EPA also broke out various sensitive population groups in its analysis, which I pick up later in the chapter in the discussion of equity effects.

EPA could have taken a different tack in estimating changes in mothers' exposure by starting with existing estimates of mercury in mothers' blood and umbilical cords. This approach, although more direct and including all sources of mercury, would still have required a mapping of pregnant women to locations with given mercury levels. EPA also could have relied on some estimates suggesting that fetal blood mercury levels exceed that of the mother. EPA's assumptions about fish consumption rates could also be challenged. At the same time, some choices, such as the use of average values for a population rather than an at-risk population, are appropriate for estimating benefits to the entire population and would have no effect on total benefits.

Overall, much effort in the RIA was expended on developing defensible estimates of exposure. Indeed, EPA's analysis includes various sensitivity analyses and the use of Monte Carlo techniques to partially address some of the above issues as well as the natural variability of response in the at-risk population. Nevertheless, more effort to address parameter uncertainties might have had a higher payoff in terms of the robustness and credibility of the conclusions reached.

Exposure–Response Functions

In quantifying benefits, EPA focused entirely on estimating the effects of mercury in blood on intelligence, as measured by the IQ score. The functions making this link come from several studies used in most analyses of such benefits: the Faroe Islands (Grandjean et al. 1997), New Zealand (Kjellström et al. 1989; Crump et al. 1998), and the Seychelles Islands (Davidson et al. 1998; Myers et al. 2003). In the Faroe Island study, slight neurological effects in children of mothers who ate whale meat and blubber (which are high in mercury and selenium) were found. In New Zealand, children did less well on performance tests when their mothers had higher hair mercury levels during pregnancy. In the Seychelles, no adverse effects were found. Because of these diverse results, EPA did the first-ever meta-analysis with these key studies, using weights that account for the variance of the estimates, a standard approach. At least one of the New Zealand studies does not count highly in the resulting estimate because the variance of its estimate is larger than those of the other studies. EPA did not assume any threshold below which changes in mercury would have no effect.

In critiquing EPA's approach, some issues lead to an underestimate of benefits and some lead to an overestimate. Considering those causing underestimates first, note that a recent study in the United Kingdom found a relationship between lower child IQ and presence of attention deficit hyperactivity disorder (ADHD) on the one hand and mothers eating less than 12 ounces of fish per week during pregnancy on the other. If EPA had known about and used this study in its meta-analysis, the resulting exposure–response coefficient would have been larger. Second, EPA did not estimate benefits for reduced myocardial infarction in adults, claiming that the epidemiological evidence is too uncertain. However, other studies of mercury reduction benefits have used the available literature to estimate such benefits, generally based on Salonen et al. (1995), who found that Finnish men with the greatest hair mercury concentrations had a two-fold increased risk of acute myocardial infarction relative to the other groups studied. A more recent study, Virtanen et al. (2005), follows up on Salonen et al.'s sample over a 13-year period, finding that high mercury content in hair increased the risk of cardiovascular morbidity and mortality and attenuated the beneficial effects of fish oils on cardiovascular health.

Indeed, none of the benefit studies in the literature noted above addressed all of the health effects associated with mercury exposure, such as reduced parasthesia (prickling, tickling, and itching sensations), reduced ADHD effects, reduced coordination problems, and lower blood pressure in adults (although this endpoint may be an input to overall cardiovascular health).

Finally, EPA's analysis notes that the IQ gains from the CAMR would be less than a single point on average. This statement is misleading because some in the population would have a greater gain; this is apparent simply by accounting for confidence intervals on the exposure–IQ coefficient, let alone by accounting for population exposure and response heterogeneity.

EPA may have overestimated benefits, however, by assuming no threshold in its exposure–response functions. Jakus et al. (2002), for instance, found in the literature an IQ effect threshold at 6 mg/kg maternal hair mercury concentration, as well as 3.3 mg/kg child hair mercury, and a child neuropsychological development threshold at 72 μg mercury/L blood.

Monetizing Endpoints

EPA attached monetary values to two endpoints of interest to this review: mortality risk reduction (not through mercury reductions, but as an ancillary benefit of fine particulate matter, or $PM_{2.5}$, reductions) and IQ improvements.

Mortality Risk Reduction

EPA uses its standard approach to monetize this health endpoint. The agency uses the results of a meta-analysis of 26 studies in the literature on the value of a statistical life (VSL)—which consists of mostly wage-risk studies and a handful of stated preference studies—to multiply by thuμμe number of deaths delayed. This figure of $5.5 million in 1999$[7] is discounted at 3 percent and 7 percent rates for the time it will take for the mercury reduction to have an effect (called a *cessation lag* by EPA) and adjusted upward to account for income growth in the population over the period in which the policy will have effects, assuming an income elasticity of willingness to pay of 1.2.

The standard welfare-theoretic model underlying modern neoclassical economics leads to an endorsement of EPA's general approach—in other words, that it is appropriate to capture preferences for reducing mortality risk in monetary terms and that these preferences can be captured through the use of revealed preference or stated preference techniques.

The specifics of EPA's approach are controversial. First, there is growing discontent among economists and other analysts for basing the VSL on wage-risk studies when the VSL is being applied in a health context. The VSLs from the wage-risk literature are based on avoidance of accidental death risks, not health-based risks, and cover a population subgroup (workers, mostly men) that may be less risk-averse, younger, and healthier than those in the population at risk from pollution.[8] The direction of bias from relying on wage-risk studies is unclear, however.

Second, EPA's use of discounting to address the cessation lag is problematic. How people discount delays in health effects has not been much studied; but the studies that have looked into this issue—by asking questions about the WTP today for a risk reduction beginning today and the WTP today for a risk reduction that starts some years in the future (see Alberini et al. 2006; Cameron and DeShazo 2004; and NRC 1991 as examples)—show rates of discount higher than 3 percent.

Thus, EPA is overestimating benefits when it uses a 3 percent rate in this example, although the 7 percent discount rate may be more accurate.

Third, the use of the income elasticity of WTP of 1.2 does not accord with the analyses in most of the literature. Consensus for this elasticity, which admittedly comes from examining preferences across people of different income levels rather than how their preferences change over time with changes in income (there are no studies in this vein), suggests that it is about 0.5 (Robinson 2007). Thus, by using the 1.2 figure, EPA overestimates this upward adjustment of benefits.

Fourth, economists increasingly believe that the same VSL should not apply to all people. One of the manifestations of this belief is to apply a different VSL to different age groups (called the senior discount in the literature) based on studies showing that the WTP for mortality risk reductions differs depending on age (see reviews of this literature, such as Krupnick 2007 and Aldy and Viscusi 2007). Applying different VSLs to different income groups is currently thought to violate ethical standards, particularly if it means discounting life loss of the poor, and is not practiced within developed countries, although it is routinely practiced when transferring VSL estimates to developing countries (using elasticities ranging from 0.4 to 1.0; see, for example, the Mexico Air Quality Management Team 2002). For the CAMR, it might have been appropriate to use VSLs for the groups that would be the biggest gainers, such as Native American tribes living near the Great Lakes. Needless to say, such estimates do not exist, so EPA cannot be faulted for that.

Finally, there is controversy among stakeholders and academics over whether the reduction in mortality risk should be expressed in terms of deaths delayed or life years gained. EPA's practice is to use deaths delayed for its main analysis, although occasionally, life years gained estimates have appeared in supplementary materials. Recently, NRC (2008) concluded that the latter metric is a better fit to the epidemiological literature, while acknowledging that the use of a constant value of a statistical life year, irrespective of the age at which effects are experienced, is implicitly discounting seniors' lives and does not accord with the relevant valuation literature. Therefore, the recommendation by NRC (2008) is to use a value of a statistical life year that varies with the age of those affected. The literature is currently insufficient for providing such estimates, however.

IQ Points

EPA's estimate for monetizing a gain in IQ is based on a cost-of-illness approach where the estimated relationship between increased earnings for people with higher IQs is used to derive a benefit of $8,807 per point increase (in 1999$). Also in this calculation is an offset to this gain from the costs of the additional education shown to occur for people with higher IQs (actually, EPA's analysis is in terms of losses in earnings being partly offset by savings in education costs).

Four studies in the literature provide an estimate to monetize IQ gain from mercury reductions. Griffiths et al. (2007) and Rice and Hammitt (2005) estimate lifetime earnings loss at $7,121 for men and $5,268 for women (in 2000$) and $16,500 (2000$) without differentiating between men and women, respectively. Thus, EPA's estimate splits these two studies. Gayer and Hahn (2005) provide estimates of the WTP to avoid chelation therapy ranging from $1,300 to $2,200 (2004$). Finally, Palmer, Burtraw, and Shih (2007) use a $10,420 per IQ point. It appears that the Rice and Hammitt (2005) estimate of lifetime earnings loss is so high relative to other estimates because of discounting assumptions and different estimates of baseline lifetime earnings.

Yet the EPA estimate would be larger if greater education costs from people with higher IQs were not subtracted from lost earnings. Is this subtraction appropriate? Common practice with mortality valuation suggests that it is not. If someone dies in an auto accident, we do not credit against the value of avoiding this death the medical cost savings that would have been incurred had that person lived a longer life. Ethical arguments may caution against this approach as well. Another benefit (which EPA did not add but should have added) to earnings gains is the remedial education cost savings from higher IQ.

Benefits and Costs Compared

Figure 7.1 is a comparison of the benefit (and cost) estimates from the RIA with those of the four studies that purport to measure benefits and costs under the CAMR using the CAIR as a baseline. Differences in these studies' estimates reflect to some degree all of the choices made that are discussed above.

EPA's benefit estimate is too small to even show up in the figure. Benefit estimates for the other studies are small if shown for IQ only, but are far larger when the ancillary mortality benefits are included (except for Gayer and Hahn 2005). Although Gayer and Hahn's (2005) estimates of benefits are quite small, they are interesting on the cost side because they contrast a command and control approach to mercury emissions reductions (which would apply under Title V's MACT provisions) with those under the cap-and-trade approach proposed in the CAMR. Costs under a trading approach are only half of those using a MACT approach—a directionally expected result because the trading system encourages the allocation of emissions reductions to the cheapest sources. Note also how low EPA's cost estimates are compared with those of Gayer and Hahn (2005). Thus, although EPA's benefit estimates are low compared to those of other studies, its costs are more than proportionally lower, making its benefit-to-cost ratio more favorable than those of Gayer and Hahn (2005), although still far less than 1.0.

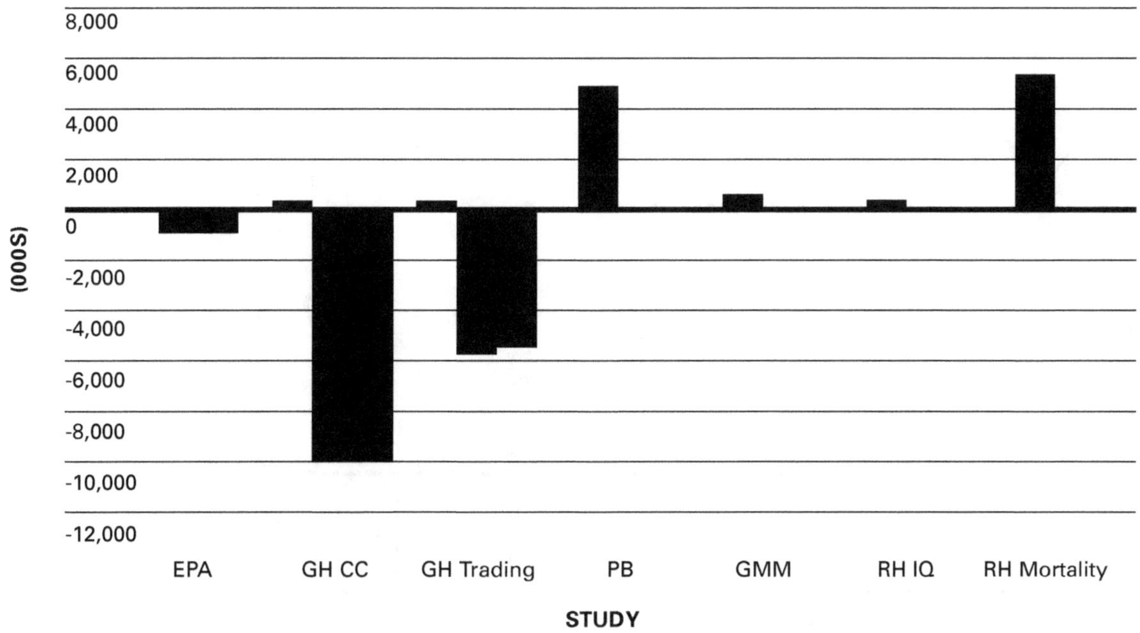

Figure 7.1
Findings of cost-benefit studies of the CAMR beyond the CAIR
(2004$)

■ Benefits
■ Cost
■ Net

Sources: EPA, U.S. EPA(2005); GH CC, Gayer and Hahn (2005) command and control approach; GH Trading, Gayer and Hahn (2005) cap-and-trade approach; PB, Palmer et al. (2007); GMM, Griffiths et al. (2007); RH, Rice and Hammitt (2005).

Distributional Analysis: Hot Spots

One of the most contentious issues leading up to the CAMR was whether a trading program would afford sufficient protection from mercury hot spots—areas of high mercury concentration. Critics argued that emissions from utilities that purchase mercury allowances under a cap-and-trade scheme could increase over some baseline and relative to an alternative approach to regulating mercury, such as technology-based standards set for each plant under the Clean Air Act's Title V MACT rules (O'Neill 2004).

This issue is exceedingly murky because commenters on both sides rarely are transparent about their baseline or counterfactual states with which they compare the policy option state. The most direct metric for these states is exposure of people to mercury. If exposure in the baseline state is less than in the policy state, then it can be said that a hot spot is being created by the policy, depending on how the baseline state is defined (see below). Concentrations in fish in the vicinity of people who eat them could be a somewhat more indirect metric. And a still more indirect, but much more easily measured and estimated, metric is the emissions from sources near people who might be exposed directly or through the fish they eat. EPA provides another definition (see EPA 2005b, Figure 4): a place where deposition contributed by a given source is enough to raise mercury in fish tissue above the safe level for consumption. Note that this definition does not refer to the change in mercury associated with a policy, so is less applicable to an RIA than the other above metrics.

Figure 7.2 provides a diagram of the situation for the emissions metric.

The graph represents a timeline for reductions in emissions at a particular plant. I assume that, absent mercury controls, either directly through the CAMR or indirectly through the CAIR, mercury emissions will increase over time with rising power production. This assumption is not critical, however. The CAIR leads to immediate reductions in 2010 and significant further reductions by

Figure 7.2
Hot Spots for Alternative Baselines

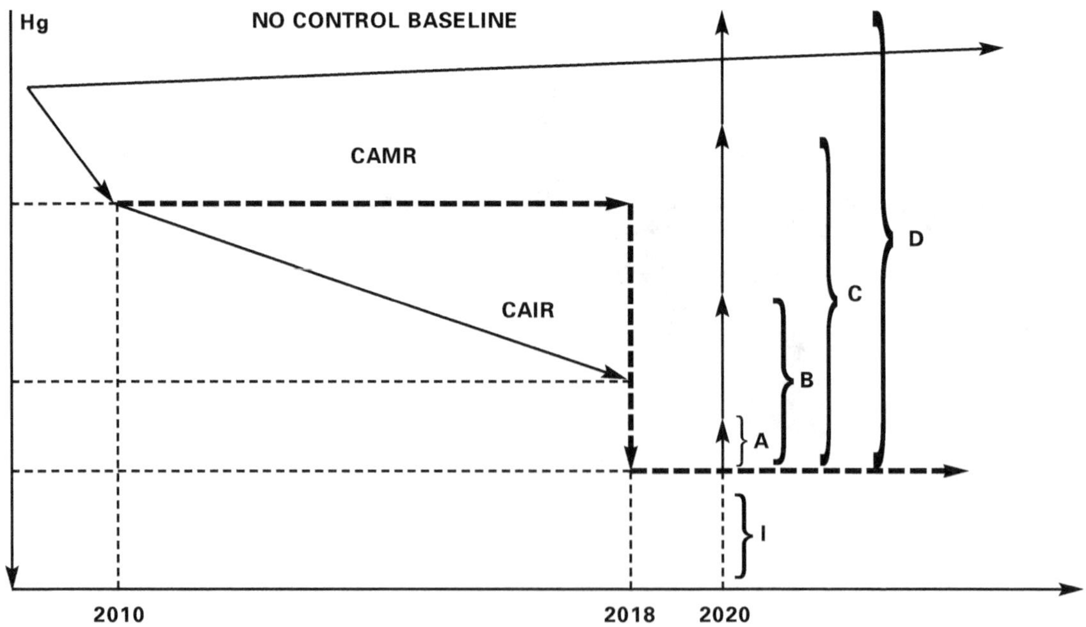

2018 through the issuance of a declining number of allowances. The CAMR's incremental cap does not become binding until 2018, at which point additional reductions are required. The graph assumes, for convenience, that trading will occur at 2020, but this assumption is not in any way critical. Line segment A represents a plant purchasing a small number of mercury allowances, and therefore emitting mercury equal to its initial allocation (line segment I) plus the purchased allowances (A). Similarly, it is possible that the plant purchases more allowances, such as that represented by line segment B, or even more, as represented by line segment C or D.

Which, if any, of these behaviors should be defined as a hot spot? Note first that emissions increase over the initial allocation at 2018 (line segment I) in all cases. But using the 2018 allocation as the baseline (or counterfactual) gives no credit for the reductions in emissions as a result of the declining caps. Thus, situation A leaves emissions above what they would have been with the CAMR and the CAIR and no trading, but below what they would have been with the CAIR alone and no trading. Situation B leaves emissions above what they would have been with the CAIR alone and no trading; situation C leaves emissions above the CAMR before 2018 with no trading, but below the no-control baseline; and situation D results in higher emissions with any of the trading programs than without any regulation at all.

Purchases of D allowances would likely make those near the plant worse off than without regulation and would be rightly considered as contributing to a hot spot (short of actually measuring exposures of people to mercury). Purchases of C, B, or A provide progressively less compelling cases for a hot spot as people are better off relative to no control in these cases.

A final conceptual issue is the consideration of trading versus MACT standards. Figure 7.3 depicts this case, repeating Figure 7.2 but adding a line for a MACT approach. Because MACT standards are rate based, increases in output can lead to increases in mercury emissions (which does not happen with trading) while still meeting the standard. Thus, as an example, the MACT line is drawn to be initially lower than the lowest emissions under the CAMR, but then rises with output to exceed

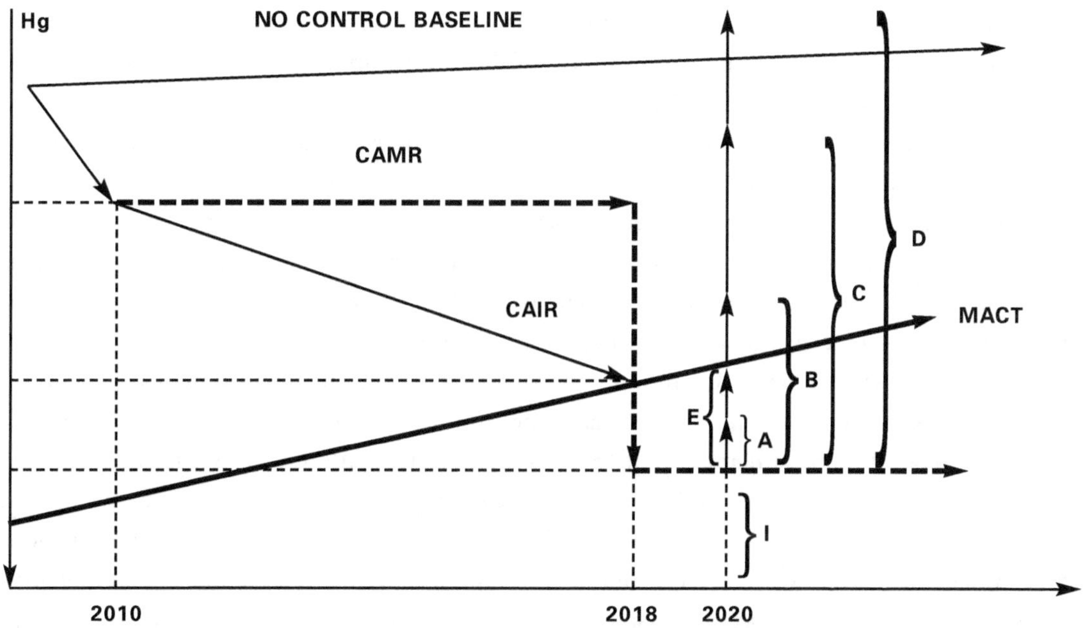

Figure 7.3
Hot Spots: Trading versus MACT

emissions under the CAMR. Which trading situation qualifies as a hot spot in this case? The most stringent definition of a hot spot would be any emissions exceeding the MACT line, but again, this definition gives no credit to the trading program for reducing emissions. Indeed, one could argue that a MACT approach creates a hot spot (equal to the length of line segment E above the CAMR baseline) if the baseline is defined as the CAMR with no trading. Even with trading, as drawn, purchase of A allowances under trading would still be preferable to a MACT approach.

The issue EPA takes up is whether the CAMR will make any areas (including those not currently hot spots) worse than they already are without any regulation (corresponding broadly to line D in Figure 7.2 or 7.3) or in any of the other situations discussed above. EPA addresses this issue by examining whether fish tissue mercury levels will be boosted above EPA's criterion of 0.3 mg mercury/kg in any fishable water bodies solely by utilities participating in the trading program. The agency uses its air quality and utility behavior models and many assumptions to make these estimates.

EPA first notes that coal-fired power plants contribute less than 5 percent of the total mercury deposition in the United States. But locally, the impacts of coal-fired power plants vary from as little as 0.05 percent to as much as 85.9 percent. Then EPA finds no "utility-attributable" hot spots after implementation of the CAMR's mercury trading program. The agency also finds that 7 to 45 percent of mercury is deposited within 50 kilometers of an "eastern site" and relatively high deposition of mercury in the Ohio River Valley; but again, not enough to violate the fish mercury standard solely from utility emissions under the CAMR.

Beyond this analysis, EPA examines aggregate emissions changes for evidence of presumptive hot spots. The RIA provides a graphic (EPA 2005b, 186) of state-by-state emissions showing improvements in mercury emissions from the CAMR relative to the CAIR. Basically, Wisconsin shows minor improvements, whereas the other two Great Lakes states examined (Michigan and Minnesota) show no improvements.

Defining hot spots in terms of state-level emissions is far too aggregate to address local concerns about hot spots, however. It is more appropriate to consider emissions at the plant level, as shown conceptually in Figures 7.2 and 7.3. EPA actually did the modeling to address this issue but did not report the results in the RIA. For this information, the reader is only referred to some runs of the Integrated Planning Model (IPM), a model of the electric utility industry that also forecasts emissions in response to various policies.

This omission is serious for the RIA. I obtained the relevant IPM runs to examine plant-by-plant data and simulation results for the key Great Lakes states (Wisconsin, Minnesota, and Michigan). Data generated from EPA's IPM were used to compare plant-by-plant[9] mercury emissions projections in the year 2020 under three different policy scenarios (EPA 2008). These scenarios are (a) a base case scenario assuming the absence of any policy, (b) a scenario assuming implementation of only the CAIR, and (c) a scenario assuming implementation of both the CAIR and the CAMR.

The IPM files list 277 unique generation stations in the states of Michigan, Wisconsin, and Minnesota. Comparisons among the three scenarios are summarized in Tables 7.2 and 7.3. Basically, projections assuming implementation of the CAIR alone show that mercury emissions would be lower at 40 of these generators and higher at 8 in 2020, compared with no policy. In the CAIR-only scenario, reductions range from a maximum of 0.01 tons to a minimum of 2.0E–4 tons of mercury. Projections assuming the implementation of both the CAIR and the CAMR show that mercury emissions would be lower at 52 power stations and higher at 2 in 2020, compared with no policy.

Table 7.2
Differences in Mercury Projections (by 2020) Under Two Policy Scenarios

	Mercury projections (tons)		
	Aggregate	Difference from baseline	Difference from the CAIR
Baseline	3.31	—	—
CAIR alone	3.26	–0.06	—
CAIR + CAMR	3.10	–0.21	–0.15

Table 7.3
Effects on Mercury Emissions from 277 Generating Plants in Michigan, Minnesota and Wisconsin (by 2020) Under Two Policy Scenarios

	Number of plants		
	Increase in mercury emissions from Aggregate	Reduction in mercury emissions baseline	Increase in mercury emissions the CAIR
CAIR alone	8	40	—
CAIR + CAMR	2	52	2

Source for tables 1–2: "IPM Analysis for the Clean Air Mercury Rule (CAMR)," last updated August 25th, 2008 at www.epa.gov/airmarkets/ progsregs/epa-ipm/camr/index.html Assessed August 5th, 2008.

After the CAMR comes into force, total reductions in mercury emissions in 2020 for the Great Lakes states are 2.5 times greater than they are with the CAIR alone.

From this analysis, it appears that only 2 of 277 plants increase emissions over the no-control level when both the CAMR and the CAIR are implemented, using the IPM simulation model (case D in Figure 7.2).

No runs were conducted with a MACT approach as the baseline. This is unfortunate. However, as noted above, EPRI (2004) has found that a cap-and-trade program actually leads to lower emissions than a MACT approach in the high-mercury regions.

Other Distributional Issues

EPA also looked at distributional issues by doing a special analysis of the exposures of high fish-eating subpopulation groups living around the Great Lakes, specifically the Hmong and Chippewa. However, the agency did not consider a subset of these groups—spearfishers. In Chapter 6 of this report, O'Neill charges that, in contrast to indigenous people's consumption of 20 grams of fish per day, this subgroup consumes 200 grams of fish per day. However, others (Great Lakes Indian Fish & Wildlife Commission 2000) note that, in response to learning about high mercury levels in fish in some areas, these groups have moved their fishing grounds to safer areas. While feeling compelled to change fishing sites is costly to be sure, health risks to these individuals and others in their group are being reduced by this behavior. I also note that the RIA does not specify relative mercury levels for African Americans and Mexican Americans whose participation in fishing and in the consumption of fish may be greater than that of the general population.

Discounting

Discounting a future stream of monetary payments or receipts to the current year—in other words, calculating the present discount value of such a stream—is the standard approach to expressing and comparing complex time streams of monetary values using a single number; this method is used by businesses and governments around the world and is taught in all universities and even in many high schools. When applied to CBAS in an RIA, it is necessary to compare a time stream of benefits of a particular regulatory option to the time stream of its costs. As long as all options are treated in the same way, the comparison of net present discounted benefits across each option permits the identification of the most efficient regulatory option analyzed, irrespective of the timing of its benefits and costs.

Although the concept of discounting benefits and costs is generally uncontroversial, the appropriate discount rate for use in RIAS has been an object of intense discussion for decades. Academic treatises on the topic abound, including whether, in violation of standard practice, it is appropriate to discount benefits of regulation at a lower rate than costs. In the absence of consensus, OMB requires agencies to use two different rates (3 percent and 7 percent), but to apply the same rates to both benefits and costs. As noted above, OMB also permits agencies to account for the delay in the effectiveness of regulations using these rates. In the CAMR RIA, EPA discounts over the lag period (assuming 5-, 10-, 20-, and 50-year lags) established for the health effects (IQ loss or the cobenefit of mortality effects resulting from $PM_{2.5}$ reduction); for the cost analysis, EPA uses the same rates and assumes a 30-year life of capital equipment purchased in response to the regulations. This approach is a standard response to OMB rules, which are themselves a reasonable response to the state of the neoclassical welfare economics literature.

Uncertainty

Analyses as complex as RIAS will have at least three different types of uncertainties. The first is statistical uncertainty. Whenever statistical analysis is used to relate one or more variables to another, as in regressions, the resulting estimate of that relationship—the exposure–response coefficient is a good example here—is the best estimate of that relationship, but it is not without some error; in other words, a distribution, or uncertainty bounds, surrounds that estimate (and this distribution, or bounds, is called the standard error). Thus, statistical uncertainty can be reasonably quantified in the sense that one can say, for example, that with a 95 percent probability, the value is at or above *x* and with a 95 percent probability that the value is no more than *y*.

The second type of uncertainty is model uncertainty. This type of uncertainty can arise literally from the need to choose a given model, such as the CMAQ model or IPM in the mercury RIA, to underpin an analysis. Model uncertainties also arise in the choice of exposure–response or valuation functions, where many choices may cover a range of coefficients. Such uncertainties can be partly quantified through sensitivity analysis, in which one "model" (e.g., exposure-response coefficient from one study) is swapped for another, or through meta-analysis, which is a technique for finding an average coefficient over a variety of studies, as EPA found by considering all of the mercury–IQ studies. But such analyses have their own problems, and one can never do all of the uncertainty analyses needed to develop a distribution equivalent to standard errors through the use of sensitivity analyses with hugely complex models such as CMAQ and IPM.

A third type of uncertainty—we might call it unknowns—is ubiquitous in RIAs and, by definition, cannot be quantified. An example is when one suspects, say from a conceptual model, that mercury affects the nervous system, but no studies are available to quantify the effect.

EPA's own RIA guidelines and OMB's guideline both address when and how to quantify and qualify these types of uncertainties. In addition, NRC (2002) lays out a variety of protocols to push federal agencies, such as EPA, to do a better job in describing uncertainties in their analyses. In the CAMR, EPA identified uncertainties relevant to its benefit and cost analyses. Some of them are quantified and others are not. EPA addressed a few uncertainties using sensitivity analysis and also indicated the impact of several uncertainties on the results. In other cases, EPA provided explanations of how it handled the uncertainties or, if necessary, simply listed the areas of uncertainties. For costs, EPA assumed some technological developments on control technologies, as incorporated into the IPM electricity model, but still felt that costs were being overestimated. The agency addressed uncertainties in fuel price and electricity demand using sensitivity analysis. Overall the agency indicates that it made conservative (high) cost assumptions. On the benefits side, EPA examined the effects of a variety of key parameters with sensitivity analyses and provided many discussions about the uncertainties underlying its estimates.

On the whole, EPA's treatment of uncertainties was unsophisticated—particularly with respect to the treatment asked for by NRC (2002)—but typical, conveying a generally qualitative and occasionally quantitative sense of the uncertainties, but not in any organized fashion. Understandably, data limitations were evident at every stage of the analysis and inhibited the agency's ability to fully quantify uncertainties.

Conclusion

The CAMR RIA is not atypical of RIAS at EPA. Its use of CBA methods was reasonable in light of data limitations as well as budget and time constraints. Its handling of uncertainty, discounting, and the VSL and other monetization issues is standard, albeit far from the ideal. And it used the agency's best models of utility behavior (IPM) and atmospheric movement of pollution (CMAQ) to estimate the benefits and costs of the CAMR. It showed creativity in addressing fetal mercury exposure issues, both in estimating total exposures and in a companion equity analysis, and did a reasonable job of capturing the epidemiological literature's information on the IQ–mercury linkage.

However, the RIA had some glaring deficiencies. The analysis considered few regulatory options, even within a trading program. Although the agency may have had administrative reasons for not considering the option of a MACT standard, the inclusion of this option would have been very helpful for decisionmaking and for explaining to readers why such a standard may not perform as well as billed by its supporters. The spatial (hot spot) analysis was less helpful than it should have been. The IPM runs the agency paid for could have been used more effectively. And the omission of at least the cardiovascular mortality effects is hard to justify because a literature existed at the time to support such analyses. A host of smaller deficiencies are detailed above.

Going forward, and based on the CAMR RIA, the agency's handling of RIAS could be improved in a number of areas. First, more time and resources must be built into the schedule to enable the agency to collect better studies, rely less on assumptions and more on data, and run a variety of scenarios to more fully explore the "policy option" space. In addition, the agency should adopt a more academic approach to such efforts, incorporating time for peer review and for its own literature review of relevant CBAs. More time and resources would also permit the agency to write clearer, more transparent reports. Backing for these efforts from the political leadership is also essential.

The CAMR RIA's equity analysis, though clever, seemed ad hoc. More attention should be paid to developing protocols for such analyses so the reader has more confidence in the conclusions. Indeed, the agency needs to more carefully justify what an equity analysis is and is not and how it relates to efficiency.

Nevertheless, in my view, RIAs are primarily about identifying efficient policy options—that is, those with the largest net benefits to society as a whole. The CBA's reliance on a utilitarian philosophy should be a prominent part of CBA "boilerplate" attached to all RIAs, as should its lack of a connection to moral and ethical considerations; such issues are clearly important, but belong in an overall analysis that incorporates the RIA results, not in the RIA per se. Secondarily, RIAs ought to describe how particularly vulnerable or favored groups fare. Thus, if poor people are made worse off and rich people are made better off, or both groups gain, but more gains accrue to the rich, these outcomes ought to be prominently displayed.

In addition, the entire area of expressing uncertainties in an RIA needs to be more carefully addressed. The recent NRC (2002) study on this topic provides many ideas for how to improve the quality and display of such analyses (e.g., the portrayal of the weight of evidence and low-probability outcomes). Also, see Krupnick et al. (2007) for more ideas in this vein. But some issues related to uncertainty are more philosophical. For instance, when, if ever, should a number for which evidence is particularly weak or suspect be judged to be more useful than a blank?[10] Currently, according to OMB rules, "gray" literature cannot be used in RIAs, possibly because of a fear that to do so opens up the analysis to legal challenge. If this hurdle can be jumped, agencies should have the right to use such literature upon appropriate justification.

Finally, EPA must make every effort to stay on top of new developments in the treatment of discounting and the valuation of health endpoints and incorporate them into its analyses, perhaps as sensitivity analyses, until they attain enough consensus to become the standard approach.

■ ■ ■

Notes

1. Senior Fellow and Director of Research, Resources for the Future. The author would like to thank the participants in the workshop for their comments on the presentation that underlies this work, as well as the funder, the Smith Richardson Foundation, and the organizers, Winston Harrington, Lisa Heinzerling, and Richard Morgenstern.

2. The agency could also have considered the benefits and costs of imposing more fishing bans, better enforced bans, or better information campaigns to reduce the consumption of fish caught from polluted waters. Although these strategies are clearly not perfect substitutes to reducing mercury emissions, they would have further illuminated the options used in this country to address mercury concerns. *See* Jakus and others (2002) for a CBA of a bass fishing ban in the Chesapeake Bay related to mercury contamination.

3. *See* www.epa.gov/camr/basic.htm.

4. A higher starting cap was ruled out by considering the CAIR as the baseline because the CAIR would have brought mercury emissions down to 38 tons annually by 2010 and then 28 tons annually by 2018, even in the absence of the CAMR.

5. A generic baseline issue with CBAS is how to treat technological change in abatement or production process technologies, recognizing that a certain degree of improvements in costs or in removal efficiencies may come about exogenously to the regulatory policy, and that the policy itself may induce such changes. EPA's typical approach is to ignore the latter and consider only on-the-shelf and nearly on-the-shelf technologies for the baseline. Treating technology change as endogenous is highly challenging in an RIA; ignoring this effect tends to lead to an overestimate of costs.

6. EPA believed that effects in the western United States would be very small.

7. This value is slightly lower than the value used in some previous RIAS, primarily for technical reasons related to the interpretation of some of the underlying studies.

8. For a review of this literature and these issues, see the National Research Council (NRC) 2008.

9. Emissions projections were actually on a stack-by-stack basis, which I aggregated to the plant level.

10. Zero should never be used as a placeholder for unquantifiable uncertainties.

References

Alberini, A., M. Cropper, A. Krupnick, and N.B. Simon. 2006. Willingness to Pay for Mortality Risk Reductions: Does Latency Matter? *Journal of Risk and Uncertainty.* 32(3): 231–245.

Aldy, J.E., and W.K. Viscusi. 2007. Age Differences in the Value of Statistical Life: Revealed Preference Evidence. *Review of Environmental Economics and Policy* 1(2): 241–260.

Cameron, T.A., and J.R. DeShazo. 2004. An Empirical Model of Demand for Future Health States When Valuing Risk-Mitigating Programs. Working paper 2004–11, Economics Department, University of Oregon.

Crump K.S., T. Kjellström, A.M. Shipp, A. Silvers and A. Stewart. 1998. Influence of prenatal mercury exposure upon scholastic and psychological test performance: benchmark analysis of a New Zealand cohort. *Risk Analysis.* 18(6): 701–713.

Davidson, P.W., G.J. Myers, C. Cox, C. Axtell, C. Shamlaye, J. Sloane-Reeves, E. Cernichiari, L. Needham, A. Choi, Y. Wang, M. Berlin, and T.W. Clarkson. 1998. Effects of prenatal and postnatal methylmercury exposure from fish consumption on neurodevelopment: outcomes at 66 months of age in the Seychelles Child Development Study. *Journal of the American Medical Association* 280(8): 701–707.

U.S. Environmental Protection Agency (EPA). 1990. Clean Air Act. www.epa.gov/air/caa/index.html (accessed August 28 2008).

———. 2005a. Clean Air Mercury Rule: Basic Information. www.epa.gov/air/mercuryrule/basic.htm (accessed August 28, 2008).

———. 2005b. Regulatory Impact Analysis of the Final Clean Air Mercury Rule. Office of Air Quality Planning and Standards. epa-452/R-05-003 (March).

———. 2005c. Clean Air Interstate Rule: Basic Information. www.epa.gov/cair/basic.html (accessed August 28, 2008).

———. 2008. IPM Analysis for the Clean Air Mercury Rule (CAMR), last updated August 25, 2008. www.epa.gov/airmarkets/progsregs/epa-ipm/camr/index.html (accessed August 5, 2008).

Electric Power Research Institute (EPRI). 2004. Mercury Fact Sheet: "Hot Spots"—Mercury Emissions and Deposition Patterns. Document number 1010996 (August). Palo Alto, CA: EPRI.

Gayer, T., and R.W. Hahn. 2005. The Political Economy of Mercury Regulation. *Regulation* 28(2): 26–39.

Grandjean, P., P. Weihe, R.F. White, F. Debes, S. Araki, K. Yokoyama, K. Murata, N. Sørensen, R. Dahl, and P.J. Jørgensen. 1997. Cognitive deficit in 7-year-old children with prenatal exposure to methylmercury. *Neurotoxicology and Teratology* 19(6): 417–428.

Great Lakes Indian Fish & Wildlife Commission. 2000. *Masinaigan Supplement.* Fall. www.glifwc.org/Publications/Mazinaigan/EnjoyFish_supplement.pdf (accessed August 8, 2008).

Griffiths, C., A. McGartland, and M. Miller. 2007. A Comparison of the Monetized Impact of IQ Decrements from Mercury Emissions. *Environmental Health Perspectives* 115(6): 841–847.

Jakus, P., M. McGuinness, and A. Krupnick. 2002. The Benefits and Costs of Fish Consumption Advisories for Mercury. Discussion paper 02-55. Washington, DC: Resources for the Future. www.rff.org/rff/documents/rff-dp-02-55.pdf (accessed August 28, 2008).

Kjellström, T., P. Kennedy, S. Wallis, A. Stewart, L. Friberg, B. Lind. 1989. Physical and Mental Development of Children with Prenatal Exposure to Mercury from Fish. Report 3642. Solna, Sweden: National Swedish Environmental Protection Board.

Krupnick, A. 2007. Mortality-Risk Valuation and Age: Stated Preference Evidence. *Review of Environmental Economics and Policy* 1(2): 261–282. doi:10.1093/reep/rem016.

Krupnick, A., R. Morgenstern, M. Batz, P. Nelson, D. Burtraw, J.S. Shih, and M. McWilliams. 2006. *Not a Sure Thing: Making Regulatory Choices under Uncertainty.* Washington, DC: Resources for the Future, February 2006.

Mexico Air Quality Management Team. 2002. Improving Air Quality in Metropolitan Mexico City: An Economic Valuation. World Bank Research Working Paper 2785. http://econ.worldbank.org/external/default/main?pagePK=64165259&theSitePK=469372&piPK=64165421&menuPK=64166093&entityID=000094946_02022604025131.

Myers, G.J., P.W. Davidson, C. Cox, C.F. Shamlaye, D. Palumbo, E. Cernichiari, J. Sloane-Reeves, G.E. Wilding, J. Kost, L.S. Huang, and T.W. Clarkson. 2003. Prenatal methylmercury exposure from ocean fish consumption in the Seychelles child development study. *Lancet* 361(9370): 1686–1692.

NRC (National Research Council). 1991. *Rethinking the Ozone Problem in Urban and Regional Air Pollution.* Washington, DC: National Academies Press.

———. 2002. *Estimating the Public Health Benefits of Proposed Air Pollution Regulations.* Washington, DC: National Academies Press.

———. 2008. *Estimating Mortality Risk Reduction and Economic Benefits from Controlling Ozone Air Pollution.* Washington DC: National Academies Press. Executive summary available: www.nap.edu/catalog/12198.html (accessed August 28, 2008).

North American Commission for Environmental Cooperation (CEC). 2003. Mercury Hot Spots of North America. www.cec.org/files/pdf/economy/hotspots_en.pdf (accessed August 28, 2008).

O'Neill, C.A. 2004. Mercury, Risk & Justice. White Paper 405. Edgewater, MD: Center for Progressive Regulation. www.progressiveregulation.org/articles/Mercury_2004.pdf.

Palmer, K., D. Burtraw, and J.-S. Shih. 2007. The Benefits and Costs of Reducing Emissions from the Electricity Sector. *Journal of Environmental Management* 83(1): 115–130.

Rice, G., and J. Hammitt. 2005. Economic Valuation of Human Health Benefits of Controlling Mercury Emissions from U.S. Coal-Fired Power Plants. Northeast States for Coordinated Air Use Management (NESCAUM), Boston, MA. www.eenews.net/Greenwire/Backissues/images/032305gwr2.pdf.

Robinson, L. 2007. How U.S. Government Agencies Value Mortality Risk Reductions. *Review of Environmental Economics and Policy* 1(2): 283–299. doi: 10.1093/reep/rem018.

Salonen, J.T., K. Seppänen, K. Nyyssönen, H. Korpela, J. Kauhanen, M. Kantola, J. Tuomilehto, H. Esterbauer, F. Tatzber, and R. Salonen. 1995. Intake of mercury from fish, lipid peroxidation, and the risk of myocardial infarction and coronary, cardiovascular and any death in Eastern Finnish men. *Circulation* 91:645-655.

United States Court of Appeals for the District of Columbia Circuit. Argued March 25, 2008. Decided July 11, 2008. No. 05-1244. *State of North Carolina Petitioners v. Environmental Protection Agency Respondent and Utility Air Regulatory Group et al., Intervenors on Petitions for Review of an Order of the Environmental Protection Agency.*

———. Argued December 6, 2007. Decided February 8, 2008. No. 05-1097. *State of New Jersey et al. Petitioners v. Environmental Protection Agency Respondent and Utility Air Regulatory Group et al., Intervenors on Petitions for Review of the Final Action of the Environmental Protection Agency.*

Virtanen, J.K., S. Voutilainen, T.H. Rissanen, J. Mursu, T.P. Tuomainen, M.J. Korhonen, V.P. Valkonen, K.Seppänen, J.A. Laukkanen, and J.T. Salonen. 2005. Mercury, Fish Oils, and Risk of Acute Coronary Events and Cardiovascular Disease, Coronary Heart Disease, and All-Cause Mortality in Men in Eastern Finland. *Arteriosclerosis, Thrombosis, and Vascular Biology* 25: 228–233.

■ ■ ■

CHAPTER 8

The Cooling Water Intake Structures Rule

WINSTON HARRINGTON

Regulations to control industrial withdrawal of cooling water from surface waters are governed by Section 316(b) of the Clean Water Act (CWA),[1] which states that

Any standard established pursuant to section 301 or section 306 of this Act[2] and applicable to a point source shall require that the location, design, construction, and capacity of cooling water intake structures reflect the best technology available for minimizing adverse environmental impact.

Two phrases in this section are so important and so often repeated in regulatory documents that acronyms are used almost universally: *best technology available* (BTA) and *adverse environmental impact* (AEI).

The U.S. Environmental Protection Agency (EPA) began working on regulations for cooling water intake structures (CWIS) in the late 1990s, in response to a 1995 consent decree entered into with Riverkeeper, an environmental advocacy group. Prior to 1995, local permit writers wrote discharge permits without the effluent guidelines that governed permit writing in almost every other industrial category. Although in principle the absence of applicable effluent guidelines allowed the permits to be based on best professional judgment (BPJ) and tailored to local conditions and the specific configurations of local facilities, in practice it led to extensive wheel reinvention and gave rise to concerns that local permit writers could be made to respond to pressure from industrial sources, who were also local employers and taxpayers.

In accordance with the consent decree, EPA prepared and issued the cooling water intake regulations in three phases. Issued in 2002, Phase I regulations targeted new facilities. Phase II (issued in 2004) regulated large, existing steam-electric plants, those that withdraw more than 50 million gallons per day and use at least 25 percent of the water for cooling purposes. Existing plants were defined as those for which construction commenced before January 17, 2002. Phase III (issued in 2006) regulated certain other existing facilities as well as new coastal and offshore oil and gas extraction facilities. In all phases, the rules were very complex, and the discussion here is limited to the Phase II regulation of existing steam plants.

Almost all sections of the CWA regulate discharges of pollutants into rivers, lakes, and other receiving waters. Section 316(b) is unusual if not unique in that it regulates not discharges into protected water bodies, but withdrawals of water from them. Its purpose is to limit the damage done to marine organisms when strong intake flows pin them against screens (*impingement*) or sweep them into the cooling system (*entrainment*) and subject them to physical trauma, fluctuating temperatures, and toxic chemicals. Not only does impingement and entrainment (I&E) result in potentially high rates of mortality of directly affected species, it also has indirect effects on the entire ecosystem. The species most likely to be drawn into the cooling system are the small organisms at the bottom of the food chain and the juveniles of all species. These impacts, direct and indirect, can be substantial, as the cooling system requirements of a large power plant can be a substantial fraction of the total streamflow or the total volume of water in a reservoir.

In addition to the relevant sections of the CWA, EPA was subject to the requirement, found in a series of executive orders (most recently Clinton's Executive Order [EO] 12866) to prepare a regulatory impact analysis (RIA) and ensure that the benefits of a regulation justify the costs. When these directives conflict, the statute takes precedence. But as an executive branch agency, EPA must still endeavor to find a way to write regulations that conforms to the requirements of the executive orders without violating the statute.

In this case, the main conflict arises because Section 316(b) required technology-based (TB) standards, which require a plant to meet a standard of performance achieved by its peers in the industry. TB standards depend only on the performance (usually, pollutant discharge rate) of the abatement technology, without specific regard for the environmental consequences that would attend meeting the performance standards. TB standards are usually uniform and can lead to very different environmental outcomes at different locations. In various contexts, a TB standard could be defined as the performance of the very best plant, the best plant of its vintage cohort, or the average of similar plants. Typically, EPA identifies an abatement technology capable of achieving the desired level of performance (usually because it is the technology employed at the exemplary plants). A plant subject to a TB standard is not usually required to install a particular technology; it has to meet a standard of performance. However, adopting the designated technology could ease the path to compliance.

In an attempt to reconcile the BTA standard with the RIA requirement in the proposed regulation, EPA argued that the language of the statute did not bar the agency from considering the relationship between plant performance and environmental impacts in the design of the cooling water intake rule. EPA clearly stated its intention to depart from the usual approach to TB standards:

> *EPA believes that section 316(b) affords EPA such discretion because unlike the sections authorizing technology-based effluent limitations guidelines and new source performance standards for the discharge of pollutants, section 316(b) expressly states that its objective is to require best technology available for minimizing adverse environmental impact. EPA believes this language affords the Agency discretion to consider the environmental effects of various technology options (EPA 2002a).*

As discussed further below, this was a very unusual, perhaps unprecedented, step to take in the promulgation of a TB standard. Of course, to consider environmental effects of pollution abatement is much easier to say than to do, and despite EPA's efforts to estimate environmental effects, questions were raised about its success in doing so.

A larger issue was also lurking in the background: the issue of thermal enrichment. Steam-electric plants are very large users of water for cooling, so much so that they are nearly always located very near oceans, lakes, reservoirs, or major rivers. In most circumstances, the least costly cooling technology is *once-through* cooling, in which water withdrawn from the water body is circulated through the heat exchanger, and the heat extracted from the spent steam from the turbines is sent directly back to the water body. Among environmentalists, thermal enrichment is in bad odor because it can affect aquatic ecology, eliminating species sensitive to heat and perhaps replacing them with more heat-tolerant but exotic species that may not be "natural" in their new habitats. There may be some divergence here between the preferences of environmentalists and recreationists, however. On balance, the species that thrive in a thermally enriched environment are often desirable game fish, and often the best fishing sites are near the outfalls of electric generating plants.

Thermal discharges from electric utility plants are governed by Section 316(a) of the CWA. Again, no effluent guidelines exist for thermal discharges; instead, National Pollutant Discharge Elimination System permit writers rely on BPJ. The preferred technology of environmentalists is closed-cycle cooling, primarily through the use of cooling towers. Closed-cycle cooling minimizes the effect of thermal releases and of I&E. (The cycle is not completely closed because evaporated water must be replaced, but withdrawals are far smaller than with other technologies.) Thus, the requirement to develop regulations to minimize the impact of CWIS raised hopes among environmentalists for a regulation that minimized the effects of cooling water altogether.

The Phase II CWIS Rule

The Phase II CWIS rule is one of the most complicated and innovative regulations ever proposed and promulgated for a TB standard.[3] It had to be complicated to accommodate the degree of flexibility that EPA was attempting to build into the rule. Among the important flexibility-enhancing features of the rule were the following:

- The performance standards were based on ecological and environmental considerations, not just on technology.
- One of the compliance options permitted a plant to weigh its specific costs of compliance against EPA estimates of the costs or against the benefits of installing and operating the compliance technology at the site.
- Among the compliance options that a plant could consider was restoration, either at the site or elsewhere in the reach.

These innovations will be discussed further after a description of the rule.

As required by EO 12866, EPA submitted to the Office of Management and Budget (OMB) a draft of the proposed rule,[4] which, among other things, identified closed-cycle cooling as the BTA for 59 plants. OMB recommended that this requirement be lifted and also suggested an additional compliance option that permitted a comparison of benefits and costs on a site-by-site basis.

In addition to the RIA requirement, discussed in the preceding section, recent executive orders also imposed new requirements on rulemaking processes. These processes are most explicitly laid out in EO 12866. The basic structure of the rule was already laid out in the proposed rule, a trial balloon that is the required first public step in this type of rulemaking. In particular, the proposed rule already contained the provision defining the performance standards based on environmental impact. In fact, the main difference between the proposed and final rule is that the final rule contained two additional compliance options that were defined in terms of the technology only and did not rely on environmental impact. These options emerged during the comment period, and on March 19, 2003, EPA issued a notice of data availability (NODA) announcing the new options.[5]

The final rule had five essential components: (a) a calculation baseline, against which performance was to be measured; (b) the definition of a performance standard that all plants subject to the rule had to meet; (c) the identification of a technology that would meet the performance standard; (d) a set of compliance alternatives by which the performance standards could be met, including the designated technology; and (e) comprehensive demonstration study requirements for each compliance alternative.

Calculation Baseline

The calculation baseline is the impingement mortality and the entrainment associated with a shoreline intake facility with once-through cooling and no impingement mortality or entrainment controls. The baseline intake technology uses screens and barriers to prevent the components of the cooling system from being clogged with debris, but not to prevent losses to aquatic biota. The conventional traveling screen consists of a set of intermittently rotating screens at right angles to the flow into the intake. After remaining in position in front of the intake for a period of time, the screen moves off and is washed with a high-pressure flow (80–120 psi) to dislodge the accumulated debris while a clean successor screen takes its place. Mortality to organisms results either from being trapped motionless on the screen for too long or from exposure to the high-pressure flow. Although fish and shellfish are most affected by the screens, diving birds are also occasionally killed on the screens. Small organisms, larvae, and eggs are more likely to become entrained than caught on the screens.

Performance Standard

EPA defined two performance standards for the CWIS rule, one for impingement mortality and the other for entrainment. The impingement performance standard requires an 80 to 95 percent reduction in impingement mortality compared with the calculation baseline. The entrainment performance standard requires a 60 to 90 percent reduction in entrainment. To take into account the ecological significance of organisms at difference ages, quantities of fish and shellfish are nor-

Table 8.1
Final CWIS Rule Performance Standard Requirements

Water body type	Capacity utilization rate	Design intake flow	Performance standard
Freshwater river or stream	< 15%	NA	I[a] only
	≥ 15%	≤ 5% MAF[c]	I[a] only
		> 5% MAF	I & E[b]
Tidal river, estuary, or ocean	< 15%	NA	I only
	≥ 15%	NA	I & E
Great Lakes	< 15%	NA	I only
	≥ 15%	NA	I & E
Lakes or reservoirs	NA[d]	Must not disrupt thermal stratification	I only

Notes:

a. Impingement mortality performance standard (80–95 percent reduction in mortality from the calculation baseline).

b. Entrainment performance standard (60–90 percent reduction in entrainment).

c. Mean annual flow.

d. Not applicable.

malized to age-one equivalence, based on survivorship probabilities. The application of these performance standards depended further on the type of receiving water body, the plant's capacity utilization rate, or the design of the intake flow, as shown in Table 8.1.

Plants with low-capacity utilization factors were exempted from the entrainment standard because those plants typically operate in the hottest part of summer and the coldest part of winter, when power demand is highest. Neither is a period when juvenile fish or shellfish are most vulnerable to uptake in cooling systems.

Identification of Compliant Technology

EPA identified two technical approaches that are effective at reducing aquatic CWIS losses from the calculation baseline. The most effective approach was to use a closed-cycle cooling system. Even in the absence of specific controls, these systems can reduce aquatic losses by 90 percent or more simply because their water consumption requirements are so much lower than the baseline technology (EPA 2002b, Chapter 3).[6] The amount of water consumed by a closed-cycle system is just the amount of evaporation plus the blowdown, an amount bled from the system to avoid buildup of dissolved solids in the cooling water.

In addition, a cylindrical wedgewire screen is mentioned explicitly as an approved technology for a CWIS located on a freshwater stream. This designation was added by the NODA. during the rulemaking process.

Compliance Alternatives

Recognizing that the costs and biological effectiveness of abatement technologies for CWIS are very site-specific, the rule allows plants several compliance options:

- Demonstrate that the flow into the intake structure has been reduced either to a flow comparable to closed-cycle cooling or to an intake velocity of no more than 0.5 feet per second. Meeting either of these requirements will be deemed to have achieved the I&E performance standards. This compliance alternative was one of two options added by the NODA.
- Demonstrate that the plant's current CWIS meets the applicable performance standards shown in Table 8.1.
- Demonstrate that the plant's current structure—together with new construction, changes in operation, or restoration measures—will meet the impingement mortality and entrainment performance standards.
- Demonstrate that the plant has installed and is properly operating an approved technology specified in the rule. Thus, plants with closed-cycle cooling systems, together with plants on freshwater streams with cylindrical wedgewire screens, were presumed to be in compliance with the rule. This alternative was also added by the NODA.[7]
- Demonstrate that the facility is entitled to a site-specific determination of compliance technology because the cost of adopting the EPA-designated technology would be significantly greater than (a) the costs estimated in the rule or (b) the expected benefits at the site.

Comprehensive Demonstration Study Requirements

Because the performance standards are defined by biological metrics, demonstrating that technology is in compliance can be considerably more complicated than is the case with the more conventional TB standard. For some compliance options, it is not difficult at all. A plant that meets the flow or intake velocity values comparable to closed-cycle cooling is deemed in compliance without further demonstration. Only slightly more difficult is the demonstration for a designated technology, such as wedgewire screens, in which case a plant has only to develop plans for installation and monitoring. For a plant trying to meet the impingement mortality and entrainment performance standard, however, the required studies included a proposal for information collection, source water body flow information, an impingement mortality or entrainment characterization study, technology and compliance assessment information, a restoration plan (if appropriate), and a verification monitoring plan. The totality of these studies could be demanding. In fact, the onerousness of these demonstration requirements was one of the reasons that EPA added the additional options whereby compliance could be demonstrated quickly and cheaply (*see* EPA 2004b, 41595). Ironically enough, EPA increased the flexibility of the final rule by adding two purely TB options.

Flexibility and Innovation in the Final Rule

As noted above, the desire to bring environmental effects into the rulemaking led EPA to a very unusual definition of performance standards. In almost every other rule, EPA defined performance standards in terms of the performance of the technology itself. For example, pollutant-reduction technologies for air or water pollutants were defined in terms of the percentage reduction compared to no treatment, or perhaps the discharge of pollutants per unit of inputs used or outputs produced. For the cooling water rule, the standard was written in terms of the effects on natural organisms and, in particular, the percentage reduction in mortality from impingement and the percentage reduction in entrainment. These definitions of performance opened the door to providing considerable flexibility in compliance.

Such flexibility was desired for several reasons. Perhaps the main reason was that the effectiveness of these technologies is highly variable, depending on local configurations and conditions. Effectiveness also depends on which species is under consideration, with some species just much more susceptible to damage than others. With so many species to consider, even defining the percentage mortality reduction is subject to interpretation.

The inherent variability in the interaction of the technology and biological systems also gave rise to the expansive range of alternative technologies that could be considered by plants complying with the rule. For the most part, these technologies work by reducing intake velocities (say, by increasing the cross-sectional area), by using gentler methods for washing the screens, or by taking advantage of hydrological principles or fish behavior to discourage organisms from approaching the screens. What works in one setting may not work in another. This variance is fairly well documented with respect to the effects of the technologies on impingement mortality. Some technologies approach the performance of closed-cycle cooling, and others provide somewhat more modest performance or are suitable in some situations but not others. For entrainment, the number of demonstrations is much smaller, so there is less data on the performance.

Still another reason EPA wished to build flexibility into the rule was the extremely high inherent variability of aquatic environments. The vulnerability of individual species or entire ecosystems to CWIS and to the various compliance technologies varies greatly, as do the values that humans attach to the affected resources.

Explicit incorporation of environmental considerations directly into TB rules is probably the most important innovation in the CWIS rule, but two others are equally interesting. First, among the unusually wide range of alternative paths to compliance, under some circumstances the CWIS rule permitted plants to engage in "restoration" activities in lieu of compliance with the standard. That is, under some circumstances, the plant could compensate for failing to meet the CWIS performance standards by investing in activities to improve or restore wildlife habitat elsewhere. To show compliance using restoration measures, a plant had to demonstrate that the use of intake technologies and operation measures alone is "less feasible, less cost-effective or less environmentally desirable" than meeting standards partly or wholly through the use of a restoration plan. In addition, the plant had to demonstrate that the restoration measures would produce ecological benefits that are substantially similar to what would have been achieved through compliance with applicable performance measures. Restoration plans are limited to one permit term but may be renewed upon reapplication. The availability of restoration plans as an alternative for compliance with the rule was highly desired by the electric utility industry, according to sources at EPA. As they have shown by their reaction to wetlands banking, environmentalists tended to be skeptical.

A second innovation was the expanded scope for cost comparisons and cost–benefit analysis. Allowing a comparison of actual to estimated costs of compliance is somewhat unusual, but it gets at a common concern in rulemakings like this, namely whether the rule will impose too great a burden on individual facilities. The comparison of costs to expected benefits is even more novel, however, and exceeds even the executive orders that require RIAS. EO 12866 asks that the total benefits of the rule justify the total costs, whereas this rule allows that, in particular instances, a plant may apply for regulatory relief if the costs of compliance are out of all proportion to the benefits. Thus, the rule not only includes a comparison of total benefits and total costs, but also the potential for marginal benefit-to-cost comparisons. This is much closer to economists' conception of how benefit and cost information should be used. The site-specific comparison of benefits and costs in this rule bears a resemblance to the vehicle-by-vehicle approach taken in the corporate average fuel economy standard for light trucks in 2005, but to my knowledge it is its first use by EPA.

Costs and Benefits

Estimating the costs and benefits of this rule was complicated by its extreme site-specificity. On the cost side, it was uncertain at many plants whether the compliance alternative preferred for its low cost would actually meet the performance standards; thus, a more robust and costly technology had to be assumed at a number of plants. The benefits of reducing these sorts of actions in biological systems are inherently uncertain and not well understood, but the uncertainty is exacerbated by the lack of complete knowledge on which technologies would be installed where and how well they would work.

Costs

As part of the background information to support the rule, EPA conducted either a short technical questionnaire or a detailed questionnaire of potential regulated plants, collecting general information on plant operations as well as specific technical information on each plant's CWIS. EPA also used data from other sources; for example, satellite photographs of plants were used to determine whether space or flow conditions at any plant's intake might preclude the adoption of a particular technology.

EPA's cost study developed 14 technology modules that can be used to reduce I&E losses. At each plant, EPA estimated the cost of applying each technology, based on the information from the surveys, and chose the technology that appeared to achieve the best performance. Table 3.2 in Chapter 3 of the TDD (EPA 2004b) shows the technology that the agency assumed would be used at each plant, together with an estimate of the costs. (Plants were identified by a code number assigned in the questionnaire, so that the identity of each plant in the table is known only to its owner.) These cost estimates had at least two purposes. First, they were to be used to build the aggregate cost estimate in the RIA. Second, the agency cost estimate was to be used by any plant seeking to employ the fifth compliance alternative above—that is, permission to make a cost-to-cost comparison to justify seeking to avoid expenditures significantly greater than the EPA estimate.

These aggregate estimated total annualized costs are shown in Table 8.2. Note that the estimated total annualized costs of the CWIS rule increased by more than 30 percent between the proposed and final stages. The change in cost estimate could be for one or more of the following reasons:

- a change in the number of plants requiring regulation;
- a change in the rule's requirements—in other words, its stringency;
- a change in the unit costs of abatement equipment;
- a change in the expected effectiveness of the abatement equipment; or,
- the addition or removal of particular categories of costs.

For the CWIS Phase II rule, the last two categories above appeared to be responsible for the increase in the cost estimate.

There were two large increases in one-time costs: capital costs increased by $41 million and connection outages increased by $33 million, whereas the cost of the initial permit application fell by more than half. With respect to the connection outage, it is apparent that those costs were not even estimated as part of the proposed rule; EPA apparently received information on those costs during the comment period.

The remaining cost changes—to both one-time and recurring costs—are connected to the expansion of the compliance alternatives available to include those that are purely technology-based. But there is an apparent paradox: when a plant is provided with more compliance options, the costs of compliance should not increase, yet in this case they do, in the whole if not in the parts. As shown in Table 8.2, the annualized capital and operating costs each increased by about 50 percent, whereas the costs associated with permitting, recordkeeping, and reporting declined slightly. Overall, costs increased from $182 million to $250 million per year, an increase of about 38 percent.

Table 8.2 Annualized Cost of the Proposed and Final CWIS Rule

	Proposed regulation (2001$)	Final regulation (2002$)
One-time costs	**121.3**	**176.1**
Capital costs	83.5	124.5
Connection outage	0.0	32.6
Initial permit application	37.8	17.0
Pilot study	—	2.0
Recurring costs	**61.1**	**73.5**
Operations and maintenance	19.0	30.6
Monitoring, recordkeeping, and reporting	34.1	30.7
Permit renewal	8.0	12.2
Total annualized costs	**182.4**	**249.5**

Source: EPA 2002a, 2004b.

In fact, between the proposed and final rules, EPA changed the decisionmaking assumptions in the cost estimation model. In the proposed rule, EPA assumed that the plant was a cost minimizer. In the final rule, the agency recognized that there would be some uncertainty regarding the exact performance at a particular site, and that, at many sites, more than one technology might be under consideration. In modeling the technology choice at those sites, EPA selected the technology expected to perform the best, rather than the least-cost technology.

Moreover, the best performing technology concept, when necessary to apply, relied on assigning technologies about a median cost, with some choices above and below. Therefore, for each model facility (or intake), in order to ensure that the technology on which costs were based would in fact achieve compliance at that model site, the agency could not rely on a one-size-fits-all, least-cost approach.[8]

The technologies added with the NODA were high-cost, high-performance technologies that would very likely achieve compliance at every site. The addition of this element of performance uncertainty into the model ensured that these high-cost technologies would perforce be selected at some plants. The change in modeling explains why capital and operating costs increased—and by approximately the same percentage. It also explains why the costs associated with permitting, recordkeeping, and monitoring declined because the amount of proof that the technology would actually work required by EPA was much lower for the added technologies.

Quantifying I and E Losses and the Effects of the CWIS Rule

Although EPA has always produced estimates of the costs of its regulations, the requirement to produce benefit estimates originated with RIAS. In principle, the benefit study must show how the regulation will affect the physical and biological environment, it must quantify those changes, and it must estimate in monetary units how much people value those changes. However, some categories of benefits put forward often resist the final valuation step, and some cannot even be quantified, at least given present methods and data. For the CWIS rules, the claims for these nonmonetized benefits are unusually large relative to the monetized benefits.

The CWIS rule is supposed to generate benefits by reducing impingement mortality and entrainment of aquatic species. Thus, EPA's task in the RIA is (a) to explain quantitatively how the regulation will reduce the mortality of species affected by CWIS I&E and (b) to estimate the welfare effects of these changes—in other words, estimate the willingness to pay.

To demonstrate (a), Chapter 4 of the Technical Development Document (TDD) (EPA 2002b) presented data on experiments at selected power plants around the United States showing that impingement mortality was substantially reduced by the installation of each of the recommended technologies. Total entrainment was also reduced, although the evidence base was smaller and more variable. In addition, EPA had little information on entrainment mortality. In the proposed regulation, EPA assumed that entrainment mortality was 100 percent. This assumption was a holdover from an early guidance document issued by the agency in 1977 to guide permit writers in the absence of a regulation. It was also the assumption used in the permitting programs of at least 12 states. After the proposed rule appeared, this assumption was challenged by the electric utilities, many of which presented studies of entrainment mortality that purported to show that it was often less than 100 percent. Prior to issuing the final rule, EPA reviewed 37 such entrainment survival studies and concluded that they were so variable in species covered, quality, and approach, that they at best only had local validity (Regional Studies Chapter A7) (EPA 2004c). For the calculation of national benefits, EPA disregarded these studies and continued to assume 100 percent entrainment mortality.

By reducing the mortality of aquatic species, the CWIS rule is designed to lead to healthier aquatic ecosystems with larger populations of valued fish and shellfish. Although mature individuals of these primary species are too large to be directly affected by CWIS, fishing yields are affected by losses to the young of those species and to the forage species that comprise their diet. EPA developed two simple models to relate losses of immature primary species and forage species to reductions in yield or future production of primary species. These models used a yield-per-recruit (YPR) approach to modeling fish populations. The models assumed that the survivor function—that is, the probability that an individual of a species would survive from age one to age two—was independent of the population density of that species, resulting in a constant yield of mature fish for any recruit of a given age. The procedure for relating forage species was somewhat more involved because it took into account the importance of various forage species in the diet of the primary species and the conversion of biomass from one to the other. Table 8.3 shows EPA's expected reductions in I&E by region, as well as the improvements in biological endpoints. EPA reported the results in three metrics, first converting mortality at each life stage to "age one equivalents," which were in turn converted to forgone fishery yield and forgone biomass production.

During the rulemaking process, EPA convened several peer-review panels to examine various aspects of the Phase II rule. One of the panels was meant to peer review EPA's biological model. The YPR model came under some criticism, both internally and from the panel, because its linearity assumption disregarded the possibility of *density dependence,* which holds that the probability of survival is a function of the density of the species in the water body and, therefore, in the current context, I&E mortality (e.g., *see* Newbold and Iovanna 2007a, 2007b). EPA used the model nonetheless, adding a disclaimer that the YPR model was most suitable to the case when I&E losses constituted a small part of total mortality. And indeed, the peer-review committee noted that the most serious issue with the biological modeling was the inadequate empirical data on population and population dynamics.

Table 8.3
Expected Reduction in I&E, Final Rule

Source: Economic Benefits Analysis Chapter C3, Table C3-1 (EPA 2004d).

Region	Reductions in impingement	Reductions in entrainment
California	31%	21%
North Atlantic	44	29
Mid-Atlantic	54	48
South Atlantic	44	17
Gulf of Mexico	59	32
Great Lakes	52	40
Inland	47	16

Table 8.4
Summary of Estimated CWIS Net Benefits, Final Rule (Exclusive of Nonuse Benefits)

Source: Economic Benefits Analysis Tables D1-4, C3-A-1 (EPA 2004d).

Region	Commercial	Recreation	Total use benefits	Costs	Net use benefits
California	\$0.5	\$2.5	\$3.0	\$31.7	(\$28.7)
North Atlantic	\$0.1	\$1.4	\$1.4	\$13.3	(\$11.9)
Mid-Atlantic	\$1.7	\$43.4	\$45.0	\$62.6	(\$17.5)
South Atlantic	\$0.2	\$6.9	\$7.1	\$9.0	(\$1.9)
Gulf of Mexico	\$0.7	\$6.2	\$6.9	\$22.8	(\$15.9)
Great Lakes	\$0.2	\$14.0	\$14.1	\$58.7	(\$44.6)
Inland	NA	\$3.0	\$3.0	\$170.1	(\$167.2)
Total	**\$3.5**	**\$79.3**	**\$82.9**	**\$389.2**	**(\$306.3)**

Monetary Benefits

Having estimated the biological losses from CWIS, to value these avoided losses, EPA considered five categories of benefits. A summary of the kinds of items in each category, data needs, and approaches is given in Table 8.5. Although five categories were considered, benefits were actually monetized in only two: commercial and recreational fisheries.

Direct use benefits for market goods. The benefits of commercial fishing to producers and consumers are referred to as direct use benefits for market goods. EPA's analysis of the affected fisheries markets in Chapter 10 of the Regional Analysis (EPA 2004c) found that the reduced I&E losses would result in small increases (between 0.03 and 2.99 percent) in harvests in the six regions. EPA asserted that these increases in output would result in negligible changes in prices, so that the benefits to consumers would be zero. Thus, EPA's analysis focuses on benefits to producers only. To determine the commercial fishing losses attributable to I&E, EPA assumes proportionality. Suppose S is the total stock of a harvested species and M is the production lost to I&E. If kS is the amount

Table 8.5: Benefit Categories

	DIRECT USE, MARKETED GOODS	
▪ Increased commercial landings ▪ Fishing tournament with entry fee and prizes	▪ Estimated change in landings of specific species ▪ Estimated change in total economic impact	▪ Market-based approach using data on landings and the value of landing data from the National Marine Fisheries Service (NMFS) ▪ Based on facility-specific I&E data and ecological modeling ▪ Based on available literature
	INDIRECT USE, MARKET GOOD	
Increase in market values: ▪ Equipment sales, rental, and repair ▪ Bait and tackle sales ▪ Increased consumer market choices ▪ Increased choices in restaurant meals ▪ Increased property values near water ▪ Ecotourism (charter trips, festivals, other organized activities with fees such as riverwalks)	▪ Estimated change in landings of specific species ▪ Relationship between increased fish/shellfish landings and secondary markets ▪ Local activities and participation fees ▪ Estimated numbers of participating individuals	▪ Not estimated for the final Section 316(b) rule analysis due to data constraints
	DIRECT USE, NONMARKET GOODS	
Improved value of a recreational fishing trip: ▪ Increased catch of targeted/preferred species	▪ Increased incidental catch ▪ Estimated number of affected anglers ▪ Value of an improvement in catch rate	▪ Regional RUM analysis ▪ Benefit transfer (inland region)
Increase in recreational fishing participation:	▪ Estimated number of affected anglers or estimate of potential anglers ▪ Value of an angling day	▪ Regional RUM analysis (not estimated for the California and inland regions)
	INDIRECT USE, NONMARKETED	
Increase in value of boating, scuba-diving, and near-water recreational experience: ▪ Enjoying observing fish while boating, scuba-diving, hiking, or picknicking ▪ Watching aquatic birds fish or catch aquatic invertebrates	▪ Estimated number of affected near-water recreationists, divers, and boaters ▪ Value of boating, scuba-diving, and near-water recreational experience	
Increase in boating and near-water recreational participation:	▪ Estimated number of affected boating and near-water recreationists	▪ Value of a near-water recreational experience
Increase in nonuse values: ▪ Existence (stewardship) ▪ Altruism (interpersonal concerns) ▪ Bequest (interpersonal and intergenerational equity) motives ▪ Appreciation of the importance of ecological services apart from human uses or motives (e.g., eco-services interrelationships, reproductive success, diversity, and improved conditions for recovery)	▪ I&E loss estimates ▪ Primary research using stated preference approach (not feasible within EPA constraints) ▪ Applicable studies upon which to conduct benefit transfer	▪ Site-specific studies or national stated preference surveys ▪ Benefit transfer, including meta-analysis of applicable studies ▪ Restoration-based values of common and/or endangered species

Source: EPA 2004c

harvested each year, then kM is the amount lost to I&E. Dockside prices are applied to this quantity to get the revenue losses resulting from I&E for a species.

Chapter 10 of the Regional Analysis (EPA 2004c) contains a table summarizing the results of four studies containing estimates of producer surplus in nine fisheries markets. In these studies, producer surplus ranges between 0 and 37 percent of total revenue.[9] In converting revenues to surplus, EPA uses these results to bracket surplus as 0 to 40 percent of revenue.

Direct use benefits for nonmarket goods. To estimate the enhancement in recreational benefits associated with the rule, EPA estimated a set of econometric models, including a random utility model to relate recreation site characteristics and travel costs to an individual's probability of visiting the site. This model predicts the average utility of the site. A second model estimates trip frequency as a function of this average utility plus individual characteristics. The main data source is National Marine Fisheries Service.

Indirect use benefits. Indirect use benefits are those welfare improvements that accrue to individuals who benefit from the enhancement of the fishery without enjoying the fish or the enhanced fishing opportunities. For example, birdwatchers might benefit if the fishery attracts more birds to the site. Another example is provided by increased populations of forage fish, which are not consumed directly, but which contribute to the primary stock. Thus, indirect use benefits can be either recreational or commercial.

Nonuse benefits. Nonuse benefits must be estimated by stated preference methods. EPA considered commissioning its own survey to elicit nonuse benefits of I&E losses, but the agency was not able to secure the required approval from OMB. EPA then attempted to use benefit transfer methods to apply studies of nonuse values at particular sites, but the analysts were uncomfortable scaling up to a national benefit level. Therefore, EPA decided not to estimate nonuse benefits.

Summary of monetary benefits. Table 8.4 shows the direct commercial and recreational benefits estimated for the Phase II rule and compares them to the costs. As shown, costs exceed benefits by about a factor of five. Recall that benefits are monetized for only two of the five categories.

The Legal Challenge

After the final rule was issued in 2004, it was challenged by states, environmental groups, and advocates for the utility industry in the U.S. Court of Appeals for the Second Circuit. This court had also heard the appeal of the Phase I rule, so it was already well educated on the substance of the rule and the situation. The state and environmentalist arguments were similar in asserting that EPA did not comply with the requirements of the CWA and in several ways wrote regulations that gave far too much discretion to utility plants in how to comply with the standard. They also argued that EPA made changes in the final rule that were not supported by the docket. The most important utility industry arguments were as follows: CWA Section 316(b) does not apply to existing facilities; the agency did not justify its definition of AEI or its assumption of zero entrainment survival; and the requirement of evaluation of qualitative nonuse benefits in site-specific cost–benefit studies was improper. The individual appeals were merged into a single case, argued on June 8, 2006, and decided on January 25, 2007 (*Riverkeeper, Inc. et al. v. EPA*).

The result was a clear win for the states and the environmental groups. Most of the aspects of the rule most troubling to environmentalists—which consisted of most of the rulemaking innovations identified and discussed above—were either rejected outright or remanded to EPA for further clarification, whereas the industry arguments found little favor with the court. To begin, the court opinion indicated that the use of cost–benefit analysis by EPA in this rule was an incorrect reading of the statute. The BTA performance standard precluded the balancing of benefits and costs, and indeed the court stated that the only legitimate use of cost information in this rule was to determine whether the cost of meeting the performance standard was something the industry could "reasonably bear." With that in mind, the court went on to observe that several utility plants did have closed-cycle cooling, and EPA's failure to identify that technology as the BTA could not be based on a cost–benefit test, but rather on what the industry can reasonably bear. It remanded this part of the rule to EPA to clarify its reasoning on this point.

One of the arguments made by the environmental groups was that the use of ranges for performance standards was impermissibly vague; these groups asked the court to require a point standard. Although the court sympathized with EPA's position that the mortality ranges were often site-specific and in any case not fully within the control of the plant, it required EPA to tighten up the rule so that a plant could not get away with achieving the minimum performance in the range when more could have been done. The key point was that the plant should do its best, not the minimum.

The court also rejected the use of restoration as a compliance alternative, ruling that restoration was impermissible "compensation" for environmental impacts, rather than "minimization," and that, in any event, restoration was not a technology as defined in the statute.

Finally, the court remanded the site-specific compliance alternatives—the cost–cost test and the cost–benefit test. The latter was eliminated for the same reasons as the more general cost–benefit tests in the rule. The former failed in part because the agency changed the basis of calculating costs in the final rule, violating the procedural requirements of informal rulemaking by giving inadequate opportunity for comment.

Since this decision, EPA has had the following statement on its Section 316(b) website:

- EPA is suspending the Cooling Water Intake Structure Regulations for existing large power plants. This suspension is in response to the 2nd Circuit Court of Appeals decision in *Riverkeeper, Inc., v. Environmental Protection Agency.*
- EPA is still studying the court decision and has not yet decided how to revise the rule to comply with its requirements. Also, the industry petitioners have appealed the decision of the 2nd Court of Appeals to the U.S. Supreme Court.

Conclusion

Few types of regulation resist the use of cost–benefit analysis as much as TB regulation does. Such regulation requires EPA to identify a technology that meets some concept of "best" performance, a concept that is ordinarily established in the legislation. The agency then must set a performance standard for all users that achieves the performance of the designated technology, regardless of costs, the environmental improvements expected, and the value of those improvements. This approach is the antithesis of cost–benefit analysis, for which cost, environmental improvement, and value are very important.

EPA justified its attempts to go beyond the usual definition of TB standards by statutory language that called for minimization of adverse environmental impact. The agency pointed out the difficulty of defining performance standards without mention of I&E mortality, which in any case were highly variable depending on species and site characteristics. These efforts were to no avail as the court struck down virtually all of the features of the regulation that allowed plants more flexibility than the limited amount usually permitted by TB standards.

Even without the TB requirement, this regulation illustrates the great difficulty of basing regulatory decisions on the likely consequences because the knowledge base for determining those consequences was sadly deficient. This showed up in both the quantification of the physical effects of the regulation and the valuation of those effects. EPA's YPR model for linking I&E mortality to population effects was exposed in the ecological peer review as very questionable. And, without a comprehensive ecological model taking into account losses to both forage and primary species, there was no way to account for the interaction of I&E effects on both food supply and juveniles of primary species. As a result, EPA's benefit studies only included direct use benefits—and those were based on an ecological model that EPA's own peer review had called into question. Nonuse benefits were considered but abandoned when OMB refused EPA permission to conduct a survey of nonuse benefits.

All of this is not to say that a conventional TB standard would perform any better. It is difficult to determine whether this pair of court decisions—first to require EPA to write effluent guidelines for CWIS and then to limit the flexibility of plants in complying—improved matters or made them worse.

■ ■ ■

Notes

1. 33 U.S.C. Chapter 26.

2. Sections 301 and 306 discuss the regulatory requirements for existing and new point sources of pollution, respectively.

3. Supporting documentation prepared by EPA for the rule can be found in the *Federal Register* notices for the proposed rule (67 FR 17121, April 9, 2002), the final rule (69 FR 41576, July 9, 2004), and a notice of data availability (68 FR 13522, March 19, 2003), which appeared during the comment period and contained information on additional compliance options available to regulated plants. In addition, the proposed and final rules were each supported by three technical reports prepared by the agency and its contractors. These are the Technical Development Document (TDD), which describes the alternative technologies for controlling impingement and entrainment and provided examples of applying those models to plants in several settings. The TDD also contains the micro-level compliance cost models. The Regional Analysis (RA) defines relevant outcome measures, provides information on the effects of the estimates on impingement mortality and entrainment in the baseline and with the technology options, and describes the micro-level benefit estimation studies. Finally, the Economic and Benefits Analysis (EBA) aggregates site-specific estimates of the benefits and costs of the rule and determines net benefits. These documents can be found at www.epa.gov/waterscience/316b/phase2/.

4. EO 12866 requires the rulemaking agency to submit a draft of the proposed rule to OMB for review. If OMB's review raises issues that cannot be resolved by negotiation between OMB and the issuing agency, the matter is re-

ferred to the president. If an issue cannot be resolved either by agency agreement or presidential intervention, the rulemaking agency cannot issue a proposed rule in the *Federal Register.*

5. EPA (2003), Phase II Cooling Water Intake Structures Proposed Rule: Notice of Data Availability (NODA). *Federal Register* 68 13522. March 19.

6. The development document also describes a dry cooling option, which works like a car radiator and essentially cuts water consumption to zero. However, EPA did not include dry cooling as a compliance option after finding it to be too costly.

7. Interestingly, the alternatives added by the NODA appeared to be more in line with the conventional meaning of TB standards. They defined compliance based on technological performance entirely; biological effects were not relevant.

8. EPA 2004b at 41650.

9. Another table providing information on numerous studies of normal profit rather than producer surplus (i.e., they do not include owners' opportunity costs) shows a much wider range of values for this ratio.

References

Newbold, Stephen C., and Rich Iovanna. 2007a. Ecological Effects of Density-Independent Mortality: Application to Cooling-Water Withdrawals. *Ecological Applications* 17(2): 390–406.

———. 2007b. Population Level Impacts of Cooling Water Withdrawals on Harvested Fish Stocks. *Environmental Science and Technology* 41: 2108–2114.

Riverkeeper et al. v. U.S. EPA, U.S. Court of Appeals, 2nd Cir., Argued June 8, 2006, decided January 25, 2007.

U.S. Environmental Protection Angency (EPA). 2002a. National Pollutant Discharge Elimination System—Proposed Regulations to Establish Requirements for Cooling Water Intake Structures at Phase II Existing Facilities; Proposed Rule. Office of Water. *Federal Register* 67: 17121, April 9.

———. 2002b. *Technical Development Document for the Proposed Section 316(b) Phase II Existing Facilities Rule.* EPA 821-R-02-003. Office of Water, April 2002. Washington, DC: EPA.

———. 2003. Phase II Cooling Water Intake Structures Proposed Rule: Notice of Data Availability (NODA). *Federal Register* 68: 13522, March 19.

———. 2004a. *Technical Development Document for the Final Section 316(b) Phase II Existing Facilities Rule (Final). EPA 821-R-04-007.* Office of Water, February 12. Washington, DC:EPA.

———. 2004b. National Pollutant Discharge Elimination System—Final Regulations to Establish Requirements for Cooling Water Intake Structures at Phase II Existing Facilities; Final Rule. *Federal Register* 69: 41576, July 9.

———. 2004c. *Regional Studies, Phase II CWIS Final Rule.* Office of Water. Washington, DC: EPA.

———. 2004d. *Economic and Benefits Analysis for the Final Section 316(b) Phase II Existing Facilities Rule.* Washington, DC: EPA.

■ ■ ■

CHAPTER 9

Improving the CWIS Rule Regulatory Analysis: What Does an Economist Want?

SCOTT FARROW

Proponents, opponents, and some who aren't too sure have spilled much ink on the merits of cost–benefit analysis in support of government decisions, including the development of regulations. In this chapter, I present a proponent's view in the specific context of establishing requirements for cooling water intake structures (CWIS) at existing power plants (EPA 2004a). The policy context, regulatory proposal, and environmental and economic evaluation of this regulation are summarized by Harrington in Chapter 8 of this report. Voluminous comments were filed on particular aspects of the estimation procedures, and one more commenter on such details is unlikely to add much value. In its response to public comments, the U.S. Environmental Protection Agency (EPA) used 5,143 pages; of those, about 1,200 focused on benefits related to economics (EPA 2004e). Consequently, the focus of this chapter is on the consistency of the CWIS cost–benefit analysis with quality criteria to which the agency might have been expected to adhere and significantly less on specific details of the existing analysis.

In the remainder of the chapter, I focus on four topics. First, I discuss criteria for evaluating the economic content of the rule and whether the rule met those criteria. Second, I investigate criteria and outcomes with respect to decision rules for the design of the regulation. Third, I address the challenge faced by agency analysts because of the frontier nature of research linking ecological and economic impacts. Finally, I provide suggestions for improvement.

Criteria and Evaluation for the Economic Review of the CWIS Rule

What are the most appropriate economically based criteria for review of the regulatory impact analysis (RIA) conducted by EPA and how did the analysis perform on those criteria? Two sets of criteria appear relevant. The first set refers to analytical standards for the content of the analysis. The second set refers to decisionmaking standards that are at the boundary of economic and other risk management approaches. Although important nuances can be found in the text, I conclude that the RIA met minimum necessary analytical economic standards and may have met additional professional analytical standards. The analysis further met some aspects of economic-based decisionmaking standards but failed a critical one based on the law. I discuss each conclusion in turn.

What an Economist Wants—1: Analytical Standards for the Content of the Analysis

In contrast to accounting, in which Generally Accepted Accounting Principles (GAAP) are issued by professional organizations, no standards from a professional society are available for cost–benefit analysis (GAO 2005; Bray et al. 2007). Instead, reviewers generally refer to two sources—guidance provided by government agencies and the published literature. These sources are equivalent to the two lowest-ranking sources in a hierarchy of standards for accountants and auditors (GAO 2005). For EPA regulations, the Office of Management and Budget (OMB) and EPA have developed guidance (OMB 1992, 2003; EPA 2000) based primarily on authority derived from executive orders related to cost–benefit analysis and regulatory development. Based on the OMB guidance, Hahn and Dudley (2007), Belzer (1999), and GAO (2005) have developed scorecards for the basic quality of an analysis. These scorecards provide a way to determine whether the analysis was consistent with elements of OMB guidance and essentially are a type of analytical process check. Cost–benefit analysis substantially lacking these elements would almost certainly be of poor quality. For instance, questions on the scorecard refer to whether benefits are stated, quantified, or monetized; whether discounting was used and at what rate; whether alternatives were evaluated; and whether uncertainty was incorporated. However, while consistency with the guidance may be viewed as a necessary condition for a good-quality analysis, it is not sufficient. In particular, the analysis may have been done incompletely or incorrectly, in which case the result would be of poor quality.

I provide a slightly modified version of the scorecard used by Hahn and Dudley (2007) in Table 9.1. In this modification, I have deleted items specifically related to health, safety, or an executive summary as these were immaterial for the case at hand. The right-hand column provides my subjective assessment of the CWIS rule. Note, however, that some items are difficult to answer with a "yes" or "no" and the record was quite extensive. For instance, a reviewer might wish to know whether all or nearly all of the material beneficial impacts of the regulation had been considered, but this is difficult to ascertain without additional information and judgments. In the CWIS case, some commenters believed and I concur that nonuse and some types of fishery stock effects were potentially large and should have been included; consequently a "no" is recorded for "monetized all or nearly all benefits" in contrast to "monetized some benefits." However, this is a matter of judgment on items where information is lacking.

At this analytical process level, the EPA CWIS RIA passes virtually all of the steps of the scorecard, in contrast to many regulations (Hahn and Dudley 2007). The RIA[1] monetized at least some costs and benefits, estimated monetized net benefits, considered alternatives, used a prescribed discount rate, and so on. The RIA earns poorer marks on the clarity of analysis to justify trade-offs made in the regulation and the completeness of the benefits estimation. For instance, the RIA provides no logical or conceptual model up front to convey the sequence of steps, and important components are spread across multiple documents including the Economic and Benefit's Analysis (EPA 2004b), Regional Analysis (EPA 2004c), and Technical Development Document (EPA 2004d). Some of the analyses are quite involved, such as the econometrically estimated Random Utility Model, and others are simpler. Although summary tables of benefit and cost results are provided in both the final *Federal Register* notice and the Economic and Benefits Analysis (EPA 2004b), these tables provide minimal caution about the analytical steps and the degree of precision. The latter

Table 9.1
Scorecard Evaluation of CWIS Cost–Benefit Analysis

Note: Evidence may be in supporting documents and not in summary documents.

Source: Scorecard format based on Hahn and Dudley (2007); some item numbers have been omitted because of their original focus on health and safety. CWIS evaluations are by the author.

Item number	Variables	CWIS
	ESTIMATION OF COSTS	
1	Stated costs exist	Yes
2	Quantified at least some costs	Yes
3	Monetized at least some costs	Yes
4	Monetized all or nearly all costs	Yes
5	Provided point estimate of total costs	Yes
6	Provided range for total costs	Yes
7	Associated costs w/ federal government	Yes
8	Associated costs w/ nonfederal government	Yes
9	Associated costs with producers	Yes
10	Provided best estimate and range for total costs	Yes
	ESTIMATION OF BENEFITS	
11	Stated benefits exist	Yes
12	Quantified at least some benefits	Yes
13	Monetized at least some benefits	Yes
14	Monetized all or nearly all benefits	No
15	Provided point estimate of total benefits	Yes
16	Provided range for total benefits	No
17	Provided best estimate or range for total benefits	Yes
18	Provided best estimate and range for total benefits	No
	COMPARISON OF COSTS AND BENEFITS	
19	Calculated net benefits	Yes
20	Provided a point estimate of net benefits	Yes
21	Provided a range for net benefits	No
22	Calculated cost-effectiveness	Somewhat
23	Provided a point estimate of cost-effectiveness	Yes
24	Provided a range for cost-effectiveness	No
25	Had positive net benefits	No
26	Calculated net benefits or cost-effectiveness	Yes
27	Calculated net benefits and cost-effectiveness	Somewhat
28	Calculated both point estimate and range for net benefits	No
29	Calculated either point estimate or range for net benefits	Yes
	EVALUATION OF ALTERNATIVES	
30	Gave at least one alt. standard/level	Yes
31	Gave at least one alt. method	Yes
32	Quantified alternatives (costs)	Yes
33	Monetized alternatives (costs)	Yes
34	Quantified alternatives (benefits)	Somewhat
35	Monetized alternatives (benefits)	No
36	Calculated cost-effectiveness of alternatives	Somewhat
37	Calculated net benefits of alternatives	No
38	Calculated net benefits or cost-effectiveness of alternatives	Somewhat
39	Considered some alternatives	Yes
40	Clarity of presentation	Average/poor
	CONSISTENT USE OF ANALYTICAL ASSUMPTIONS	
41	Identified dollar year	Yes
42	Used consistent dollar year	Yes
43	Identified discount rate	Yes
44	Used consistent discount rate	Yes
45	Discount rate = 7 percent	Yes
46	Used consistent costs and benefits	Yes
47	Identified and consistently used discount rate and dollar year	Yes

issue also relates to the treatment of uncertainty in the final *Federal Register* notice itself, which does not convey some of the extended uncertainty analyses carried out in various supporting documents. Similarly, although implicit throughout the analyses, EPA provides no direct reporting of an average cost per adverse environmental impact (AEI) for different alternatives. The alternatives analysis, with its implicit wet cooling tower benchmark technology, was not clearly brought into the analysis as a likely "default" technology against which other regulations were measured. This is an example of an instance in which the usual "do nothing" alternative of cost–benefit analysis would not seem to be appropriate given the clear direction to "do something." Consequently, although many of the necessary aspects identified in the guidance were done well, other aspects—such as those identified in Table 9.1—were touched upon but could have been improved.

What of the role of a review standard based on the professional literature that goes beyond the necessary aspects identified in guidance and scorecard approaches? This criterion is more nebulous because of the ambiguity and vastness of the professional literature as applied to a specific problem. Variations may also exist for standard practice, best practice, and frontier application. However, this concern for the quality of the analysis is related to other guidance from OMB (2002) based on the Data Quality Act. In this guidance document, quality is composed of utility, objectivity, and integrity. Procedural steps are identified through which agencies can achieve quality; these procedures include the use of peer review panels, whereby the review is conducted by people with a professionally equivalent or advanced understanding of the problem investigated by the agency. However, if an analysis appears in a peer-reviewed publication, then it may—but need not—have met the data quality standards. In the case of a peer review panel, peer reviewers are not blind to the identity of the author and are selected by the reviewing authority, whereas in the case of a peer-reviewed publication, the reviewers are generally "blind" and are selected by the editors of the publishing outlet. To some extent, the availability of a document for public comment can be viewed as a nonblind review process in which the selection of the reviewers is based on self-interest, which can include payment from any parties. The OMB review of RIAS might be viewed by some as another type of peer review and, in fact, OMB appears to hold that opinion.[2] Three tests might be identified based on these different processes. First, did the agency conduct a peer review of the analysis and, if so, what was the outcome? Second, how did professional commenters respond during the public comment period? And third, is the analysis based on a peer-reviewed publication or, somewhat weaker, is it likely that a peer-reviewed publication would publish a paper based on the analysis? I briefly address these three checks on quality in turn.

Did EPA assemble a peer review panel of the cost–benefit analysis and, if so, what were the results? Partially—a peer review panel was convened for the ecological and fisheries aspects of impingement and entrainment, and a separate panel was convened on nonuse values, although the latter panel addressed a later phase of the regulation. As is often the case, the review panels developed a variety of comments and suggestions for improvement (EPA 2002; RTI International 2005). The reviewers raised a number of issues about broader ecological impacts and fisheries dynamics, although EPA appears to express substantial concern with linking such measures with economic valuation (Stratus Consulting 2004; EPA 2004e). With regard to nonuse values for the later regulatory phase, the panel appeared critical of the particular way in which EPA estimated nonuse values but supported EPA's efforts to provide a nonuse value estimate. OMB reviewed both the proposed and final rules and completed its actions by determining that the rule was "consistent with changes" (Regulatory Information 2002, 2004).

How did professional commenters respond publicly to the final rule? Unfortunately, this is basically unknowable because the public comment period applies only to the proposed rule. However, several professional economists criticized analytical elements of EPA's proposed rule and, later, data availability, on a variety of fronts.[3] Further, at least two of the economists, Frank Ackerman and Robert Stavins, disagreed with each other (see Chapter 10), while the other economists presented a somewhat more common methodological view. At the risk of ignoring other economists not obvious in the record, I counted four economists whom I interpret as having similar methodological interpretations and one with a different interpretation.[4] Some of the methods opposed by the larger group of economists were removed from the final analysis, whereas movement of the monetized cost–benefit analysis toward the views of the economist with the minority interpretation appeared to be slight. Because no comment period existed for the final rule, the extent of economic commenters' agreement on the quality of the final rule is unknown.

Finally, was the completed analysis published in a formal, externally peer-reviewed source? No. Could it potentially be published in such a source and thus by demonstration meet some level of professional standards? Possibly. Although this question has many facets—including the length of the CWIS analysis, which is more suited to a book than to a journal article, and the lack of a journal currently devoted to cost–benefit analysis—my sense is that a journal article devoted to the CWIS case could appear in an applied, peer-reviewed journal. In particular, the topic is at the frontier of integrating ecological and economic analysis, extensive information is provided in the analysis on technological alternatives, and a variety of quantitative analyses— including some relatively advanced econometric analyses of recreational choices—are provided. I conclude that a paper based on the CWIS rule could potentially appear in a peer-reviewed publication.

Although the evidence that the CWIS rule met the ambiguous "professional quality" standard is somewhat weaker, the fact that EPA carried out some peer reviews, that its analysis moved in the direction of the majority of professional commenters, and that it may be publishable in a peer-reviewed journal, indicates that it probably meets guidelines provided by the Data Quality Act and the more ambiguous auditing standard of consistency with professional norms. The difficulty in determining the quality of a controversial analysis may also indicate the ambiguity of criteria and the potential usefulness of work in this area.

What an Economist Wants—2: Decisionmaking Standards

Much of the controversy surrounding the CWIS rule appears to involve not only the analytical methods, but also the decisionmaking standards and the role played by OMB's Office of Information and Regulatory Affairs (OIRA; Heinzerling 2006). The choice of a decisionmaking standard by decisionmakers is outside of the role of economists, although a large body of public policy and economic literature, including cost–benefit analysis, suggests normative decision rules that could be followed in making a government decision. The Federal Water Pollution Control Act as amended in this case states that "[a]ny standard established…shall require…the best technology available for minimizing adverse environmental impact" (33 U.S.C. s. 316(b)). In addition, agencies are directed by presidential executive orders, primarily Executive Order (EO) 12866 (1993), (a) not to use the RIA to displace the agencies' authority or responsibilities as authorized by law (EO 12866, s. 9); (b) to assess both the costs and benefits of an intended regulation and propose or adopt a regulation only upon a reasoned determination that the benefits of the intended regulation justify its costs (EO 12866, s. 6); (c) to identify and assess alternative forms of regulation that, to the extent feasible, specify performance objectives rather than a specific behavior or manner of compliance (EO 12866, s. 8); and (d) to tailor regulations to impose the least burden on society, including various societal elements (EO 12866, s. 11).

Consider first the decisionmaking standard under the law. Although a technology-based standard may appear simpler to define than an economics-based one, the apparent simplicity may belie complex issues in implementation and interpretation (Freeman 1980). Importantly in this case, EPA chose never to explicitly define the AEI (EPA 2004a, 41612). Even with the simpler focus on impingement and entrainment used by the EPA, substantial uncertainty about environmental impact remained because of a paucity of what may seem like basic data, such as the natural mortality rate of a species (EPA 2004a, Regional Studies, A6–5). Consequently, it is unknown whether EPA chose the technology that would minimize that impact. Further, the cost of technologies was identified as a possible, though secondary, element of the decision (EPA 2004a; *Riverkeeper, Inc. v. EPA* 2007).

Several weaknesses result from the lack of definition of the AEI, uncertainty about environmental impacts, and secondary use of cost information. Regardless of the definition of the AEI, it has multiple dimensions. EPA at least identified impacts, including the following: effects on various commercial, recreational, rare, sensitive, exotic, and disruptive species; disruptions of ecological relationships and public satisfaction; organic carbon, nutrient, and energy transfer; and decreased local biodiversity (EPA 2004a, 41662). Determination of the AEI would require a weighting of many factors. Analysts would be forced to come up with weights of relative importance, to defend equal weighting, or to explicitly acknowledge impacts that have zero weight in quantitative or monetized analyses but that may be discussed in the text. Decisions would have to be made about which impacts are material to the decision. Economists would generally seek to use observed or inferred functions of prices as weights; alternatively, other weighting approaches could be devised. In any event, the benefit of the regulation is essentially the reduction in the weighted AEI with a standard economic benefit analysis using monetized weights.

Because any AEI developed based on the record would contain substantial uncertainty about impacts on the environment, it is extremely unlikely that one technology would be unambiguously best, an issue made even more complex by the multiple environments being considered, such as estuaries, freshwater lakes, and rivers. Thus, following the direction of the statute appears to allow for the potential for a benefit evaluation, measured by uncertain AEI reduction, and some evaluation of cost as a secondary element.

OMB guidance (2003) further indicates that in situations of uncertainty, expected (mean) values should be used as the foundation for analysis; however, other treatments of uncertainty could be used, possibly including probabilistic AEI and cost analysis. Consequently, it is conceivable that EPA could have used an analysis that included elements of AEI reduction (benefits) and costs in a manner more consistent with the statute than developing a cost–benefit analysis that did not depend on a definition of AEI.

In general, however, the determination of a decisionmaking standard that complies with the law is a legal question that is not answerable by economists. In hindsight, the actual analysis carried out by EPA and the regulation failed the decisionmaking standard of both the law and EO 12866 (*Riverkeeper, Inc. v. EPA* 2007), although that conclusion is under review by the Supreme Court at the time of this writing.

What of the additional requirements of EO 12866? Did EPA assess the costs and benefits and choose an approach in which the benefits justify the costs? Yes and no. EPA assessed costs and benefits in the final CWIS regulation; on the basis of that analysis, the monetized benefits did not justify the cost (EPA 2004a; EPA 2004b, D1–4). A break-even analysis was provided to illustrate how large the nonmonetized benefits must be for the rule to break even on a monetized net benefit basis. One possible inference is that statutory responsibilities trumped this element of the executive order because, on a monetized cost–benefit basis, the country would be better off without the rule in the standard interpretation. Alternatively, EPA may have determined that, taken together, the monetized and nonmonetized benefits justified the costs, although they do not appear to explicitly make such a statement in the final regulation (EPA 2004a, 41663).

Did EPA specify a performance objective instead of the manner in which regulated entities must comply? Yes, to a large degree. An important aspect of the CWIS regulation was precisely the effort to identify a performance standard—the degree of mitigation provided by cooling towers—and to specify technological alternatives to meet that standard, which also varied by environmental conditions. The regulation specified a set of technologies that meet a range of impingement and entrainment performance reductions as well as some alternative means of compliance. Although this could be seen as an effort to acknowledge the uncertainty in both the AEI and the performance of technology, the integration of uncertainty and performance could have been better clarified in the determination of the standard.

Did EPA choose an approach that generated the least burden on society? Yes, to a large degree. This is the decisionmaking element closest to the discussion of the appropriateness of cost-effectiveness in *Riverkeeper, Inc. v. EPA* (2007). EPA clearly chose an approach that imposed a significantly lighter burden on society than the cooling tower option that formed the performance basis for the regulation. EPA estimated that a cooling tower option would have an annualized post-tax compliance cost of $2,316 million (EPA 2004a, proposed Economic and Benefits Analysis [EBA], B7–10, 2001$) and a total social cost of $3,507 million computed with a 7 percent real discount rate. This compares with an estimated annualized post-tax compliance cost of $250 million (EPA 2004a, final EBA, D1–3, 2002$) and a social cost of $389 million for the final regulation.

Although commenters debated many aspects of these estimates (EPA 2004e), it appears that EPA did incorporate the economic burden on society in its determination. The record provides substantial evidence that the agency considered a lower-cost alternative to meeting a standard with the potential to save approximately $3 billion in annualized dollars or approximately $40 billion in present value.[5] To put this in a different context, the GAO would probably score an approximate $40 billion dollar nongovernmental cost savings if the EPA made the regulatory design change away from cooling towers in response to a recommendation by the GAO6.

In summary, with regard to what decisionmaking standards were applied, the results are mixed, but EPA's failure to comply with statutory requirements appears to trump other aspects of the analysis at the time of this writing. Clearly, however, EPA did not apply a strict monetized cost–benefit decision rule, although cost-effectiveness information was applied with the potential for materially reducing the burden on society.

The Frontier of Linking Ecological and Economic Systems

In what ways does this case study illustrate weaknesses in linking ecological impacts with a cost–benefit approach? Consider that economists can be preoccupied with the monetary valuation stage and the normative, economics-based decision rules that constitute their area of comparative advantage. However, nothing can be monetarily valued without a change in quantity (or quality) in the environment broadly considered. Here lie the key difficulties in the case study. The environmental impacts of CWIS affect freshwater and saltwater ecosystems in a variety of ways, some of which are poorly understood. In short, the AEI was incompletely specified, not an unusual occurrence for ecological impacts. Even determining what is adverse requires some value judgment. For instance, increased recreational fishing or the congregation of an endangered species such as the manatee near thermal outlets may be viewed as adverse from one perspective if it is a deviation from an environmental baseline. In fact, EPA did not define the objective function of its regulation, the AEI (EPA 2004a, 41612).

From the standpoint of economically valuing impacts, the challenge is to find the material ecological outputs or services that people value and to find ways to measure those impacts and values. For instance, impingement and entrainment do not easily translate into dimensions that people value. As a consequence, the links that connect the ecological and economic impacts are difficult to measure. Commercially and recreationally landed fish provide the most concrete linkage between the ecological and economic measurements in the RIA, but other linkages proved problematic.

Governmental practitioners and consultants have been asked to resolve such basic research challenges. As indications of this frontier challenge, consider that the Committee on Valuing the Protection of Ecological Systems and Services (C-VPESS) of the EPA Science Advisory Board is drafting a report on the topic that may result in future guidance (EPA C-VPESS 2007, 2008). In addition, a National Research Council report has appeared since the regulation was finalized (National Research Council Committee on Assessing and Valuing the Services of Aquatic and Related Terrestrial Ecosystems (NRC C-AVSARTE) 2004). Would the CWIS analysis have been substantially assisted by having these documents available at an earlier date?

First, both the C-VPESS draft report and the NRC C-AVSARTE report (NRC C-AVSARTE 2004) discuss the difficulties in modeling ecological and economic systems and in linking the impacts. The C-AVSARTE report focuses primarily on economic approaches to valuation, embracing both use and nonuse values. The C-VPESS draft report considers an expanded set of valuation methods useful at various stages of the regulatory process but states that, in the case of RIAS, the economic component is to be "conducted in accordance with the methods and procedures of standard welfare economics" (EPA C-VPESS 2008, 122). The draft report included a survey of methods for social-based valuation approaches that are at the frontier of research; for instance, it posits that people may have different values as citizens than as consumers. However, at least some economists would probably be concerned about the decisionmaking context in which individuals are placed to elicit values for cost–benefit analysis (Spash 2007). For instance, a person placed in the experimental context of a citizen decisionmaker with a group of peers facing a relatively unknown problem may be influenced by the social context, the formation of the group, and the hypothetical nature of the setting, perhaps including issues related to the absence of actual budget limitations or the scope of choices being considered. List et al. (2004) recently found that a lack of social isolation may create a bias roughly equivalent to that created by the hypothetical nature of surveys that elicit economic values. These authors interpret social settings for value elicitation as inducing "respondents to include any number of utility-enhancing values that come from publicly advertising one's own goodwill. But, since these 'externality-type' values are not germane to the good in question, rather to a class of goods, it is incorrect to lump them with any particular good's value" (List et al. 2007, 749).

The C-VPESS citable draft report (the latest version is not citable per its webpage) has a special section on valuation for national rulemaking (EPA C-VPESS 2007, s. 6.1). Examples of draft guidance (which may change) include the following:

- an early conceptual model of the ecological and economic system being analyzed (s. 6.1.2.1);
- early identification of socially important impacts that may not be limited to economic methods;
- early interaction of ecologists and economists to inform the prediction of biophysical changes in value-relevant terms (but the draft report notes, in the concentrated animal feeding operation example used, that "the combination of variation complexity, and gaps in information and understanding make it difficult for the Agency to assess the ecological impacts of its actions, particularly at the national scale," 122); and
- likely use of benefit transfer methods and quality checks in the development of monetized valuation measures.

Had this draft document become final, the advice may have served early on in the CWIS process to help frame and direct research. However, as is the case with general cost–benefit guidance from OMB, it is unlikely to have been a substantial help in resolving the difficulties in defining ecological impacts and linking those impacts to economic valuation measures.

Conclusion

Economists want decisionmakers to consider economic trade-offs based on credible information. We want a cost–benefit type of analysis to analyze our definition of efficiency while recognizing that cost–benefit analysis will not unambiguously identify a socially preferred policy (Arrow et al. 1996). If legal or other constraints exist, we typically want a cost-effectiveness analysis if the law is immutable or a cost–benefit analysis if we are considering changes in the law. We want price or functions of prices as societal measures of value from the interaction of many people in the marketplace. If markets are lacking, we want an experiment conducted that generates numbers as if a market existed. We want sufficient precision to distinguish positive from negative net benefit values or to test a specific hypothesis or question. Economists want this information as they seek the largest economy, broadly considered, that is consistent with people's preferences, technology, and environmental conditions. The largest economy, broadly considered, includes leisure time, the provision of environmental amenities, and nonmarket as well as market activity. If the distribution of goods and services that result from market forces in this largest economy is deemed inequitable, then economists currently look to the political process for distributional adjustments. Economists, and some other stakeholders, want an economic analysis to conform to norms of the discipline, which may be difficult to infer. Economists do not want decisionmakers to be provided only with distributional information.

Table 9.2
Research Recommendations

Research recommendations by the National Research Council's Committee on Assessing and Valuing the Services of Aquatic and Related Terrestrial Ecosystems (2004, 258), referred to as "overarching research needs," are as follows:

Although much is known about the services provided by aquatic ecosystems and methods for valuing changes in these services exist, the committee believes that there are still major gaps in knowledge that limit our ability to incorporate adequately the value of ecosystem services into policy evaluations. Drawing from the preceding major conclusions and overarching recommendations provided above, the committee has identified the following research needs. The committee believes that funding to address these needs is necessary if progress toward improving the use of ecosystem valuation in policy decisions is to be made, and it recommends that such funding be a high priority.

- Improved documentation of the potential of various aquatic ecosystems to provide goods and services and the effect of changes in ecosystem structure and functions on this provision
- Increased understanding of the effect of changes in human actions on ecosystem structure and functions
- Increased interdisciplinary training and collaborative interaction among economists and ecologists
- Development of a more explicit and detailed mapping between ecosystem services as typically conceived by ecologists and the services that people value (and hence to which economic valuation approaches or methods can be applied)
- Development of case studies that show how these links can be established and templates that can be used more generally
- Expansion of the range of ecosystem services that are valued using economic valuation techniques
- Improvements in study designs and validity tests for stated-preference methods, particularly when used to estimate nonuse values
- Development of "cutting-edge" valuation methods, such as dynamic production function approaches and general equilibrium modeling of integrated ecological-economic systems
- Improved understanding of the spatial and temporal thresholds for various ecosystems, and development of methods to assess and incorporate into valuation the uncertainties arising from the complex dynamic and nonlinear behavior of many ecosystems
- Improvements in the methods for assessing and incorporating uncertainty and irreversibility into valuation studies.

Other stakeholders in the regulatory process may want a different type of analysis. For instance, a package of reports might be associated with an analysis. In the context of the CWIS rule, the ecological impacts are important to some stakeholders in their natural units; impacts in natural units also form a first step in an economic analysis. Some other type of noneconomic valuation—such as energy or other modeling of the ecosystem—may gain credence. One can easily imagine, however, a set of summary tables that proceeds from qualitative impacts, to quantitative impacts in their natural units, to valuations, and finally to a cost–benefit table. The regulatory requirement for a cost–benefit or cost-effectiveness analysis, however, identifies an aspiration to report somewhere in the document a specific type of professionally recognized analysis.

What might EPA do to proceed, both in the specific CWIS case and for its economic regulatory evaluations in general? Regarding CWIS, ultimately, EPA created a complex regulation without a transparent message and analysis. An analysis of technologies that considers cost-effectiveness by defining a weighted AEI, taking explicit account of uncertainty and using cost as a secondary consideration, appears to be supported in the legal record and appears to contain many, if not all, of the elements that an economist would want. In particular, an explicit discussion of weighting an AEI may be one way of investigating alternative methods to capture society's preferences.

From a broader regulatory evaluation perspective, creating new interdisciplinary science is a high hurdle for a decision-support document like an RIA. EPA and other agencies, such as the National Science Foundation, might choose to foster additional frontier work so that models better linking ecosystems and the environment may be available commercially the next time a regulation calls for such analysis. Both the C-AVSARTE report (2004) and the EPA C-VPESS draft report (2008) contain recommendations for further research (those of the C-AVSARTE report are provided in Table 9.2). To these research issues, I would add the following topics: an examination of instances in which it is better to use a number in place of a default of zero, an investigation of faint behavioral trails for nonuse value, and research that more explicitly recognizes uncertainty in the risk-management decision that can lead to new valuation measures.

Regarding the use of default values of zero, cost–benefit analysts often are unable to find a number—whether related to quantity, value, or cost—that is exactly designed for the location or other context of the regulatory setting. Analysts often substitute estimates but, not infrequently, may choose to use a zero in the monetized cost–benefit computation because of imprecision or other reasons. In the context of the CWIS analysis, the final monetized cost–benefit analysis used zero values for fish that were not captured or that were not an input into commercial or recreational fishing, as well as for nonuse value. In another setting, the U.S. Army Corps of Engineers chooses not to monetize the probability of a loss of life, although such impacts are typically discussed in the text of corps documents (GAO 2005). In environmental applications, one aspect of this general issue has been discussed as *benefits transfer* (Freeman 2003), in which means have been sought to improve the accuracy of benefits that are estimated in one location and "transferred" to an analysis for another location. This issue is central to many debates about omitted impacts in which the analyst chooses to use zero in the monetized estimate instead of an estimate transferred from a related study or estimated by other means. Research and improved guidance might exist to help analysts determine when zero is a better estimate than another value. For instance, one can test for a value different from zero both in a statistical sense—via standard statistical testing—and in a decision-analytical sense. One could investigate questions such as, How far away from the true value does an estimate have to be before a value of zero is a better estimate? A preliminary result is that, given a mean squared error loss function, a value of zero is a worse estimator than any number less than two times the (unknown) true value (Farrow 2005).

The value people ascribe to a resource that they may never use—its nonuse value—is also a difficult impact to monetize. The authoritative panel that provided guidance on the use of survey-based (contingent) valuation methods also suggested a search for a faint behavioral trail of revealed behavior (Arrow et al. 1993). In my view, little work has been conducted relevant to this suggestion compared with the work to extend the methods of contingent valuation. For instance, researchers might pursue the thin trail that links observed news gathering behavior to follow-on activity, such as charitable donations or changes in consumer purchases.

Finally, and probably for ease of implementation, regulatory reviews focus on a basic approach requiring benefits to justify costs. However, the basic decision rules can change substantially in the presence of uncertainty, irreversibility, and an ability to obtain more information (Dixit and Pindyck 1994). In cases where a property right is under dispute and the goal is to maintain a particular level of environmental services, it may be appropriate to spend significantly more than expected benefits (Farrow and Morel 2001). Analytically, such approaches suggest additional, difficult to measure, elements of the problem that are related to uncertainty. These elements are seldom considered, although they are mentioned in OMB guidance (2003).

Economists teach that wants are insatiable. What an economist wants to improve the regulatory process is probably insatiable as well, without consideration of the constraints on agency resources and the value of the economic information in the debate. Finding the analytical and communication level that is as simple as possible but no simpler—as was famously said by Einstein regarding natural science models—remains an art, not a science. Detail that is unlikely to change a decision should not be analyzed, but one should continue to ask, What policy alternative will improve the welfare of society? Economists have evolved their approach for more than 150 years, and regulations like the CWIS rule continue this evolution. Important questions can be easy to ask and hard to answer.

■ ■ ■

Notes

1. The RIA is here taken to be the final notice in the *Federal Register* and the final version of supporting documents. Where important, distinctions among documents are noted in the text.

2. OMB guidance appears to exempt regulatory analyses from peer review, saying "[t]his Bulletin covers original data and formal analytical models used by agencies in Regulatory Impact Analyses (RIAS). However, the RIA documents themselves are already reviewed through an interagency review process under E.O. 12866.... In that respect, RIAS are excluded from coverage by this Bulletin..." (OMB 2005, 2674).

3. Commenters with economics Ph.D.s included Frank Ackerman, Thomas Grigalunas and James Opaluch (together), Robert Stavins, and Ivar Strand. Economists may have been elements of other teams providing comments.

4. It is true that science is not a democratic, majority process, and the author could add yet one more view on particular analytical aspects of the regulation. However, the challenge I trace here is the difficulty of determining an acceptable level of "quality" given the mandates placed on the agency.

5. These are approximations without correcting for the one-year difference in the value of the dollar, differences in discount rates used by EPA for costs and benefits, and assuming a 7 percent real discount rate over an infinite time horizon. The difference between the infinite time horizon and a shorter one is V/r*(1 - e - rT) where V is

the annual value, r is the discount rate, and T is the terminal time. For instance, a 25-year time horizon at 7 percent would reduce the infinite time horizon value by 17 percent. If the alternative 3 percent discount rate were used, the cost savings would be approximately $100 billion.

6. This did not occur; instead I provide this as an illustration of "scoring" cost savings such as used by the GAO (2008).

References

Arrow, K., M. Cropper, G. Eads, R. Hahn, L. Lave, R. Knoll, P. Portney, M. Russell, R. Schmalensee, V.K. Smith, and R. Stavins. 1996. Is There a Role for Benefit–Cost Analysis in Environmental, Health, and Safety Regulation? *Science* 272 (April): 221–222.

Arrow, K., R. Solow, P. Portney, E. Leamer, R. Radner, and H. Schuman. 1993. *Report of the NOAA Panel on Contingent Valuation.* Washington, DC: National Oceanic and Atmospheric Administration.

Belzer, R. 1999. Project on Regulatory Oversight: Appendix C, Evaluative criteria for compliance: EO 12866. Washington University in St. Louis. http://wc.wustl.edu/csab/regulation/Regulation.htm (accessed December 29, 2008).

Bray, C., S. Farrow, and T. Guinane. 2007. Microeconomic Performance: Uses, Gaps, and Opportunities in Government Program Evaluation. Vol. 23, *Research in Law and Economics.* Amsterdam: Elsevier.

Dixit, A., and R. Pindyck. 1994. *Investment Under Uncertainty.* Princeton, NJ: Princeton University Press.

Executive Order 12866. 1993. Regulatory Planning and Review. *Federal Register* 58, no. 190 (October 4): 51735–51744. www.whitehouse.gov/omb/inforeg/eo12866.pdf (accessed December 29, 2008).

Farrow, S. 2005. When is some number better than no number. Paper presented at the annual meeting of the Association for Public Policy Analysis and Management, Washington, DC, November (available from the author).

Farrow, S., and B. Morel. 2001. Continuation Rights, Options, and the Precautionary Principle. *Risk, Decision, and Policy* 6: 145–155.

Freeman, M., III. 1980. Technology-Based Effluent Standards: The U.S. Case. *Water Resources Research* 16(1): 21–27.

———. 2003. *The Measurement of Environmental and Resource Values.* 2nd ed. Washington, DC: Resources for the Future Press.

Hahn, R., and P. Dudley. 2007. How Well Does the U.S. Government Do Cost–Benefit Analysis? *Review of Environmental Economics and Policy* 1(2): 192–221.

National Research Council, Committee on Assessing and Valuing the Services of Aquatic and Related Terrestrial Ecosystems (NRC C-AVSARTE). 2004. *Valuing Ecosystem Services: Toward Better Environmental Decision-Making.* Washington, DC: The National Academies Press.

Heinzerling, L. 2006. Statutory Interpretation in the Era of OIRA. *Fordham Urban Law Journal* 33(4): 101–120.

List, J., R. Berrens, A. Bohara, and J. Kerkvliet. 2004. Examining the Role of Social Isolation on Stated Preferences. *American Economic Review* 94(3): 741-752.

OMB (Office of Management and Budget). 1992. *Guidelines and Discount Rates for the Benefit–Cost Analysis of Federal Programs.* Circular A-94. Washington, DC: OMB.

———. 2002. Guidelines for Ensuring and Maximizing the Quality, Objectivity, Utility, and Integrity of Information Disseminated by Federal Agencies. *Federal Register* 2 no. 67 (February 22): 8452.

———. 2003. *Regulatory Analysis.* Circular A-4. Washington, DC: OMB.

Regulatory Information. Historical Reports, Various Years. RIN 2040-AD62. www.reginfo.gov (select EPA for 2002 and 2004; accessed December 29, 2008).

Riverkeeper, Inc. v. U.S. Environmental Protection Agency (EPA), 475 F.3d 83, 90 (2d Cir. 2007), also known as Riverkeeper II.

RTI International. 2005. *Report of the Peer-Review Panel Analysis of Nonuse Benefits for 316(b) Phase III Rule.* RTI project number 09393.000, May. Research Triangle Park, NC: RTI International.

Spash, C. 2007. Deliberative Monetary Valuation: Issues in Combining Economic and Political Processes to Value Environmental Change. *Ecological Economics* 63(4): 690–699.

Stratus Consulting. 2004. *Summary of Peer Review of I and E Methods,* regulations.gov. Document ID: EPA-HQ-OW-2002-0049-0985.

U.S. Environmental Protection Agency (EPA). 2000. *Guidelines for Preparing Economic Analysis.* EPA 240-R-00-003. Washington, DC: EPA. http://yosemite.epa.gov/ee/epa/eed.nsf/webpages/Guidelines.html (accessed December 29, 2008).

———. 2002. Peer reviews available at regulations.gov. Document IDs: EPA-HQ-OW-2002-0049-1402 through 1406.

———. 2004a. Final Regulations to Establish Requirements for Cooling Water Intake Structures at Phase II Existing Facilities. *Federal Register* 69 (July 9): 41576–41693. www.epa.gov/waterscience/316b/ (accessed December 29, 2008).

———. 2004b. Economic and Benefits Analysis. EPA-821-R-04-005, www.epa.gov/waterscience/316b/phase2/econbenefits/final.htm (accessed December 29, 2008).

———. 2004c. Regional Analysis. www.epa.gov/waterscience/316b/phase2/casestudy/final.htm (accessed December 29, 2008).

———. 2004d. Technical Development Document. www.epa.gov/waterscience/316b/phase2/devdoc/final.htm (accessed December 29, 2008).

———. 2004e. Response to Public Comment March 29. www.epa.gov/waterscience/316b/phase2/comments/index.html (accessed December 29, 2008).

U.S. Environmental Protection Agency Science Advisory Board, Committee on Valuing the Protection of Ecological Systems and Services (EPA C-VPESS). 2007. Valuing the protection of ecological systems and services (draft report, September 24). http://yosemite.epa.gov/sab/SABPRODUCT.NSF/81e39f4c09954fcb85256ead006be86e/C6814A495B15D3228525736A004C4680/$File/cvpess_draft-09-24-07.pdf (accessed December 29, 2008).

———. 2008. Valuing the protection of ecological systems and services (draft report, March 11). http://yosemite.epa.gov/sab/SABPRODUCT.NSF/81e39f4c09954fcb85256ead006be86e/F4771258F94FDA8C85257409006711186/$File/C-VPESS+Draft-03-11-08.pdf (accessed December 29, 2008).

U.S. Government Accountability Office (GAO). 2005. *Economic Performance: Highlights of a Workshop on Economic Performance Measures.* GAO-05-796SP. Washington DC: GAO.

———. 2008. *Performance and Accountability Highlights: Fiscal Year 2007.* GAO-08-2SP. Washington, DC: GAO. www.gao.gov/new.items/d082sp.pdf (accessed May 30, 2008).

CHAPTER 10

Fish Tales

DOUGLAS A. KYSAR

Before it was amended in 1972, the Federal Water Pollution Control Act, commonly known as the Clean Water Act (CWA), required the setting of various ambient water quality standards that, in turn, were to afford a basis for determining acceptable levels of pollution in interstate navigable waterways. As the Supreme Court has noted, federal and state regulators operating under the early CWA approach found it "very difficult to develop and enforce standards to govern the conduct of individual polluters."[1] This difficulty was caused in substantial part by informational demands, scientific uncertainties, and valuation questions that accompanied the task of establishing acceptable water quality standards. Thus, "prompted by the conclusion of the Senate Committee on Public Works that 'the Federal water pollution control program has been inadequate in every vital respect,'"[2] Congress dramatically overhauled the CWA, requiring the placement of maximum effluent limitations on point sources of water pollution in addition to the achievement of water quality standards, and creating the National Pollutant Discharge Elimination System (NPDES) as a means of enforcing effluent limitations. Primary authority for implementing the CWA and for issuing and overseeing NPDES permits was vested in the U.S. Environmental Protection Agency (EPA), although states could continue to play a substantial role if they developed EPA-approved permitting programs.[3] Significantly, effluent limitations were to be determined by regulators with reference not to ambient water quality standards, but to the level of pollution reduction that could be achieved through the application by industry of identified high-performing technologies.

This new technology-based approach proved much more effective than the earlier standard-based approach;[4] nevertheless, few would contend that the CWA has been a complete success. For instance, nonpoint sources of water pollution have proven extremely difficult to subject to regulatory control, and the health of aquatic ecosystems continues to be impaired by dams and channelizations, withdrawals and transfers, and other disruptions. In addition, EPA has struggled mightily to implement the CWA's requirement that the agency regulate cooling water intake structures at point sources to reduce the structures' adverse environmental impact.[5] Electricity generating plants and other industrial facilities often depend on the withdrawal of water from rivers, lakes, and other waterways to manage excess heat generated during their production processes. The amount of water required is vast, on the order of 50 million gallons or more per day for a large plant and totaling billions of gallons of water per day for all national facilities. The resulting environmental impact is also dramatic: EPA estimates that over 3.4 billion fish and shellfish (expressed as "age 1 equivalents") are killed by cooling water intake operations each year, either from being trapped against components of the cooling water intake structure and therefore suffering "exhaustion, starvation, asphyxiation, and descaling," or from being drawn into the cooling water system and therefore suffering "physical impacts," "pressure changes," "sheer stress," "thermal shock," and "chemical toxic effects."[6] These two mortality threats, referred to as *impingement* and *entrainment*, affect not only the various fish and shellfish species for which EPA was able to generate quantitative estimates, but also certain endangered, threatened, and other special status species, such as sea turtles, Chinook salmon, and steelhead, as well as immeasurable quantities of phytoplankton and zooplankton at the base of aquatic food chains. Moreover, impingement and entrainment are only the most obvious and measurable adverse effects of cooling water intake operations on aquatic ecosystems. Other impacts include "diminishment of a population's compensatory reserve; losses to populations including reductions of indigenous species populations, commercial fisheries stocks, and recreational fisheries; and stresses to overall communities and ecosystems as evidenced by reductions in diversity or other changes in system structure and function."[7] For all of these myriad effects, regulators only can generate rough predictions of their likelihood and magnitude, given that "[p]opulation dynamics and the physical, chemical, and biological processes of ecosystems are extremely complex."[8]

Cognizant of these kinds of informational difficulties, Congress in Section 316(b) of the CWA mandated that "the location, design, construction, and capacity of cooling water intake structures [must] reflect the best technology available for minimizing adverse environmental impact."[9] EPA's first effort to implement this statutory provision was remanded on procedural grounds following an industry challenge in 1977,[10] after which the agency formally withdrew the regulation in 1979.[11] In a hopeful gesture, EPA reserved space in the Code of Federal Regulations for future cooling water intake rules.[12] Nevertheless, amidst the changing political climate of the 1980s and the agency's enormous backlog of regulatory responsibilities under other federal environmental statutes, EPA essentially abandoned its Section 316(b) rulemaking efforts, leaving regulation of cooling water intake structures instead to the case-by-case decisionmaking of NPDES permit issuers.[13] Eventually, facing a legal challenge by environmental groups, EPA agreed in 1995 to a consent decree that required the agency to establish cooling water intake rules in multiple phases.[14] Phase I, involving new facilities, was completed by the agency on December 18, 2001, and generally required facilities to achieve environmental performance standards based on what is known as *closed-cycle cooling technology*, a process in which cooling water is recycled and only periodically replenished from neighboring waterways, rather than continuously withdrawn. Although environmentalists had argued on behalf of *dry cooling technology*, an even more protective approach that did not require the withdrawal of water at all, the Second Circuit in 2004 accepted EPA's conclusion that the expense of this technology rendered it not reasonably available to industry.[15] The court did, however, remand other portions of the Phase I rules, holding that EPA had impermissibly allowed facilities to use environmental restoration efforts to meet part of their compliance obligations, despite the CWA's clear mandate that facilities prevent, rather than compensate for, the environmental degradation associated with cooling water intake.[16]

Phase II involved the much more politically nettlesome category of large existing power plants. EPA's final Phase II regulations were issued on July 9, 2004, and involved a complicated array of compliance options that were built around a set of impingement and entrainment performance standards that required only certain facilities to engage in reductions and that, for those covered facilities, only formally required 80 percent and 60 percent reductions, respectively, in impingement and entrainment. These performance standards were particularly notable because they marked a refusal by EPA to use closed-cycle cooling technology as the benchmark against which other proposed protection measures might be evaluated. Despite acknowledging that impingement and entrainment provide the "primary and distinct types of harmful impacts associated with the use of cooling water intake structures,"[17] and that "closed-cycle, recirculating cooling towers . . . can reduce mortality from impingement by up to 98 percent and entrainment by up to 98 percent,"[18] EPA nevertheless adopted standards that were based on the performance of less effective technologies. It did so because it chose to "interpret CWA section 316(b) as authorizing EPA to consider not only technologies but also their effects on and benefits to the water from which the cooling water is withdrawn."[19] More specifically, the agency chose to interpret Section 316(b) to allow performance standards that required not the best technology available for minimizing impingement and entrainment, but rather the technology that best equalizes the marginal ecological benefits of reducing impingement and entrainment with the marginal economic costs of doing so.[20] This efficiency-oriented approach had a dramatic effect: EPA estimated in its regulatory impact analysis that 125 facilities would adopt no impingement and entrainment controls at all under the Phase II rules.[21] Moreover, rather than up to 98 percent reduction in impingement and entrainment, most facilities would only achieve a 30.9 to 59.0 percent reduction in impingement and a 16.4 to 47.9 percent reduction in entrainment.[22] In short, the "best technology available for minimizing adverse environmental impact" became merely the technology that produces an acceptable benefit-to-cost ratio, irrespective of its overall level of environmental benefit.

EPA never quite got around to fully disclosing and accounting for this maneuver in the preamble to its final regulations. At one point, the agency noted that it "believes it is reasonable to vary performance standards by the potential for adverse environmental impact in a waterbody type."[23] However, attempting to correlate the stringency of performance standards with the ecological benefit to be gained from doing so does not seem to be a proper reading of the statutory mandate to minimize adverse environmental impact. EPA essentially admitted as much in the preamble to its proposed Phase II regulations when it noted that states might choose to use their own authority to impose more stringent requirements, such as performance standards based on dry cooling technology, to garner "additional protection above the levels provided by these technology-based *minimum* standards."[24] The statute requires maximum levels of environmental protection, not minimum. Put another way, it may well be that the appropriate minimizing technology will vary by water body type, but those technologies will never be equivalent to ones that optimally balance costs and benefits by water body type. Interestingly, when offering the proposed Phase II rule, EPA actually stated that it was "concerned about the cumulative overall degradation of the aquatic environment as a consequence of . . . intakes located with or adjacent to an impaired waterbody,"[25] because the goal of the CWA is to steadily restore the biological integrity of the nation's water bodies, even those that are quite severely impaired at any given moment. Allowing the degraded condition of a water body to reduce the level of protection that it receives—as EPA did—therefore seems inconsistent with the CWA's more dynamic, long-term goal of progressive improvement.

If at times it appeared that EPA was of two minds on the issue of how to establish performance standards, it was because, in fact, the agency's deliberations had been heavily influenced by an outside force. As Lisa Heinzerling has explained, EPA's hand was essentially forced by the Office of Information and Regulatory Affairs (OIRA) within the Office of Management and Budget

(OMB).[26] Documents released during the rulemaking process indicate that EPA's Phase II approach shifted dramatically following initial review by OIRA. In particular, EPA was pressured to abandon its original plan to impose standards based on closed-cycle cooling in favor of the eventual suite of compliance options, the most important of which were founded, either directly or indirectly, on a comparison of the demonstrated ecological benefits of protection technologies with their accompanying costs. The result of being forced to resort to cost–benefit analysis (CBA) in setting the Phase II performance standards was that much of EPA's analysis seemed to be at odds with the rules that it adopted. For instance, at one point the agency awkwardly observed that closed-cycle cooling would be granted a "streamlined" approval process under the Phase II rules because use of "this highly effective technology . . . would always achieve the performance standards."[27] In an attempt to disclaim the obvious inference to be drawn from this concession, the agency quickly added that closed-cycle cooling had been deemed "not economically practicable for many existing Phase II facilities."[28]

In part because of this conceptual awkwardness and dissonance that appeared throughout the Phase II rulemaking documentation, a Second Circuit appeals panel remanded the Phase II regulations almost in their entirety in 2007. The basic defect of the rules, in the panel's view, was EPA's apparent decision to use regulatory CBA to identify the performance standard that could be attained by "the best technology available for minimizing adverse impact." As the court observed, *best technology available* (BTA) requirements long have been understood to preclude reliance on cost–benefit balancing by EPA.[29] Because under a BTA requirement "Congress itself [already has] defined the basic relationship between costs and benefits," EPA's responsibility is simply to identify the most environmentally protective technology available at a cost that can be "reasonably borne" by the regulated industry.[30] In determining the ultimate standard, EPA may take account of "other factors, including cost-effectiveness, to choose a less expensive technology that achieves the same results as the benchmark."[31] When identifying the initial BTA, however, the agency must simply ask what level of environmental protection can be achieved by the *best technology available* at "not the average plant, but the optimally operating plant, the pilot plant which acts as a beacon to show what is possible."[32]

On April 14, 2008, the U.S. Supreme Court accepted three consolidated petitions for review arising out of EPA's rulemaking, with the court's consideration limited to the sole question of "whether Section 316(b) of the Clean Water Act, 33 U.S.C. 1326(b), authorizes the [EPA] to compare costs with benefits in determining the best technology available for minimizing adverse environmental impact at cooling water intake structures." To many observers, the court's decision to hear the case was surprising: the Second Circuit's opinion had lacked a dissent; there was no obvious split among the federal courts of appeals as to how to interpret Section 316(b); and the U.S. solicitor general had argued on behalf of the federal government against granting the petition for writ of certiorari. Moreover, although the statutory language of Section 316(b) is slightly ambiguous—in that EPA is left to determine what, in the context of cooling water intake structures, is the relevant "adverse environmental impact" to minimize, unlike the effluent reduction context, which simply demands minimization of pollutant outflows—few observers would have regarded EPA's eventual Phase II rule as a plausible interpretation of the long-familiar BTA language contained within the provision. Thus, the Supreme Court's willingness to consider overturning the Second Circuit's remand of the Phase II rules seemed to many observers to represent another chapter in the long campaign, waged through both political and judicial channels, to subject environmental laws to the discipline of CBA, irrespective of their statutory language.

Regardless of the ultimate outcome of the case, the controversy thus far has imposed substantial administrative costs: the Phase II rulemaking docket included some 2,805 documents, many of which are hundreds of pages long; and by itself, the agency's final response to public comments on the rulemaking extended for 5,143 pages.[33] Much of this paperwork and analysis was generated by EPA to pursue a course of environmental impact assessment and cost–benefit balancing that was not required by the statute and that may well have been prohibited by it. So deeply dependent on CBA were the Phase II rules that EPA was forced, after the court's decision, to simply suspend operation of the rules entirely, reverting instead to the same case-by-case judgment that the agency had already agreed was inadequate to satisfy the mandate of Section 316(b).[34] Thus, unless the Supreme Court intervenes as a deus ex machina to rescue the aggressively creative Phase II rules, EPA, the regulated community, and the billions of life forms consumed by cooling water intake structures each year all will find themselves in the same dubious legal position that they have occupied for the previous 35 years. This would be an especially frustrating result given that, as the Second Circuit noted, the CWA plainly required EPA to begin adopting cooling water intake regulations *concurrently* with effluent limitation standards under Sections 301 and 306, which is to say, decades ago.[35]

The Phase II rulemaking provides a valuable opportunity to assess, in context, the competing merits of regulatory CBA and more traditional technology-based approaches to environmental law and policy. As a result of EPA's unprecedented approach to the Phase II rulemaking, scholars now have available for study both an expansive regulatory CBA and the technology-based standard that would have been adopted by EPA had the agency followed the straightforward technology-based approach that it took in the Phase I rulemaking. As I argue in this chapter, EPA's Phase II rulemaking illustrates several limitations of regulatory CBA that have yet to be overcome in practice and that, in some cases, cannot be overcome even in principle. Most notably, the practical challenges that prompted Congress to amend the CWA in 1972 remain just as insurmountable today. Although many advances have been made in the understanding of ecosystem functioning and in methods of monetizing environmental impacts, the process of reducing environmental policymaking to empirical technique remains deeply flawed. To be clear, no one denies the wisdom of acquiring information regarding the consequences of regulation as part of the policymaking process. But the wisdom of hinging regulatory outcomes on how well that information can be made to fit the form of a CBA exercise has not been demonstrated. Instead, renewed critical attention is required regarding whether and how to use environmental law to realign burdens of proof for the demonstration of harm, whether the staunch value monism of welfare economics adequately represents the significance and diversity of environmental "goods," and whether the unyielding atomism of environmental economic valuation techniques is consistent with the demands of democracy. Because proponents of regulatory CBA often incorporate debatable views regarding how these issues of information, valuation, and process should be resolved—while still contending that "questions about the correct measurement of benefits and costs" can be kept separate from "philosophical" or "ethical" discussion[36]—they end up offering cost–benefit accounts that, like fish tales, are best heard with skepticism.

Information

The choice of phrasing that Congress used in setting out the requirements of Section 316(b) of the CWA—"best technology available for minimizing adverse environmental impact"—was not incidental. The CWA includes a dizzying array of technology-related standards, each carrying a subtly, but significantly, distinct meaning. Section 301(b) of the CWA, for instance, requires EPA to establish initial effluent limitations for existing sources based on the "best practicable control technology currently available" (BPT). By 1989, EPA was to replace those standards with more stringent ones based on the "best available technology economically achievable." Section 306, in contrast, requires effluent limitations for new sources to be based on the "best available demonstrated control technology [offering] the greatest degree of effluent reduction." As the Second Circuit noted, these varying standards carry significantly different implications for the permissibility of CBA use by EPA. Under the initial BPT approach, regulatory standards could be premised on an explicit comparison of compliance costs to environmental benefits. For the setting of second-generation existing source standards and new source standards, however, EPA generally is prohibited from engaging in such comparison.

Because the language of Section 316(b) closely resembles the language of these more stringent standards, EPA's Phase II rules probably should not have been premised on cost–benefit balancing. Instead, the agency should have focused simply and directly on the affordability of increasingly efficacious environmental control technologies, recognizing that Congress itself already had determined that the benefits of regulation are sufficiently vast and difficult to quantify that only the "best" control technology will suffice. That approach would have mandated the achievement of performance levels equivalent to closed-cycle cooling because EPA earlier had acknowledged that "[c]losed-cycle cooling systems . . . are the most effective means of protecting organisms from [impingement and entrainment],"[37] and the imposition of standards based on closed-cycle cooling would have been affordable to the industry as a whole. Nevertheless, following interventions by OIRA and industry commenters, EPA, in its issuance of the final Phase II rules, sought to bootstrap its way into a CBA regime by reading the word *practicable* into the CWA's cooling water intake structure provision: "Section 316(b) authorizes consideration of the environmental benefit to be gained by requiring that the location, design, construction, and capacity of cooling water intake structures reflect the best economically *practicable* technology available for the purpose of minimizing adverse environmental impact."[38] This attempt by EPA to transform the BTA standard of Section 316(b) into a BPT standard—an attempt that failed to survive judicial review under the Second Circuit's analysis, but that may yet receive the blessing of the Supreme Court—ignored the pragmatic wisdom of Congress's effort to prescribe stringent technology-based standards irrespective of the agency's ability to precisely demonstrate the benefits of doing so. The decision to adopt such standards—which, on the surface, may appear to be one-sided or irrational because of the rejection of cost–benefit maximization—is one that pays due respect to the concerns of practical realizability that must accompany any lawmaking exercise.

Regulatory CBA, in contrast, carries an implicit assumption that the policy space within which EPA operates is informationally rich and probabilistically sophisticated, such that the agency easily can identify courses of action that maximize expected social welfare outcomes. This assumption is unwarranted: the scientific and economic information necessary to fulfill textbook efficiency analysis is almost always lacking in the environmental context. Thus, when proponents of regulatory CBA argue that only the cost–benefit test "will—by definition—lead consistently to decisions which make the world better off,"[39] they fail to acknowledge that CBA only does so "by definition," that is, by an a priori assumption that the information needed to satisfy the form of the optimization exercise is both attainable and costless. In the case of cooling water intake structures, this assumption was even less warranted than in the typical environmental policymaking task. To meet the form of regulatory calculus demanded of it by CBA, EPA reduced a complex and highly uncertain decision to a question of how much to invest in "reductions in impingement and entrainment as a quick, certain, and consistent metric for determining performance."[40] Increased fish survival became the primary determining factor because, at least for those fish that are commercially or recreationally valuable, that factor offered an ecological benefit that was readily quantifiable and monetizable.

As the agency acknowledged, however, the potential impact of cooling water intake structures is much broader and more complex than direct mortality effects (which are, themselves, only roughly estimable).

In addition to their importance in providing food and other goods of direct use to humans, the organisms lost to [impingement and entrainment] are critical to the continued functioning of the ecosystems of which they are a part. Fish are essential for energy transfer in aquatic food webs, regulation of food web structure, nutrient cycling, maintenance of sediment processes, redistribution of bottom substrates, the regulation of carbon fluxes from water to the atmosphere, and the maintenance of aquatic biodiversity. Examples of ecological and public services disrupted by [impingement and entrainment] include:

- *decreased numbers of ecological keystone, rare, or sensitive species;*
- *decreased numbers of popular species that are not fished, perhaps because the fishery is closed;*
- *decreased numbers of special status (e.g., threatened or endangered) species;*
- *increased numbers of exotic or disruptive species that compete well in the absence of species lost to [impingement and entrainment];*
- *disruption of ecological niches and ecological strategies used by aquatic species;*
- *disruption of organic carbon and nutrient transfer through the food web;*
- *disruption of energy transfer through the food web;*
- *decreased local biodiversity;*
- *disruption of predator–prey relationships;*
- *disruption of age class structures of species;*
- *disruption of natural succession processes;*
- *disruption of public uses other than fishing, such as diving, boating, and nature viewing; and*
- *disruption of public satisfaction with a healthy ecosystem.*[41]

These various ecological and public services received no monetary value in EPA's economic analysis. Indeed, as the agency candidly admitted, even its focus on impingement and entrainment losses was highly incomplete, as it only accounted for losses insofar as they impacted commercial and recreational fish harvest; hence, the agency "was not able to monetize benefits for 98.2% of the age-one equivalent losses of all commercial, recreational, and forage species for the section 316(b) Phase II regulation."[42]

Even for the limited data on cooling water impacts that it did have available, EPA warned that "[b]ecause of... methodological weaknesses, EPA believes that studies... should only be used to gauge the *relative* magnitude of impingement and entrainment losses."[43] EPA also acknowledged that "only 150 out of 554 Phase II facilities have indicated... that they have ever performed an impingement and entrainment... study."[44] Yet these methodologically weak and incomplete studies provided the raw material for the agency's economic analysis that rejected dry cooling and closed-cycle cooling technologies in favor of weaker performance ranges. Nor was the agency's counsel to use impingement and entrainment studies only to gauge "relative magnitude" the only unheeded disclaimer to appear in the rulemaking record. The agency also warned that "[t]o rely only on estimated use values would substantially undervalue the benefits of the final section 316(b) rule"; that "[t]he organisms that remain unvalued in the analysis provide many important ecological services that do not translate into direct human use"; and that "[t]o the extent that the latter are not captured in the benefits analyses, total benefits are underestimated."[45] Elsewhere, the agency offered the sage advice that "[a] comparison of complete costs and incomplete benefits does not provide an accurate picture of net benefits to society,"[46] and that "there is a real possibility that ignoring non-use values could result in serious misallocation of resources."[47]

Despite these multiple and seemingly sincere disclaimers, the agency ultimately could not resist claiming that its "proposed rule has the largest estimated net benefits, $452 million, of the five regulatory options analyzed."[48] Thus, one important level of objection to the cooling water intake rulemaking focuses on EPA's decision to allow regulatory CBA to heavily influence the ultimate selection of environmental performance standards when the CBA itself was woefully incomplete and uncertain. With so many effects remaining off the balance sheet, regulators actually had little reason to be confident that the conclusions offered by CBA were welfare-maximizing. Nonetheless, as in other contexts, the promise of an "objective" quantitative analysis seemed difficult to resist in the face of a heavily politicized, deeply uncertain, and morally fraught decision. This cognitive lure of CBA was especially evident in the public comments of one prominent environmental economist, who suggested that regulators should seek to identify the technology "which protects the target resources... up to the point where the incremental benefit from increased protection *just equals* the incremental cost of increased protection."[49] Because this standard of empirical sophistication is never met in the area of environmental regulation, important policy judgements must be made regarding how to handle information gaps, scientific uncertainties, system complexities, and other quantitatively intractable features of regulatory decisions such as the cooling water intake rules.

These kinds of policy judgments became especially evident in a difference of view that arose among public commenters regarding whether nonuse values attributable to saved organisms should be estimated by EPA. Beginning with the background precautionary assumption that not everything valuable about the environment can be discerned, dissected, and quantified by present human modes of understanding, one commenter advocated the use of admittedly imperfect attempts to quantify and monetize the variety of ecological impacts that go beyond simple reductions in commercially and recreationally valuable fish mortality. For instance, citing a literature review of use and nonuse value studies in the environmental context, the commenter noted that agencies could rely on a simple presumed ratio between use and nonuse values to provide an approximate numerical estimate of the latter values, conservatively estimated at a multiple of two dollars of nonuse benefit for every one dollar of use benefit.[50] This ratio was offered in contrast to EPA's customary "50 percent rule," which had been used to set nonuse benefits at half of the use benefits, but which this commenter argued was based on outdated studies. Regardless of the ultimate heuristic chosen, however, the commenter's most fundamental point was that EPA should "avoid placing an effective value of zero on categories of value that the Agency does not have time or resources to analyze in detail."[51] Any other approach would be inconsistent with the environmentally precautionary tilt of the CWA.

Another commenter, in contrast, regarded the literature on use-to-nonuse value ratios to be inadequate to support quantitative estimations in the immediate policymaking context. Because the studies underlying the literature review cited by the first commenter addressed a variety of environmental and natural resources issues other than the impacts of cooling water intake structures on aquatic ecosystems, this commenter argued that the studies provide "no evidence of why, given the *specific* environmental improvements associated with the proposed regulations, non-use value should be of any *specific* magnitude."[52] In the policymaking context, this standard of exactitude carries vastly different implications than in an academic one. By demanding original, detailed, and unambiguous valuation studies of nonuse benefits and other ecological impacts, and by refusing to assign any nonzero value in the absence of such studies, the latter commenter's approach assigned a burden of proof to the agency that predictably biases decisions against environmental protection. Such exactitude might be appropriate within the ivory towers of the university, where scholars aim to bolster the scientific credentials of welfare economics by portraying the discipline as one of "objective implementation of benefit–cost analysis, based on established economic theory and empirical research."[53] In the real world of policymaking, however, decisions must be made in advance of comprehensive knowledge. Nevertheless, in light of the intense criticism that the agency received even for its cautious use of a 50 percent ratio to estimate nonuse benefits, EPA ultimately assigned no numerical value at all.[54]

A second, and potentially more significant, level of objection focuses on the opportunity cost of conducting regulatory CBA. One of the defining characteristics of environmental law is its demand that society make choices in advance of complete and reliable information regarding the consequences of those choices. Accordingly, a primary thrust of the precautionary approach to environmental law and policy is that regulatory decisionmaking should be designed not merely to react to existing information, but to actively intervene in the processes and institutions by which information is generated. For instance, through careful assignment and management of the burden of proof, regulators may be able to marshal the information-generating resources of firms and other private actors in service of the public's environmental aspirations. By instead placing the burden on regulators to identify, quantify, and monetize potential adverse impacts of market activity, regulatory CBA does not merely construe uncertainty against the environment, it also forfeits a valuable opportunity to use incentives, penalty defaults, and other regulatory strategies directly in furtherance of informational goals. Such an opportunity is especially significant in the environmental regulatory context given that, as public choice theory would predict, the regulated community is typically better represented and better resourced than nongovernmental organizations and other representatives of the public interest during administrative rulemaking. As the Phase II rulemaking process made all too plain, identification of the social welfare–maximizing policy is a function not merely of regulatory costs and benefits—or even of policy decisions regarding how to address inevitable uncertainties in costs and benefits—but also of the power of stakeholders to invest in shaping the understanding and perception of costs and benefits.

Lest it be genuinely obtuse, environmental policymaking must be conceived of with this informational and political context in mind. Many proponents of regulatory CBA, however, treat information acquisition and management as matters that simply happen out there. One important task for the improvement of regulatory CBA is therefore to consider more fully the institutional context within which it unfolds and to expressly link its use to conditions within which it can be expected to generate reliable guidance. Regulatory CBA as currently practiced can perhaps be best understood as a method of executive control over agency decisionmaking, rather than as a genuine effort to identify welfare-maximizing uses of public resources. The Phase II rulemaking experience seems to bear this cynical view out, as one can read between the lines a sense of frustration among EPA staff members regarding OIRA's simultaneous demand for robust quantitative and monetary estimates of regulatory impacts on one hand and its refusal to grant EPA the required approvals to conduct original valuation studies on the other.[55] More broadly, the agency's Phase II rulemaking documents repeatedly stress that adequate valuations for the myriad nonuse benefits of the rule were not developed because of a lack of adequate time and resources. Again, the burden of proof becomes an issue in any real-world policy analysis such as this one, and OIRA seems to have influenced the rulemaking such that the burden fell on EPA to demonstrate harm rather than, as the CWA arguably prescribes, on those agents that adversely impact aquatic ecosystems to demonstrate the infeasibility of harm reduction.

Valuation

Even assuming that EPA did have reliable and comprehensive information regarding the myriad, complex environmental impacts of cooling water intake processes, the agency still, under regulatory CBA, needed to transform those impacts into a uniform monetary measure to enable quantitative comparison with the expected costs of protection technologies. As noted above, when designing the Phase II rules, EPA focused only on "reductions in impingement and entrainment as a quick, certain, and consistent metric for determining performance."[56] Moreover, reductions in impingement and entrainment were valued only insofar as they resulted in identifiable gains to commercially and recreationally valuable fish.[57] This choice of metric was useful for EPA because it allowed the agency to avoid intractable valuation questions that would have accompanied the effort to account for threatened or endangered species impacted by cooling water intake processes.[58] Indeed, the entire approach of focusing on reductions in impingement and entrainment as "a convenient indicator of the efficacy of controls in reducing environmental impact"[59] seemed designed to ensure that the agency could comply with the quantification and monetization demands of regulatory CBA, irrespective of the actual share of the consequentialist landscape that the agency encompassed within its calculations. Unable to measure what was important, EPA instead chose to make important what it could measure.

In earlier stages of its analysis, EPA candidly acknowledged that existing valuation techniques in the environmental economic literature tend to understate the benefits of environmental protection by focusing only on the most readily understood and quantifiable effects of human interference with ecosystems.[60] Thus, the agency sought to supplement its initial analysis with an indirect measure of the value of environmental protection, one that asks what the cost would be of replacing the variety of goods and services that are provided by a healthy, functioning aquatic ecosystem. Although conventional valuation techniques ask whether an environmental resource is worth saving based on estimates of the monetary worth of its fruits, EPA's *habitat replacement cost* (HRC) method instead asked what those fruits are worth based on how difficult and costly it would be to develop substitutes for the environmental resource that generates them. The former approach reflects a demand-side estimate of environmental value based on the amount that individuals appear willing to pay to preserve discrete environmental goods or services; the latter approach generates a supply-side estimate of environmental value based on the amount that society would need to expend to replace those same environmental goods and services in their interrelated ecological context, essentially taking as given that those goods and services must be provided. The great advantage of the latter approach, as the agency noted, is that "[t]he HRC method can be used in benefit–cost analyses to value a broad range of ecological and human services associated with [impingement and entrainment] losses that are either undervalued or ignored by conventional valuation approaches."[61]

Nevertheless, one public commenter strongly condemned HRC as "a completely illegitimate method of analysis," stating that it is "essentially oxymoronic and completely invalid," and that it commits "one of the gravest of errors in economics" by confusing environmental costs and benefits.[62] Not content with this invective, the commenter also sought to denounce as "false" all of EPA's claimed reasons for deploying HRC estimation:

[I]t is claimed that 'the HRC method can be used to value a broad range of ecological and human services losses....' False. It is also asserted that 'it can be used as an alternative to conventional approaches that are based on recreational and commercial fishing impacts.' False. And it is stated that 'in addition, HRC can supplement conventional valuation results....' False.[63]

What accounts for such strong criticism? HRC estimations essentially assume that an object of environmental protection is a unique capital resource that produces a flow of valuable goods and services, the worth of which can only be approximated by asking what it would cost to develop a substitute resource that produced those same goods and services. To some observers, this approach is wrongheaded because the very point of regulatory CBA is to ask whether the environmental resource is worth preserving at all. That is, rather than just assume that society must have clean water, biodiversity, and the variety of other goods and services that flow from intact ecosystems (or their built replacements), regulators should estimate the monetary amount that individuals are actually willing to pay to obtain those specific, disaggregated goods and services. In this view, no resource or service—not fish, not freshwater, not human life—is considered sufficiently important to avoid being subjected to instrumentalist trading.

A second commenter objected to this radically commensurated worldview and sought instead to portray environmental resources as capital items in the manner assumed by HRC. The commenter asked:

Are the natural resources that are affected by cooling water intake best thought of as long-lived capital goods—or are they more like consumer goods that people, or power plants, might choose to consume when they are hungry? If you eat the last cookie and then throw out the box, you may not have to pay the full 'cookie replacement cost.' Perhaps you are getting tired of cookies and don't plan to buy any more, so there is no need to worry about replacement cost. Something along these lines seems to be involved in the claim that HRC overstates the value of environmental resources: if we are planning to consume the ecosystem without replacement, then HRC might overestimate the values at stake.[64]

Through this skeptical metaphor, the commenter joined rather illustrious company, even among economists: in Daniel Defoe's novel *Robinson Crusoe*—a favorite literary example of economists[65] and a key progenitor of the myth of the self-reliant individual that animates much of liberal political theory (and by extension welfare economics)[66]—the legendary protagonist despaired over his declining supply of ink for writing on his island home. Although he attempted to engineer crude substitutes—such as the arrangement of large physical objects to serve as memory joggers akin to a written diary or a list—in the end, Crusoe was left sadly bemoaning the disappearance of an irreplaceable asset: "My Ink, as I observed, had been gone some time, all but a very little, which I eek'd out with Water a little and a little, till it was so pale it scarce left any Appearance of black upon the Paper. . . ."[67] Whatever Crusoe's role in furthering the economic conception of human nature, his supply of ink seemed not to have been, to him, just another box of cookies.

The debate between these two commenters illustrates a more general theoretical distinction that exists between conventional environmental economists, who tend to view all natural and human-made capital as substitutable, and ecological economists, who in contrast view many features of natural capital and ecosystem services as practically irreplaceable. In this context at least, Congress appears to have sided with the ecological economists: the goal of the CWA is "to restore and maintain the chemical, physical, and biological integrity of the Nation's waters."[68] The goal is not to view those waters as merely contingent resources, to be impaired or sacrificed at any moment for the promotion of an abstract and undifferentiated maximization of welfare. Nor is the CWA at all unusual in this regard: much of the project of environmental law aims to identify elements of the environment—for instance, particular species, habitats, ecosystems, or global atmospheric processes—that are sufficiently important to human well-being or sufficiently worthy of admiration and respect in their own right and to take them outside of the realm of instrumentalist trading; that is, to cordon them off from the market's continual demand that they demonstrate their monetary worth in order not to be consumed, developed, or otherwise sacrificed.

Such an approach should not be completely alien to welfare economists: in political theory, eminent welfarists such as Amartya Sen similarly offer an "objective list" of essential human goods or capabilities as the relevant indicia of social well-being that governments ought to pursue, rather than undifferentiated increases in overall well-being as approximated through willingness-to-pay, gross domestic product, or similar wealth-based measures.[69] Lexical orderings such as Sen's can be justified deontologically as conferring certain "rights" or "trumps" on individuals to help effectuate the liberal ideals of equality and autonomous flourishing (or, in the case of the rights of nature, to move beyond liberalism's relentless anthropocentricism). They also can be justified more pragmatically as offering better approximations of overall well-being than the preferentialism of welfare economics, given the insurmountable costs of information and deliberation that prevent preferentialist approaches from achieving their aim of comprehensive evaluation. It may be both just and efficient, in other words, to at least occasionally refuse to treat policy decisions as mere investment choices. Proponents of regulatory CBA would be wise to embrace, or at least experiment with, such alternative value criteria. Under Sen's approach, for instance, the so-called inevitable trade-off between efficiency and equity is accommodated by bringing the underlying value criterion, which ultimately determines what counts as efficient, into better alignment with the demands of equity. Likewise, by placing ecologically determined constraints on market ordering, such as through the mandated protection of endangered species or the steady renewal of the nation's water bodies and coastal areas, Congress has, in essence, made certain environmental goals foundational, such that they need not be seen as derogating from efficiency. Instead, they are put in place prior to the market operations that later will determine efficiency.

Nonetheless, to some proponents of CBA, value remains a steadfastly monistic concept. As one public commenter wrote, "[i]n economic terms, the benefits of some action are equivalent to the aggregate of the willingness to pay (WTP) by the affected human populations for that action or outcome."[70] From this perspective, everything of value in the world can be readily commensurated because, by assumption, value only takes the form of individual human preference, as manifested in measurable expressions of willingness to pay. Such a tendency toward value monism also was apparent in EPA's flirtation with a trading program as a compliance alternative for firms under the final Phase II rules. Under this radically commensurated approach, facilities would have been permitted to purchase credits from other firms representing environmental impact reductions equivalent to the purchasing facilities' regulatory obligation.[71] Interestingly, here EPA balked at the "comparability and implementation challenges" implied by such a trading program: "EPA does not believe that it is possible at this time to quantify with adequate certainty the potential effects on ecosystem function, community structure, biodiversity, and genetic diversity of such trades, especially when threatened and/or endangered species are present."[72] Those same challenges, however, also applied to the attempt to conduct a regulatory CBA of cooling water intake structure requirements. That is, the very challenges of information and valuation that caused EPA to back down from a trading program should have caused it also to withdraw from regulatory CBA.

Even assuming adequate information, the value monist approach of welfare economics still must meet the ecological economist's more fundamental conceptual objection that willingness-to-pay valuation measures confuse the categories of capital and income, stock and flow, and future and present. Characterizing natural resources as capital goods or stocks illuminates their irreducible intergenerational aspect. In the context of fish, for instance, one might say that the "optimal" management criterion for a given fishery is to seek maximum sustainable yield.[73] That intuitively appealing criterion represents, at bottom, a normative judgment regarding intergenerational distributive equity: the value of the fishery's yield is considered sufficiently large and distinct to merit preserving the underlying capital stock, such that the flow continues irrespective of opportunity costs. From the strict welfare economic perspective, on the other hand, the fishery would be managed sustainably only if the anticipated benefits of doing so happened to justify the costs. If it turned out that "liquidating" the fishery and reinvesting the monetary proceeds in other investment opportunities proved to have a higher net present value, then nothing in the value monism of welfare economics would counsel against such liquidation. As Robert Solow famously put it, from this perspective, "the world can, in effect, get along without natural resources, so exhaustion is just an event, not a catastrophe."[74]

This debate also touches on a longstanding difference of view among economists, philosophers, and others regarding the proper role of temporal discounting in regulatory CBA. Over the years, OMB has supported relatively high discount rates, often based on historical returns on private investment rather than arguably more appropriate measures, such as the real return on long-term government debt or, indeed, no discounting at all.[75] In 2003, for instance, OMB dismissed a public comment that had advocated a relatively low (e.g., 3 percent) discount rate for long-term policy evaluation, arguing that such a rate would "not be appropriate for regulations that had a strong displacing effect on capital investment."[76] Behind OMB's argument again lies an assumption that natural and human-made capital are generally substitutable, such that the proper rate of discount for the former safely can be assumed to be the rate of return given by the latter. If this assumption holds, then discounting by market rates of return will benefit future generations by ensuring that their eventual resource base has taken advantage of the best available investment opportunities during intervening time periods. If the assumption does not hold, however, future generations may be left immeasurably harmed. In fact, they may be caused *not to exist at all,* even as cost–benefit models assure them that they are somehow "better off" as a result.[77]

Despite the obvious normative issues raised in this context, OMB's defense of intergenerational discounting has been meager. In its 2003 revised Circular A-4, for instance, OMB offered two misguided arguments: (a) future generations "will be wealthier and thus will value a marginal dollar of benefits and costs by less than those alive today," and (b) the longer time horizon involved for intergenerational policy analysis implies a greater level of uncertainty regarding future costs and benefits.[78] As Derek Parfit long ago pointed out, these two defenses fail even to address their proper subject: the first defense discounts consequences because they happen to people who are better off; the second, because the consequences may not actually happen. Neither defense actually bears an unequivocal relationship to time, which is, after all, the dimension along which discounting purports to proceed.[79] In fact, none of the conventional arguments in defense of discounting in the intergenerational context withstand analysis, except potentially the defense based on opportunity costs, which, as argued above, depends on a debatable assumption of perfect substitutability among the varieties of natural and human-made capital goods. Even the defense offered by EPA in connection with the Phase II rulemaking—which simply notes that individuals in the current generation tend to reveal a rate of time preference for present over future consumption[80]—fails to grapple with the intergenerational context because it assumes that one generation's rate of time preference can simply be transposed onto another. It even confuses the relevant interest holder: it implicitly treats society as one infinitely lived individual, rather than acknowledging the need to correlate periods of discounting with the time spans during which individuals actually are alive and experiencing temporal impatience. Such conceptual slippage is symptomatic of a broader confusion regarding CBA in the intergenerational context: when future generations are at issue, there can be no revealed or stated preferentialist approach to welfare maximization. Instead, analysts must engage directly with the inevitably paternalistic project of helping to structure the environment, the possibilities, and the very identities of future individuals. Recent work by Matthew Adler and Paul Dolan attempts to take this challenge seriously while still generating valuations that can be of practical worth.[81] Their work should be of great interest to the community of academics, bureaucrats, policy analysts, and others who are invested in the philosophically appropriate use of regulatory CBA.

For the cooling water intake rulemaking, these fundamental conceptual issues manifested in the question of whether EPA had underestimated the commercial and recreational value of fish by calculating only the value of fish that eventually are landed. From one perspective, uncaught fish should be credited as valuable because they form the stock from which future fish grow, including those caught fish that later register in EPA's benefit calculations. Treating natural capital instead as a pure consumable may lead to a situation that is deemed efficient even as it fails to ensure sustainability. This issue should have played a far more prominent role in EPA's decisionmaking, given the dire state of many of the nation's fisheries and the fact that "some studies estimating the impact of impingement and entrainment on populations of key commercial or recreational fish have predicted substantial declines in population size."[82] Likewise, the value of threatened, endangered, and other "special status" species was left unquantified in EPA's regulatory impact analysis, despite ample evidence that such species are threatened by the impacts of cooling water intake. The attempts the agency did undertake to value threatened and endangered species, moreover, simply treated species preservation as an investment decision to be based on individuals' revealed preference for species survival. From this perspective, certain charismatic species, such as the loggerhead sea turtle, might fare well based on their high nonconsumptive value as objects of photography and other indirect uses, whereas more "obscure species" would be relegated to the notoriously difficult to estimate category of "pure non-use value."[83] Although preferable to EPA's eventual failure to credit endangered and threatened species impacts at all, the investment approach nevertheless fails to take seriously the awesomeness of extinction as a collective deed. Like the irreversible depletion of a once-grand fishery stock, species extinction is a deed that lies outside of secular time. Once these stocks and species are gone, they are gone. Thus, the decision to establish standards of protection entails not just a question of maximizing allocative efficiency—of squeezing the most net present value out of resources that are assumed to belong to the current generation—but a question of determining whether, as a matter of intergenerational distributive equity, they should remain available for the benefit of future generations. No matter how rigorous its techniques of valuation, regulatory CBA has no capacity to resolve the latter question, which must remain one of morality and law.

Process

Because EPA eventually felt the need to abandon both its conservative 50 percent rule for nonuse value estimation and the HRC measure of ecosystem value,[84] the agency ended up capturing only a very limited picture of the value of cooling water intake regulation. The agency did present what it styled as a "break-even analysis," which illustrated the amount of total and per capita annual willingness to pay for unvalued ecological impacts that would be required for the costs of the rule to equal its benefits.[85] This approach—which essentially allowed policymakers and their various publics to examine the value of habitat preservation implied by the CWA's precautionary mandate and to ask themselves, "Is it worth it?"—bore some relationship to another valuation technique that EPA was forced to back away from after harsh public comments, the *societal revealed preference* method.[86] Under this approach, the agency examined the level of compliance costs that had been tolerated by society in connection with previous governmental actions that sought to attain the same kinds of ecological benefits that the Section 316(b) rulemaking would promote. Also sometimes referred to as the *cost of control* method, this valuation approach appears to have been pioneered by public utility commissions in Massachusetts and other progressive states as part of an effort to monetize environmental externalities during electricity generation planning: "[t]he basic rationale for using cost of pollution control as a measure of the value of pollution reduction is that the cost of pollution controls required by the government provides an estimate of the price that society is willing to pay to reduce the pollutant."[87] Like EPA's HRC and break-even analyses, then, the societal revealed preference methodology represented an innovative way of responding to OIRA's demand for quantitative, monetized information regarding the costs and benefits of the Phase II rule, despite the great uncertainty characterizing ecological effects.

Some commenters, however, were no more pleased with the societal revealed preference methodology than with HRC; thus, EPA ultimately decided to back away from its innovative approach. As one environmental economist wrote, "[l]ike the HRC method, this approach has no foundation in economic theory, is not accepted by economists as a legitimate empirical method of valuation, and is no more than a method of cost analysis mistakenly applied to the benefit side of the ledger."[88] Such criticism of the societal revealed preference methodology was instructive because it raised in a concrete setting the basic normative questions posed by Mark Sagoff in his classic critique of regulatory CBA: What, after all, is wrong with allowing valuations of collective goods to emerge from society's willingness to act collectively as citizens to preserve a threatened good? What is wrong with simply allowing individuals to express their preferences for environmental goals through political channels, by voting in favor of environmentally inclined or disinclined politicians or by holding direct referenda on environmental policies?[89] To be sure, a monetary value of the environment will be implied after the fact by the level of resources necessary to fulfill adopted environmental policies—a value that will even be quantitatively estimable through use of the societal revealed preference methodology—but this value will not drive the initial selection of policy. Rather the monetary value will simply be an ancillary effect of a policy choice that instead is premised on social values, explicitly discussed, and mediated through democratic decisionmaking processes.

This distinction between market-determining and market-determined views of the content of law is perhaps best illustrated through analogy to a context in which the sublimation of law to welfare analysis is more obviously problematic, or at least unfamiliar. In the wake of the 9/11 World Trade Center attacks, OMB issued a call for research on how to measure the value of liberty and privacy interests that would potentially be sacrificed by new antiterrorism measures.[90] Responding to this call, researchers at Harvard University attempted to measure individuals' willingness to trade off civil liberties for safety and convenience during airport security checkpoint procedures. In particular, the researchers questioned survey respondents' willingness to accept racial profiling in exchange for reduced time spent waiting in line. Not surprisingly, white respondents were generally more willing to accept racial profiling than nonwhite respondents.[91] The principle of race neutrality did figure somewhat in the white subjects' decisionmaking: when told that only nonwhite groups would be targeted for profiling, white respondents expressed an unwillingness to support the explicitly race-based program for a meager 10-minute time savings. For a 30-minute savings, however, the white respondents overcame their moral qualms.[92] They revealed, in other words, their willingness to pay to preserve the principle of race neutrality.

By calling for empirical research of this nature, OMB sought to develop procedures for estimating the welfare impacts of civil liberties restrictions. From the welfarist perspective, the value of civil liberties was thus made to be contingent on the level of burden that individuals appeared to be willing to accept to maintain them. In a constitutional scheme of government, however, the cart is typically at the other end of the horse: civil liberties are protected by law, and their value, assuming one insists on quantification, is implied by the level of burden that individuals are made to accept to maintain them. At least on their face, environmental, health, and safety laws also frequently display this structure: employees are *entitled* to a safe workplace, endangered species to necessary habitat, citizens to clean air and water, and so on. These entitlements are meant to be inviolable; that is, they are meant to be protected by law, such that their value is a creation, rather than a determinant, of law. Just as a constitutional scheme of rights is intended to protect certain interests of individuals from incursion by majoritarian politics, democratically enacted environmental laws often seek to protect certain interests from shortsighted destruction by land development, industrialization, and other potentially harmful human activities.

The most critical question raised by regulatory CBA, therefore, is not to be found in the details of ever more refined modes of valuation, but rather in the initial staging decision regarding what is, and is not, a proper subject of instrumentalist trading. Perhaps for the welfare economist a better example than civil rights would be the classical liberal principles of strong private property rights and freedom of contract. Should the content of *these* laws be determined through regulatory CBA? If so, how would the relevant valuations of cost and benefit be derived, considering that the very laws under consideration are the same ones that give rise to the market structure and exchange activities that enable revealed preference analysis? As with civil liberties, property rights, and freedom of contract, many of the central questions and tasks of environmental law and policy also are of a foundational character and therefore cannot be adequately posed within the language of regulatory CBA. Instead, they must be answered through the more familiar discourses and procedures of democratic constitutionalism.

Opposition to methodological innovations, such as the HRC and societal revealed preference valuation techniques, may be driven at bottom by a suspicion of precisely these familiar discourses and procedures. Yet it is not clear that individual willingness-to-pay valuations are above suspicion. Critics imply that the societal revealed preference methodology is especially unreliable because there is no guarantee that individuals *"actually (and voluntarily) incur [ ] costs* to avert (or tolerate) the environmental disruption in question."[93] The very domain of regulatory CBA, however, is one in which private market activity—that of individuals *"actually (and voluntarily) incurring costs"*—is presumed to have failed to maximize social welfare. Indeed, recent revisions to the executive order mandating regulatory CBA purport to make the identification of some market failure along these lines a prerequisite to agency action.[94] By assumption, then, conventional willingness-to-pay valuation exercises must apply outside of the context that is said to have made them especially reliable as indicators of value. This point needs to be stressed: policy analysts devote a great deal of attention to the practice of *benefits transfer analysis,* whereby values derived from one economic study are transferred for use in a somewhat analogous, but nevertheless distinct, regulatory setting. They spend comparatively little attention worrying about the fact that *every* willingness-to-pay study, when used in a regulatory context, is a benefits transfer exercise entailing deep conceptual and practical difficulties.

Thus, even putting aside the foundational normative issues raised by critics like Sagoff, it would be wrong to believe that conventional willingness-to-pay methodologies provide an objective measure of benefits. For example, even individual market valuations may differ dramatically depending on whether the individual must pay to obtain an entitlement or instead can simply demand an amount of monetary compensation adequate to compel voluntary relinquishment of the entitlement. Because welfare economists hope to avoid such overtly normative questions, they typically use only willingness-to-pay valuations. This choice can be defended on the methodological ground that such valuations are better behaved than willingness-to-accept valuations because the former are disciplined by actual or hypothetically imposed budget constraints and therefore tend to give more consistent (and lower) values. The conventional privileging of revealed preference studies over stated preference studies is often defended on similar pragmatic grounds.[95] But these choices also have the convenient normative effect of forcing individuals to demand greater levels of protection only through markets, using their existing rights and resources, rather than through law. The transformative potential of law is constrained by tethering social choice valuation to the status quo—the same status quo whose problematic features gave rise to a need for law. We should hesitate to regard the fact that willingness-to-pay valuations are better behaved than willingness-to-accept valuations as evidence that society can identify the optimal level of harm through conventional regulatory CBA. Instead, we should see consistently lower willingness-to-pay valuations as evidence that workers, consumers, fish, and other imperiled entities ought to be given more power directly through law to help close the gap between the two valuations.

Also not properly characterized as objective are stated preference valuation methodologies, in which individuals are asked through hypothetical survey instruments to state their willingness to pay to protect a public good, such as an endangered species or a portion of rainforest. Such contingent valuation exercises have attracted criticism from two different extremes. To some critics, the exercises are practically unreliable because they require no actual performance of an individual's stated commitment. To others, the exercises are philosophically objectionable because they seem to force a market-based values framework onto experimental subjects, rather than approaching the environmental preservation decision as a social or collective one. Although economists have responded to the former criticism by enhancing the level of methodological sophistication found in contingent valuation experiments, they have only recently begun grappling with the latter objection. In one noteworthy study, researchers examined the role of social isolation on stated preferences by varying the degree of anonymity afforded to subject responses.[96] Their findings are significant: in addition to finding a gap between stated willingness to pay and actual contribution levels—thus confirming the complaint of the practical critics of contingent valuation studies—the researchers also documented large disparities among willingness-to-pay valuations depending on the social context within which subjects are asked to state their response amounts, thus confirming the complaint of the philosophical critics.

This apparent context-dependence of environmental valuations poses yet another theoretical challenge to the welfare economic framework: the process of preference elicitation itself seems to influence the resulting content of preferences, thereby further complicating the effort to identify preferences in a manner free from analyst bias.[97] Not that anyone should be especially surprised by this result; after all, as Amartya Sen has argued, "[t]he very idea that I treat the prevention of an environmental damage just like buying a private good is itself quite absurd.... [I]t would be amazing if the payment I am ready to make to save nature is totally independent of what others are ready to pay for it, since it is specifically a social concern."[98] During the public comment period for the Phase II rulemaking, one contributor argued that this quotation from Sen merely reflects the commonplace notion that, in the case of public goods, such as environmental conservation, individuals can be expected to free-ride, thereby driving a wedge between their true demand for public goods and their observed (unregulated) contributions to the provision of such goods. Thus, the commenter disagreed with others who viewed Sen's comment as more strongly supporting the need for alternative valuation mechanisms to conventional revealed preference methodologies.[99]

But in fact, Sen was making the stronger claim, namely that specifically social concerns demand specifically *social* methods of deliberation, valuation, and decisionmaking. As Sen noted in another context, "[w]e can have many reasons for our conservational efforts—not all of which are parasitic on our own living standards and some of which turn precisely on our sense of values and of fiduciary responsibility."[100] Because of this value pluralism, which is inherent in contemporary liberal societies, Sen emphasized the need to guarantee to individuals the "freedom to participate" in environmental policymaking. Thus, the famed economist and political theorist cast his lot, not with conventional revealed preference methodologies or even with the supposedly radical economic methodologies of HRC and societal revealed preference, but rather with the kind of democratic forums, institutions, and processes championed by the likes of Sagoff. As Sen wrote:

The relevance of citizenship and of social participation is not just instrumental. They are integral parts of what we have reason to preserve. We have to combine the basic [instrumentalist] notion of sustainability... with a broader view of human beings—one that sees them as agents whose freedoms matter, not just as patients who are no more than their living standards.[101]

In recognition of the force behind such arguments, a growing body of literature within environmental economics attempts to fuse deliberative processes with economic valuation techniques, aiming to address both the political demand for participatory legitimacy and the economic demand for disciplined monetary valuation. As Clive Spash, a leading participant in this emerging literature, noted, proponents of such deliberative monetary valuation techniques "have sought processes where there is time for reflection, potential for information gathering and group deliberation," an approach that "seems to accept that preferences are formed during a process aiming to value environmental changes and that the type of process is therefore something which needs to be openly discussed as a matter of institutional design."[102] Deliberative monetary valuation therefore occupies a middle position between the radically atomized individualism of welfare economics and the political collectivity of conventional lawmaking. Although it is unclear whether such efforts to sensitize welfare economic policy analysis ultimately will be able to meet the democratic objection of thinkers like Sagoff and Sen, they undoubtedly represent an improvement over status quo valuation techniques.

Conclusion

Despite the implicit claim of welfare economists, laws are not only instrumental tools for serving human "patients," but also repositories of cultural value and meaning that both flow from and help to form human "agents." The temptation to reduce this complexity is great. Many economic observers, for instance, contend that BTA standards devolve in practice to CBA because, as noted above, courts often permit agencies to reject a technology whose costs are "wholly disproportionate" to environmental benefits.[103] However, such limited cost consideration does not render BTA standards extensionally equivalent to CBA. Technology-based standards continue to carry expressive connotations that are vastly different from those of CBA. The determination by Congress that an environmental, health, or safety goal is sufficiently important that only society's collective "best" efforts will suffice opens an inevitable and lamentable gap between statutory aspiration and regulatory achievement. Those harms that are not prevented under a BTA standard accordingly serve as moral remainders, indicating the collective need to constantly seek ways of doing better—of further protecting life and lowering environmental impact in the future. In contrast, under regulatory CBA, those harms that are not avoided simply represent the right, the efficient, or the optimal level of harm. Precisely because it purports to account for all relevant consequences of decisionmaking, CBA must inevitably round to zero the moral remainders of risk regulation, leaving nothing further to signify the societal need for redoubled vigilance. In EPA's Phase II rulemaking, for instance, literally billions of fish each year are simply ignored in the agency's CBA, treated as if their loss is meaningless because the question of their worth has been abandoned.

Environmental decisionmaking cannot be premised solely on static, localized costs and benefits in this manner. Regulatory approaches instead must aim to alter—over time and in rather dramatic macroscale ways—the economic and technological forces that combine to structure any given regulatory context, with its microscale snapshot of cost and benefit information that seems to admit of only one optimal solution. The latter approach ignores what is rightly regarded as law's transformative potential. For instance, as the Second Circuit observed when striking down EPA's Phase II rules, "the most salient characteristic of [the CWA's] statutory scheme, articulated time and again by its architects and embedded in the statutory language, is that it is technology forcing."[104] When EPA subtly shifted from expressing concern over the impact of cooling water intake structures on already impaired water bodies to using that present impairment as an efficiency-oriented excuse for lowering levels of protection, the agency abandoned the transformative potential of the CWA. The quest to use environmental law to progressively restructure the economic and technological landscape that gives rise to any momentary depiction of costs, benefits, and optimality instead became a surrender to those momentary depictions. To effectively serve the original, more ambitious aims of environmental law, statutes and regulations must form a part of the social glue that binds a political community together in pursuit of long-term, uncertain, and often simply audacious collective goals. To serve that aim, in turn, laws and policies must have "continuity with ordinary discourse and hence with real communities, real values, and real politics."[105] By literally denying the sacredness of life, and indeed of anything, regulatory CBA fails these tests.

■ ■ ■

Notes

1. *EPA v. State Water Resources Control Board*, 426 U.S. 200, 202–203 (1976).

2. Id. at 203.

3. *See* 33 U.S.C.A. §§ 1251, 1342. *See also Weyerhaeuser Co. v. Costle*, 590 F.2d 1011, 1041–42 (D.C. Cir. 1978); *Bethlehem Steel Corp. v. EPA*, 538 F.2d 513, 514–15 (2d Cir. 1976); *Hooker Chems. & Plastics Corp. v. Train*, 537 F.2d 620, 623–24 (2d Cir. 1976).

4. *See* Oliver A. Houck, *The Clean Water Act TMDL Program: Law, Policy, and Implementation* (Washington, DC: Environmental Law Institute, 1999), 3.

5. *See* 33 U.S.C.A. § 1326(b).

6. U.S. Environmental Protection Agency (EPA), "National Pollutant Discharge Elimination System—Final Regulations to Establish Requirements for Cooling Water Intake Structures at Phase II Existing Facilities," *Federal Register* 69 (July 9, 2004): 41576, 41586 (to be codified at 40 C.F.R. pts. 9, 122–25) [hereinafter, "Final Regulations"].

7. Id.

8. Id. at 41,586.

9. 33 U.S.C.A. § 1326(b).

10. *See Appalachian Power Co. v. Train*, 566 F.2d 451, 457 (4th Cir. 1977).

11. *See Federal Register* 44: 32956.

12. *See Cronin v. Browner*, 898 F. Supp. 1052, 1056 (S.D.N.Y. 1995).

13. *See* Karl R. Rábago, "What Comes Out Must Go In: Cooling Water Intakes and the Clean Water Act," *Harvard Environmental Law Review* 16 (1992): 429, 467.

14. *See Riverkeeper, Inc. v. United States Environmental Protection Agency*, 358 F.3d 174, 181 (2d Cir. 2004) [hereinafter, Riverkeeper I]; *Riverkeeper, Inc. v. United States Environmental Protection Agency*, 475 F.3d 83, 90 (2d Cir. 2007) [hereinafter, Riverkeeper II].

15. *See* Riverkeeper I, 358 F.3d at 196.

16. "Final Regulations," *Federal Register* 69: 41583.

17. Id. at 41598.

18. Id. at 41,601. *See also* U.S. Environmental Protection Agency, *Economic and Benefits Analysis for the Final Section 316(b) Phase II Existing Facilities Rule*, EPA-821-R-04-005 (Washington, DC: U.S. EPA, 2004), A2-5 ("Closed-cycle cooling systems . . . are the most effective means of protecting organisms from [impingement and entrainment]."), available at www.epa.gov/waterscience/316b/phase2/econbenefits/final.htm [hereinafter, Final Rule EBA].

19. "Final Regulations," *Federal Register* 69: 41583.

20. *See* id. at 41603 ("[I]n determining that the technologies on which EPA based the compliance alternatives and performance standards are the best technologies available for existing facilities to minimize adverse environmental impact, EPA considered the national cost of those technologies in comparison to the national benefits . . . .").

21. Final Rule EBA, at D1-1 (predicting "no compliance action" for 200 facilities but noting that 75 of these facilities already have adopted closed-cycle cooling).

22. Final Rule EBA, at C3-2.

23. "Final Regulations," *Federal Register* 69: 41599. *See also id.* at 41,604 ("[I]n a waterbody that is already degraded, very few aquatic organisms may be subject to impingement or entrainment, and the costs of retrofitting an existing cooling water intake structure may be significantly greater than the benefits of doing so."); U.S. Environmental Protection Agency, "National Pollutant Discharge Elimination System—Proposed Regulations to Establish Requirements for Cooling Water Intake Structures at Phase II Existing Facilities," *Federal Register* 67 (proposed Apr. 9, 2002): 17122 (to be codified at 40 C.F.R. pts. 9, 122–25) ("In general, the more sensitive or biologically productive the waterbody, the more stringent the requirements proposed as reflecting the best technology available for minimizing adverse environmental impact.") [hereinafter, "Proposed Regulations"].

24. "Proposed Regulations," *Federal Register* 67: 17,168 (emphasis added). A similar slip occurred with respect to non-impingement and entrainment environmental effects, which EPA claimed would be "acceptable at the national level." "Final Regulations," Federal Register 69: 41586. Again, the standard is minimization of environmental harms, not acceptability.

25. "Proposed Regulations," *Federal Register* 67: 17136.

26. *See* Lisa Heinzerling, "Statutory Interpretation in the Era of OIRA," *Fordham Urban Law Journal* 33 (2006): 1097.

27. "Final Regulations," *Federal Register* 69: 41601.

28. *Id.*

29. Riverkeeper II, 475 F.3d at 97–98. *See also* Heinzerling, *supra* note 26, at 1102.

30. Riverkeeper II, 475 F.3d at 99 (quoting *Am. Textile Mfrs. Inst. Inc. v. Donovan*, 452 U.S. 490, 509 (1981)).

31. Riverkeeper II, 475 F.3d at 100.

32. *Kennecott v. United States EPA,* 780 F.2d at 445, 448 (4th Cir. 1985).

33. *See also* Riverkeeper II, 475 F.3d at 104.

34. *See* U.S. Environmental Protection Agency, "National Pollutant Discharge Elimination System—Suspension of Regulations Establishing Requirements for Cooling Water Intake Structures at Phase II Existing Facilities," *Federal Register* 72 (July 9, 2007): 37107.

35. Riverkeeper I, 358 F.3d at 185.

36. Robert N. Stavins, Letter to Water Docket ID no. OW-2002-0049, Re: Comments on Proposed Rule RIN 2040-AD62; Notice of Data Availability Clean Water Act Section 316(b) —National Pollutant Discharge Elimination System—Proposed Regulations to Establish Requirements for Cooling Water Intake Structures at Phase II Existing Facilities, at 4 (April 21, 2003) [hereinafter, Stavins 4/21/2003 Letter].

37. Final Rule Economic Benefit Analysis (EBA), at A2-5.

38. "Final Regulations," *Federal Register* 69: 41,603 (emphasis added). *See also id.* at 41,591 (stating that EPA compliance alternatives were designed "to ensure that the rule requirements are economically practicable"); id. at 41596 ("EPA was not able to identify a uniform set of technologies that would be available and economically practicable....").

39. Robert N. Stavins, Letter to Proposed Rule Comment Clerk—W-00-32, Re: Comments on Proposed Rule RIN 2040-AD62 Clean Water Act Section 316(b)—National Pollutant Discharge Elimination System—Proposed Regulations for Cooling Water Intake Structures at Phase II Existing Facilities, EPA ICR no. 2060.01, at 9 (July 19, 2002) [hereinafter, Stavins 7/19/2002 Letter].

40. "Final Regulations," *Federal Register* 69: 41,586.

41. U.S. Environmental Protection Agency, *Regional Analysis Document for the Final Section 316(b) Phase II Existing Facilities Rule,* EPA-821-R-02-003 (Washington, DC: EPA, 2004), A9-1, available at www.epa.gov/water science/316b/phase2/casestudy/final.htm [hereinafter, Final Rule RS].

42. Final Rule EBA, at C3-2. *See also* "Final Regulations," *Federal Register* 69: 41,661 ("The Agency's direct use valuation does not account for the remaining 98.2% of the age 1 equivalent aquatic organisms estimated to be protected nationally under today's rule.").

43. U.S. Environmental Protection Agency (EPA), *Case Study Analysis for the Proposed Section 316(b) Phase II Existing Facilities Rule,* EPA-821-R-02-001 (Washington, DC: U.S. EPA, 2002), A8-1 (emphasis added), available at www.epa.gov/waterscience/316b/phase2/casestudy/ [hereinafter, Proposed Rule CS].

44. Final Rule EBA, at A2-1.

45. Final Rule RS, at A9-8.

46. Final Rule EBA, at D1-5.

47. "Final Regulations," *Federal Register* 69: 41,660 (quoting Myrick Freeman).

48. U.S. Environmental Protection Agency (EPA), *Economic and Benefits Analysis for the Proposed Section 316(b) Phase II Existing Facilities Rule,* EPA-821-R-02-001 (Washington, DC: EPA, 2002), D1-4, available at www.epa.gov/water science/ 316b/phase2/econbenefits/.

49. Stavins 7/19/2002 Letter, at 10 (emphasis added).

50. Frank Ackerman, Comments on Proposed Rule, RIN 2040-AD62 Clean Water Act §316(b)—National Pollutant Discharge Elimination System—Proposed Regulations for Cooling Water Intake Structures at Phase II Existing Facilities, EPA ICR no. 2060.01, August 1, 2002, at 11–13 (citing Thomas C. Brown, "Measuring Nonuse Value: A Comparison of Recent Contingent Valuation Studies," in *Benefits and Costs Transfer in Natural Resource Planning, Sixth Interim Report,* ed. J.C. Bergstrom (Athens, GA: University of Georgia, Department of Agricultural and Applied Economics, 1993).

51. Frank Ackerman and Rachel Massey, Comments on Notice of Data Availability, EPA 40 CFR Part 125 Clean Water Act §316(b)—National Pollutant Discharge Elimination System—Proposed Regulations for Cooling Water Intake Structures at Phase II Existing Facilities; Notice of Data Availability, March 19, 2003, June 2, 2003, at 4.

52. Stavins 4/21/2003 Letter, at 3 (emphasis added).

53. Id. at 2.

54. "Final Regulations," *Federal Register* 69: 41624.

55. Final Rule RS, at A13-13.

56. "Final Regulations," *Federal Register* 69: 41586.

57. Id. at 41,624.

58. Id. at 41,587 (observing that threatened, endangered, and other special status species might be impacted by cooling water intake and noting, as an example, "that 3,200 threatened or endangered sea turtles entered enclosed cooling water intake canals at the St. Lucie Nuclear Generating Plant in Florida").

59. Id. at 41,612.

60. Proposed Rule CS, at A11-1 ("Conventional techniques to value the benefits of technologies that reduce [impingement and entrainment] losses at § 316(b) facilities can omit important ecological and public services.").

61. Id. at A11-2.

62 Stavins 7/19/2002 Letter, at 3, 15.

63. Id. at 25.

64. Ackerman, *supra* note 50, at 14.

65. See Ulla Grapard, "Robinson Crusoe: The Quintessential Economics Man?" Feminist Economics 1 (1995): 33.

66. *See* Bernadette A. Meyler, "Daniel Defoe and the Written Constitution," *Cornell Law Review* 94 (2008): 73 (citing Ian Watt, *Myths of Modern Individualism: Faust, Don Quixote, Don Juan, Robinson Crusoe* (Cambridge, UK: Cambridge University Press, 1996).

67. Daniel Defoe, *Robinson Crusoe,* ed. Michael Shinagel (New York, NY: W.W. Norton and Company, Inc., 1993), 76.

68. 33 U.S.C.A. § 1251(a). *See also Rapanos v. United States* 547 U.S. 715 (2006).

69. *See generally* Amartya Sen, *Commodities and Capabilities* (Amsterdam, the Netherlands: North–Holland, 1985).

70. Stavins 7/19/2002 Letter, at 17.

71. "Final Regulations," *Federal Register* 69: 41610.

72. *Id.* at 41,610.

73. *See* Douglas A. Kysar, "Law, Environment, and Vision," *Northwestern University Law Review* 97 (2003): 675, 690.

74. Robert M. Solow, "The Economics of Resources or the Resources of Economics," American Economic Review: Papers and Proceedings 64 (1974): 1, 11.

75. *See* Daniel H. Cole, "'Best Practice' Standards for Regulatory Benefit–Cost Analysis," *Research in Law and Economics* 23 (2007): 6; Lisa Heinzerling, "Discounting Our Future," *Land and Water Law Review* 34 (1999): 39.

76. Cole, supra note 75, at 15 (quoting Office of Management and Budget, *Informing Regulatory Decisions: 2003 Report to Congress on the Costs and Benefits of Federal Regulations and Unfunded Mandates on State, Local, and Tribal Entities* (Washington, DC: OMB, 2003), 176.

77. *See* Douglas A. Kysar, "Discounting... on Stilts," *University of Chicago Law Review* 74 (2007): 119.

78. Office of Management and Budget, *Informing Regulatory Decisions: 2003 Report to Congress on the Costs and Benefits of Federal Regulations and Unfunded Mandates on State, Local, and Tribal Entities* (Washington, DC: OMB, 2003), at 36.

79. *See* Derek Parfit, *Reasons and Persons* (Oxford, UK: Oxford University Press, 1986). *See also* Tyler Cowen and Derek Parfit, "Against the Social Discount Rate," in *Justice between Age Groups and Generations,* ed. Peter Laslett and James S. Fishkin (New Haven, CT: Yale University Press, 1992), 144, 144–61.

80. *See* Final Rule RS, at A14-1 ("Discounting refers to the economic conversion of future benefits and costs to their present values, accounting for the fact that individuals tend to value future outcomes less than comparable near-term outcomes.").

81. *See* Matthew D. Adler and Paul Dolan, "Introducing a 'Different Lives' Approach to the Valuation of Health and Well-Being" (University of Pennsylvania, Institute for Law & Economics Research Paper no. 08-05, March 19, 2008), available at http://papers.ssrn.com/sol3/papers.cfm?abstract_id=1112636.

82. "Proposed Regulations," *Federal Register* 67: 17,138. For instance, one study using conservative entrainment survival assumptions estimated that each year up to 20 percent of striped bass, 25 percent of bay anchovy, and 43 percent of Atlantic tom cod would be lost to entrainment in the Hudson River estuary from just three power plants. *See id.*

83. Final Rule RS, at A13-12.

84. "Final Regulations," *Federal Register* 69: 41624–41625.

85. *See id.* at 41,664 (indicating a break-even value of $5.07 annual willingness to pay per household at the national level).

86. *See* Proposed Rule CS, at E5-1 (noting that, because of data, budget, and time constraints on conventional valuation techniques, "EPA is pursuing an approach that uses actual sums of money which society has dedicated to restoring and preserving [threatened and endangered] species of fish as an indication of society's revealed preference valuation for protecting those species.").

87. Richard D. Gary and Michael L. Teague, "The Inclusion of Externalities in Electricity Generation Resource Planning: Coal in the Crossfire," *West Virginia Law Review* 95 (1993): 839, 851 (quoting Massachusetts Department of Public Utilities, Integrated Resource Management Practices, 116 Pub. Util. Rep. 4th (PUR) (Boston, MA: Mass. DPU, 1991): 67, 91. *See also* Stephen S. Bernow and Donald B. Marron, *Valuation of Environmental Externalities for Energy Planning and Operations,* Update 3-4 (Boston, MA: Tellus Institute, 1990).

88. Stavins 7/19/2002 Letter, at 3.

89. *See* Mark Sagoff, *The Economy of the Earth: Philosophy, Law, and the Environment* (Cambridge, UK: Cambridge University Press, 1988), 7–14.

90. *See* Office of Management and Budget, "Draft 2003 Report to Congress on the Costs and Benefits of Federal Regulations," *Federal Register* 68 (Feb. 3, 2003): 5492, 5499.

91. W. Kip Viscusi and Richard J. Zeckhauser, "Sacrificing Civil Liberties to Reduce Terrorism Risks," *Journal of Risk and Uncertainty* 26 (2003): 99, 104–105.

92. Id. at 105, Table 1.

93. Stavins 7/19/2002 Letter, at 25 (emphasis in original).

94. *See* President, "Executive Order no. 13,422," *Federal Register* 72 (Jan. 23, 2007): 2,763–2,765; Office of Management and Budget, "Final Bulletin for Agency Good Guidance Practices," *Federal Register* 72 (Jan. 25, 2007): 3,432.

95. *See, e.g.,* Stavins 7/19/2002 Letter, at 20 ("There is a consensus view in economics that when an appropriate and reliable revealed-preference approach is available for valuing a particular environmental amenity, then that approach should be used, rather than resorting to a stated preference approach, such as contingent valuation.").

96. *See* John A. List, Robert P. Berrens, Alok K. Bohara, and Joe Kerkvliet, "Examining the Role of Social Isolation on Stated Preferences," *American Economic Review* 94 (2004): 741.

97. Cf. Douglas A. Kysar, "Preferences for Processes: The Process/Product Distinction and the Regulation of Consumer Choice," *Harvard Law Review* 118 (2004): 525.

98. Amartya Sen, "The Discipline of Cost–Benefit Analysis," *Journal of Legal Studies* 29 (2000): 931, 949.

99. *See* Stavins 4/21/2003 Letter, at 18 (contending that Sen offered "not a commentary on peoples' valuations of public goods (indirectly estimated by valid willingness-to-pay or willingness-to-accept measures), but a commentary on actual voluntary payments for the use of public goods").

100. Amartya Sen, "Why We Should Preserve the Spotted Owl," *London Review of Books,* 26 (2004): www.lrb.co.uk/v26/n03/sen_01_.html.

101. Id.

102. Clive L. Spash, "Deliberative Monetary Valuation (DMV): Issues in Combining Economic and Political Processes to Value Environmental Change," *Ecological Economics* 63 (2007): 690.

103. *See, e.g.,* Riverkeeper II, 475 F.3d at 105, 113 n. 25. Indeed, EPA argued that judicial interpretations of Section 316(b) are consonant with its own in the final Phase II rule. See "Final Regulations," *Federal Register* 69: 41,626 (citing *Seacoast Anti-Pollution League v. Costle,* 597 F.2d 306, 311 (1st Cir. 1979). The referenced case, however, only upheld EPA's position at that time that costs need not be imposed that are "wholly disproportionate to any environmental benefit." Id. Thus, the case provides no support for EPA's more ambitious present claim that Section 316(b) authorizes full CBA.

104. Riverkeeper II, 475 F.3d at 99 (quoting *Natural Res. Def. Council, Inc. v. EPA,* 822 F.2d 104, 123 (D.C. Cir. 1987)).

105. Sidney A. Shapiro and Christopher H. Schroeder, "Beyond Cost–Benefit Analysis: A Pragmatic Reorientation," *Harvard Environmental Law Review* 32 (2008): 433 (quoting James Boyd White, "Law as Rhetoric, Rhetoric as Law: The Arts of Cultural and Communal Life," *University of Chicago Law Review* 52 (1985): 684, 701).

CHAPTER 11

What We Learned

WINSTON HARRINGTON, LISA HEINZERLING, AND RICHARD D. MORGENSTERN

This report began by noting some of the controversies surrounding the use of economic methods to evaluate the benefits and costs of new environmental regulations, including the concern by some about the excessive focus on economic efficiency criteria, the limited ability to quantify health and environmental damages, and quite fundamental questions about the monetization of these effects. While recognizing the importance of these issues, we have deliberately placed some of the broader questions beyond the reach of this volume, and have chosen instead to focus on what we believe to be the most tractable question, namely the development and use of regulatory impact analyses (RIAS) by the U.S. Environmental Protection Agency (EPA). A key goal of an RIA should be to help inform regulators, Congress, and the general public about the expected consequences—both the benefits and the costs—of pending decisions.

To provide focus, we decided to examine as case studies three recent, relatively sophisticated RIAS conducted by EPA, and to engage experts, both economists and lawyers, with diverse perspectives on the issues. Our process involved the development of in-depth critiques of the three RIAS, with an opportunity for debate among the authors and outside reviewers, including academic, private, and government experts.

At the time of case selection, each of the three rules chosen had been appealed by various stakeholders, but only one outcome had been reached (the cooling water rule had recently been invalidated by a federal appeals court). Since then, our study has achieved a rare trifecta: all three rules have been overturned by the courts, sometimes for reasons explicitly linked to the economic analyses.

In choosing to dig deeply into individual RIAS, we hoped to focus on the current practice of cost–benefit analysis (CBA) in the regulatory process and to downplay the strictly textbook or philosophical issues that sometimes surround debates about the use of the technique. At the outset, we stipulated that the objective was not to defend or attack CBA, but to improve its use in environmental decisionmaking. Thus, we assumed, as others have, that CBA is here to stay. Our goal is to improve the quality, acceptability, and usefulness of the analyses that are undertaken.

Whereas the challenge for the authors of the RIA critiques was to assess the individual studies conducted by EPA, we the editors took on the task of preparing comparisons and developing a set of recommendations for changes to current practices on which the three of us could jointly agree. At the outset, we recognized that it might not be feasible to reach consensus among ourselves and that any consensus we did reach might not represent meaningful reform.

This chapter presents the results of our work. We make no claims about the revolutionary nature of our recommendations. At the same time, we believe that they are both substantive and achievable, and that by embracing them, EPA, and possibly other agencies, could improve the overall credibility and usefulness of RIAS.

A natural starting point is to review, in brief, the assessments of the three individual RIAS, comparing and contrasting the views expressed by the various chapter authors. From there, we launch directly into our recommendations for reform.

Summary of RIA Critiques

Clean Air Interstate Rule

Nat Keohane and Wendy Wagner develop in-depth and broad-ranging assessments of the RIA prepared by EPA for the Clean Air Interstate Rule (CAIR), a regulation designed to achieve major reductions in power plant emissions of sulfur oxides and nitrogen oxides. In general terms, Keohane and Wagner both see the RIA as a quite competent example of CBA in many respects, including use of clear and consistent baselines, consideration of various categories of benefits and costs, and an innovative treatment of uncertainty. At the same time, they both strongly criticize the RIA for its failure to consider alternative options. They also see the RIA as somewhat unfocused, devoting excessive attention to the estimation of some very small benefit categories, such as emergency room visits for asthma and lower and upper respiratory symptoms in children, and virtually ignoring potentially major issues, such as ozone mortality and ecological damages.

The exclusive focus on the particular policy option selected by EPA rather than on a broader set of alternatives, as mandated in the agency's RIA Guidelines for Preparing Economic Analyses, is identified by Keohane and Wagner as the RIA's major flaw.[1] This failure to consider alternatives is all the more surprising considering that the agency had already prepared assessments of competing proposals in unsuccessful efforts to advance the Bush administration's Clear Skies legislation. Wagner goes so far as to label the RIA as principally a litigation-support document, albeit a technically sophisticated one, rather than the genuine aid to decisionmaking envisioned by RIA advocates. Keohane sees the single-option focus as an attempt to mask the greater net benefits that could probably have been achieved by a more stringent standard. Further, he notes that EPA's approach precludes development of a cost-effectiveness analysis that, ironically might have strengthened the legal underpinnings of the rule.

Keohane and Wagner both express concern that the excessive technical complexity of the document limits its usefulness for nonexperts in the field. Thus, they believe it fails in what should be one of the RIA's key objectives: providing clear, transparent information to Congress and the general public about the true societal impacts of the CAIR.

Beyond the similarities in their assessments, the issues raised by Keohane and Wagner also differ in important ways. Keohane focuses on various technical aspects of the RIA, including both benefit and cost estimations, the consideration of equity and the differential impacts among subpopulations, the discounting of delayed benefits and costs, and the treatment of uncertainty. He argues that, because even a simple assessment would reveal the presence of large net benefits from the chosen option, little is gained from the excessive detail presented, especially so late in the regulatory process. As an alternative to this false precision, Keohane favors a different focus, namely a simpler, more straightforward study that would be accessible to a broader audience. For example, he favors greater use of physical units, in addition to monetary terms, for estimating benefits.

Wagner takes a somewhat different tack in her proposals for reform. Given her assessment that the RIA is largely designed to protect the agency against legal challenge, she seeks to develop institutional incentives to make the document more relevant to actual decisionmaking. In that regard, she would try to separate the RIAS from judicial review as much as possible. For example, she would reward agencies for high-quality analyses, perhaps by attaching a strong presumption in favor of the policy choices made in the rulemaking—well beyond the "arbitrary and capricious" standard commonly used. Wagner also endorses development of a set of criteria that could help determine whether an RIA met the high-quality standard that would qualify it for more deferential judicial treatment. Further, she would require that the RIA be completed at a much earlier point in the rulemaking process. Consistent with Keohane's approach, Wagner would also encourage more qualitative assessments and a greater emphasis on estimates denominated in natural units rather than monetary terms. Interestingly, Wagner compares this emphasis on early, less technical analyses to the scoping documents prepared for environmental impact statements under the National Environmental Policy Act. Very much like Keohane, Wagner favors a more open, transparent process for decisionmaking and the development of documents to support such a process.

Clean Air Mercury Rule

Alan Krupnick and Catherine O'Neill both present detailed reviews and analyses of the RIA prepared by EPA for the Clean Air Mercury Rule (CAMR), a regulation designed to cut power plant mercury emissions via a cap-and-trade approach. Not surprisingly, Krupnick and O'Neill agree in their assessments of a number of the RIA's shortcomings. At the same time, there are also important disagreements between them on a range of technical issues as well as on the basic economic efficiency approach adopted in the analysis.

Focusing first on examples of agreement, chapter authors Krupnick and O'Neill acknowledge the daunting task of analyzing the benefits and costs of mercury emissions controls, given the complexities and uncertainties of the underlying problem and the evolving nature of the available scientific information. Nonetheless, they both highlight the very limited set of options considered in the RIA, including the sole focus on emissions trading, and the failure to analyze the costs and benefits of adopting maximum achievable control technology (MACT) standards, especially in light of the prior determination by the Clinton administration that mercury is a hazardous air pollutant as defined under section 112 of the Clean Air Act. They also highlight the failure to consider alternative baselines—for example, treating the benefits and costs of CAIR as ancillary to CAMR, rather than solely defining CAMR as ancillary to CAIR. Further, both authors chastise the EPA for the virtually contemporaneous issuance of the RIA and the underlying regulation, thus undercutting the use of the RIA as a decision tool.

Krupnick and O'Neill also agree on a number of the limitations of the exposure analysis, including the emphasis on freshwater fish consumption, and the failure to consider damages other than IQ loss. With regard to benefits monetization, both authors raise concerns about the netting out of educational costs from the estimates of reduced lifetime earnings attributable to IQ loss. On this point, O'Neill cites Professor Rena Steinzor: "...the good news is that stupider children need less school and earn just a little more money because they are working rather than sitting in a classroom."[2]

Beyond the similarities in the issues raised by Krupnick and O'Neill, their disagreements fall into two categories: technical and philosophical. From a technical perspective, O'Neill questions the selection of studies chosen for inclusion in the RIA, noting that in the Bush administration, all judgment calls went "one way." Krupnick focuses more on likely errors of omission in study selection without suggesting bias. O'Neill sees the failure to quantify or monetize certain benefit categories as a fundamental flaw, whereas Krupnick sees it more as a sign that too few resources were committed to following recent National Research Council recommendations on this issue.

Another technical difference involves their approaches to distributional issues. O'Neill focuses on the high baseline mercury blood-level concentrations among Chippewa or other ethnic populations that have a strong identification with freshwater fish consumption. She is concerned that the RIA does not explicitly consider who will bear the costs and benefits of the rule, nor whether the decision ameliorates or worsens current inequities. She is also concerned that the delays involved in implementing CAMR rather than the MACT standard will cause permanent harm to millions of children. Krupnick hones in on the issue of emissions trading, specifically whether the use of trading will create "hot spots." Using publicly available EPA data, he finds that for the vast majority of plants, there are no increases in mercury emissions compared to the no-control baseline with or without CAIR in place. At the same time, he does find that several plants are predicted to increase mercury emissions over their baselines when emissions trading is allowed. He chides the EPA for failing to exploit its own data on this highly contentious issue.

Finally, we note the not insignificant philosophical divide between Krupnick and O'Neill. Overall, Krupnick sees the CAMR RIA as a quite reasonable approach in light of data limitations, as well as budget and time constraints. In general, he sees the deficiencies—which are extensive in some cases—as inherently remediable with greater effort on the part of the agency. He recommends allotting more time and resources to enable EPA to collect better studies, to rely less on (arbitrary) assumptions and more on actual data, and to more fully explore relevant policy options. He also proposes that the agency adopt a more thorough, academic-style peer review process.

In contrast, O'Neill calls for a more interdisciplinary approach to the analysis of regulatory issues, with less emphasis on economic efficiency and without use of a single analytical approach purporting to incorporate all considerations. Importantly, O'Neill sees certain resources such as mink and loons, which serve as Chippewa clan symbols, as the type of priceless resources that should neither be ignored nor subjected to traditional cost–benefit analysis. In short, whereas Krupnick seeks greater technical sophistication to enhance the usefulness of the RIA in agency decisionmaking, O'Neill seeks to employ the tools of multiple disciplines to enhance the economic analyses and to make the RIA more accessible to the broader (nonexpert) public.

Cooling Water Intake Structures Rule

The U.S. Supreme Court is now reviewing the decision by the Court of Appeals for the Second Circuit to remand the Cooling Water Intake Structures (CWIS) Phase II rule. As Doug Kysar explains in his chapter, the sole issue before the court is whether a cost–benefit comparison will be allowed in the determinations of the best technology available (BTA) for individual permittees. It bears repeating that the retail-level comparison of benefits and costs goes far beyond the requirements imposed by Executive Order (EO) 12866 to ensure that the benefits of regulations justify their costs. This instruction applies at the rulemaking stage and refers to the comparison of total benefit and cost estimates of a new rule. But, although not required, the case-by-case cost–benefit comparison is much closer to the economists' notion that benefits and costs should always be compared at the margin.

The cost–benefit test was not the only issue discussed at the appeals-court level. Also at issue was whether the statutory language authorizing the CWIS rule supports the kind of flexibility that EPA inserted into the rule. This flexibility included not only the site-specific comparison of benefits and costs, but also allowed as abatement technologies what looked to environmentalists like impermissible compensatory measures, in particular the use of habitat restoration measures elsewhere in the river reach to compensate for ecological and environmental damages at the site of the CWIS.

The Second Circuit's opinion was pretty clear that this level of flexibility was over the line. For one thing, the same court had remanded Phase I of the rule (applying to new plants) a few years previously because it allowed more flexibility than the court considered statutorily appropriate, and Phase II had attempted to adopt an even more flexible approach to existing plants. The statutory requirement to use the "best technology available" might admit of some wiggle room with respect to the economic hardship imposed on the utility industry and its investors, but seemingly much less so with respect to the environmental consequences (at least according to the Second Circuit).

Both chapter authors Scott Farrow and Doug Kysar identified flaws in the CWIS RIA, but on the whole they seemed to take opposing positions as to the overall effectiveness and value of the document. Farrow's critique is largely restricted to the CBA itself and is generally the technical assessment of a professional economist. Because there are so many criteria for judging, Farrow subjects the RIA to a contents "checklist" developed by Robert Hahn and Patrick Dudley (2007). As Farrow points out, a "good" score on such a checklist does not necessarily indicate a high-quality RIA, but a poor score certainly indicates a bad one. The score for the CWIS RIA is quite high; it is missing or lacking in only a few elements considered essential by economists. Farrow even suggests that as a professional document, the quality of the analysis may achieve a standard required for publishing in a professional journal, were it not for the document's great length. However, Farrow has more reservations about the quality of the data supporting it.

As an aid to decisionmaking, Farrow sees the document as mixed. He argues in particular that EPA based much of the regulatory flexibility introduced into the rule on the adverse environmental impact (AEI) of the rule without ever defining the AEI. Admittedly, this would have been very difficult because the agency did identify a long list of potential impacts qualitatively; it just did not try to provide weights so that an overall assessment of the AEI could be made. Providing such weights for a comparison of benefits with costs and with other benefits is one of the principal objectives of CBA. In this case, so many categories of benefits were left nonmonetized that the cost–benefit comparisons were not meaningful. However, considering the analysis instead as an exercise in cost-effectiveness, Farrow found it to be more valuable. EPA claimed that, compared with cooling towers—a technology that achieves at least a 90 percent reduction in aquatic mortality—a wire screening technology for water intakes would, where applicable, achieve better than half the level of reductions at about 10 percent of the cost. EPA asserted that, under most circumstances, this level of protection could meet the BTA standard, even though it was possible to do much better.

Kysar's assessment of the rule is both broader and more critical. Deeply skeptical of the value of CBA to begin with, he found little in this RIA to change his mind. Three issues in particular concerned him. First, after EPA had prepared a proposed regulation that relied much more heavily on regenerative cooling as the BTA, the Office of Management and Budget (OMB) intervened, evidently encouraging EPA to provide more flexibility in the definition of BTA to reduce cost. In Kysar's view, the OMB intervention went well beyond the regulatory review required by EO 12866. In addition, the substance of OMB's comments and revisions were not made part of the public record, so it was difficult for parties interested in the outcome of the rulemaking to discover essential elements of the decision process. Moreover, these inputs were not subject to public comment during the rulemaking process.

Second, Kysar complains that the technical information made available to EPA (and made available by EPA) was not adequate to support valuation. As he puts it in his chapter, "CBA... carries an implicit assumption that the policy space within which EPA operates is informationally rich and probabilistically sophisticated," pointing to the fact that, of 554 facilities subject to the regulation, only 150 had performed impingement and entrainment studies. A defender of the regulation might point out that a sample of 150 out of 554 is frequently adequate to draw conclusions, depending on how the sample is drawn, but in fact these studies focused only on the direct and indirect mortality of the species subject to impingement and entrainment and not on the larger questions of local and global ecosystem effects. EPA acknowledges and provides a long list of such ecological effects (reproduced in Kysar's chapter), about which little is known and which are unquantifiable based on current information.

On the other hand, steam-electric plants with once-through cooling systems have been in place on the nation's water bodies for nearly 100 years and constitute only one of myriad environmental insults to aquatic systems. The fact that they are still capable, according to EPA, of destroying billions of organisms each year, along with the fact that on most water bodies a relatively small portion of the flow (of a stream) or volume (of a lake or estuary) passes through the steam plant, suggests that the aggregate biota of these water bodies is still very large. This, in turn, may suggest that those systems are not in long-term crisis, at least not from steam-electric generation.

Kysar's third point concerns valuation, and he suggests two alternatives to the individual willingness-to-pay (WTP) approach that is the intellectual backbone of CBA. One, a proposal to value environmental assets at their replacement costs is not likely to find acceptance among economists in academia or at EPA because it assumes what is at issue, namely that the threatened resource is worth saving. Yet statutes like the Clean Water Act take as their premise that natural resources—such as rivers, lakes, and streams—are indeed worth protecting. A methodology aimed at evaluating individual regulations may perform best when it respects this foundational policy determination. The other proposal is to substitute a measure of WTP that is partially or totally determined by group interactions. Some observers in political science, psychology, and similar disciplines believe that this would usefully replace the private, utility-based method with one more appropriate to the valuation of public goods by emphasizing the collective nature of these decisions. As discussed elsewhere in this chapter, group-mediated valuation need not result in an increase in valuation estimates.

Perhaps more critically, the question of valuation gets back to the question of information. If information about the physical consequences of regulatory actions is nonexistent or inadequate, what is the point of valuation? When both valuation and physical-effects data and methods are less than adequate, which offers the largest marginal improvement?

Recommendations for Reform of RIAs

In considering the three RIAS analyzed in this volume, and drawing on our experiences in the field of regulatory assessment, we have developed a series of specific reforms that we believe would enhance the overall quality and usefulness of the substantial studies that are conducted as part of the regulatory development process. We develop a dozen recommendations addressing the content of the RIAS as well as the process by which they are prepared. These recommendations cover five overarching topics:

- technical quality of the analyses;
- relevance to the agency decisionmaking process;
- transparency of the analyses;
- treatment of new scientific findings; and
- balance in both the analyses and the associated processes, including the treatment of distributional consequences.

In addition, we have developed two recommendations involving future research. Most of the recommendations could be implemented by the agency alone, although in a few cases changes in the governing executive order would be desirable. Only one of the recommendations requires statutory reform, specifically of the Paperwork Reduction Act (PRA) of 1995.

Technical Quality of the Analyses

1. Give greater consideration to meaningful alternative policy options

If an RIA is truly designed to inform and guide regulatory decisionmaking—and not, as Wendy Wagner suggests, simply to serve as a litigation support document or, in Nat Keohane's view, only to provide information about the consequences of a regulatory decision made on other grounds—then it must examine a reasonable set of alternative policy options. An RIA that only compares the proposed action to the existing regulation, such as the RIA produced for the CAIR, or that considers only very limited options, such as the one developed for the CAMR, does little to help decisionmakers determine the appropriate course to take.

As noted at the beginning of this report, CBA and the RIAS that embody it are not intended to be decisive in the regulatory process; they are inputs, or tools, rather than dispositive frameworks. Thus, even with a very high-quality RIA, regulators may well end up selecting an approach that is not the most efficient from an economic perspective, as concerns about equity or other factors may drive the decision in another direction. At the same time, given the acceptance in economic circles of the efficiency criterion and the appeal of quantitative analysis even to those outside the cost–benefit world, EPA decisionmakers may be reluctant to adopt a "second-best" approach by choosing a regulatory option that generates fewer net benefits than an alternative. The path of least resistance is to analyze only one alternative and thereby avoid explaining why a different, more efficient, choice has been rejected. However understandable this may be from a bureaucratic or political perspective, we do not believe this approach is consistent with the underlying purpose of the executive orders governing regulatory analysis. Thus, we recommend that meaningful alternative options be analyzed in RIAS. Although it may be tempting to stipulate some minimum number of alternatives to be considered, we prefer to focus on the term *meaningful,* which we define to include the full set of options deemed to be technically feasible and legally defensible.

2. Choice of baselines should reveal choices and trade-offs, not conceal them

The expected outcomes of a regulation cannot possibly be understood without reference to what would have happened in its absence. As a result, expected outcomes are routinely measured against baselines, which represent the development of an intricate set of choices made by the regulator to generate a future or a set of alternative futures that would take place if the rule were not issued. They are also known by the more revealing name of *counterfactuals.*

Constructing a baseline requires a legion of assumptions concerning such matters as future population and economic growth, rates of improvement of existing technologies or replacement by new ones, and trends in future regulation. The credible evaluation of benefits and costs is not possible without a well-constructed baseline or set of baselines. The construction and presentation of baselines are every bit as important to the estimation of net benefits as the construction and presentation of alternative regulations. RIAS should reflect that reality.

A vivid example can be found in Catherine O'Neill's case study in this report of mercury emissions from coal-fired power plants. Control of airborne mercury emissions was widely anticipated under the new MACT standards enacted as part of the Clean Air Act of 1990. Although EPA did promulgate MACT rules for two important sources (municipal and hospital waste incinerators) in the late 1990s, and began work on a third (emissions from electric power generation) in 2003, agency analysts involved in the technical and economic aspects of the utility MACT rule were instructed by top management to stop their work. Instead, they were to begin drafting a new rule based not on the MACT section of the statute, but on a cap-and-trade policy modeled after the sulfur dioxide (SO_2) trading program for fossil electric plants.

The initial regulation, like all MACT regulations, would be required by statute to achieve the emissions reduction performance of the top 12 percent of existing plants and was expected to be implemented around 2007. Its replacement, the CAMR, would only be implemented after the CAIR, a cap-and-trade program for SO_2 emissions that was to be phased in beginning in 2010, with a lower cap to be phased in beginning in 2018. The difference in the performance and timing of the two rules could hardly be more dramatic: whereas the MACT rule would require nearly a 90 percent reduction in mercury emissions by 2007, the CAMR would not achieve its objective of a 70 percent reduction until nearly 2030.

We take no position in this report on whether the abandoned MACT rule was or was not a superior rule to the CAMR, which eventually adopted. Certainly, the MACT timetable and stringency would have produced more emissions reductions and would have produced them much sooner, and thus would have produced much greater benefits. But the costs would have been much higher as well. And because the net benefits of the CAMR were negative, at least according to the EPA analysis, moving up and expanding the emissions reductions would only make things worse. Of course, many skeptics of CBA, including O'Neill, would strenuously disagree.

The point is that EPA could and perhaps should have been more informative in the CAMR about the earlier MACT analysis, perhaps including MACT implementation as an alternative to the customary "no policy" baseline. This would have provided a useful historical perspective and made it clear how much broader were the regulatory options than EPA's regulatory documents let on at the time.

3. Develop a checklist of good practices that all RIAs should have, and provide an explanation for missing items

All three of the RIAS examined in this volume violated one or more elements of EPA's Guidelines for Preparing Economic Analyses.[3] Other studies based on larger samples have reported similar findings, including a quite broad range of deviations from the approaches advanced in the guidelines (Hahn and Dudley 2007).

It is not entirely clear why there is such a gap between the agency's guidelines and current practices. Insufficient resources is an oft-cited reason, although it strains credibility to say that after spending more than $1 million to develop a major analytical effort, funds are not available to conduct one or two additional model simulations.

Robert Hahn has long advocated a checklist to assess RIA quality. In fact, it would be fairly simple for an agency to report on its adherence to some basic quality criteria, or to explain why it did not adhere to such criteria. The criteria reported on need not reflect every nuance covered in the guidelines but should focus on certain key topics. For example, they could include some or all of the issues suggested by Hahn and Dudley (2007), as described by chapter author Farrow. Perhaps the Economics Subcommittee of the agency's Science Advisory Board (SAB) could offer guidance on the "top ten" elements to include in an RIA. The EPA administrator could voluntarily report the checklist as a means of strengthening his or her hand with the public, OMB, and the courts, and could present the checklist results in the preamble to a rule, in concert with the actual presentation of the RIA findings. Alternatively, the president, acting through OMB, could require the checklist. In the absence of a sound peer review process, a high score on such a checklist would not provide complete assurance of RIA quality; however, a low score would be a sure indicator of failure. In her chapter in this report, Wendy Wagner proposes that RIAS deemed to be of high quality be given special deference by the courts.

Beyond the use of a checklist, other approaches could be used to encourage quality improvements in RIAS. For example, one could establish a formal review process involving outside experts, based either at EPA or at OMB, to more directly grade or otherwise evaluate RIA quality. Although appealing at many levels, however, such procedures would probably introduce further delays into an already lengthy regulatory development process. Thus, we propose the development of a checklist, with initial implementation to be carried out by EPA, presumably in consultation with the SAB.

Relevance to Agency Decisionmaking Processes

4. Be more strategic about devoting agency resources to the estimation of the benefits and costs of regulation

The value of regulatory analysis, with or without monetary estimates of benefits, is limited by the absence of coverage of important benefit categories. It is also limited by the precision and accuracy of the estimates of the physical effects of regulation. Although these observations may seem obvious, they are sometimes overlooked by both advocates and skeptics of CBA. Sometimes this can result in an overemphasis on certain scientific and economic issues that may not be entirely relevant to the decision. In other cases, the key issues may be underemphasized.

In several of the RIAs considered in this volume, the focus on precision for some relatively low-value benefit categories at the expense of even a rudimentary scoping of other, potentially higher-value categories is inexplicable. For example, both Nat Keohane and Wendy Wagner note the extensive details in the CAIR RIA on emergency room visits for asthma and lower and upper respiratory symptoms in children, and the absence of analysis of major issues such as ozone mortality and ecological damages.

At the same time, except for air quality management for criteria pollutants, most of the research effort into benefits assessment goes into the estimation of WTP for a given environmental improvement. Thus, whereas the models connecting a regulation to its effects are often fairly rudimentary, the WTP estimates are increasingly sophisticated. Certainly among economists, the professional rewards for developing better methods and data for estimating WTP for nonmarket goods exceed the rewards for linking regulation to physical outcomes. Similarly, whereas the incentives of natural scientists are to link causes to physical outcomes, they often ignore or devalue the effects that the behavioral responses of firms and individuals to regulation can have on regulatory outcomes. Research into physical effects usually involves interdisciplinary research combining natural and behavioral scientists. As anyone knows who has tried to do it, such research is quite difficult to do.

Skeptics of CBA can be as indifferent to the physical effects of regulation as they are to the monetary benefit estimates. For the skeptics' preferred regulatory alternative—best-technology standards—it often doesn't matter very much what the effects of regulation are.

In our view, the usefulness of RIAs would be enhanced if, at the outset of the rulemaking, an explicit judgment is made regarding the best way to allocate resources toward examining the consequences of the regulation. Regulators rarely have all the information they would like about either physical outcomes or their valuation. But not all information has the same value at the margin, and additional forethought about where the biggest payoffs are would probably be well rewarded. In addition to the current intra-agency review of the analytical plans for RIAs, it might be appropriate to send them to the SAB for review, possibly to a special subcommittee established for such a purpose.

5. Make key aspects of the RIAs available to decisionmakers earlier in the regulatory development process

Under current agency procedures, draft RIAs are required to be circulated to top decisionmakers three weeks in advance of final agency review. This applies equally for proposed and final regulations. Reportedly, these deadlines are often not met. However, even when the internal deadlines are met, important opportunities for constructive use of the RIA results in rule development may be missed.

Typically, key elements of rule design are decided fairly early in the regulatory development process, sometimes by midlevel staff. Based on those early decisions, work is begun on monitoring, data collection, development of enforcement strategies, and related issues. If the RIA subsequently finds that the preferred approach is not the most efficient one, strong internal pressures discourage change.

Accordingly, we propose that agency procedures be modified to require that a preliminary RIA be prepared at least six months in advance of final agency review of proposed and final regulations. Understandably, a preliminary RIA may be incomplete and subject to greater uncertainties than the full study. At the same time, this preliminary RIA would characterize the full set of options being analyzed and would provide at least rough estimates of the benefits and costs of each option. It would also provide an opportunity to assess whether the most important benefit (and cost) categories are being assessed, as in recommendation number four. As noted by Wendy Wagner in her chapter on the CAIR, in some respects, a preliminary RIA would be similar to the scoping analysis conducted under the National Environmental Policy Act.[4]

Transparency of the Analyses

6. Include in RIAs detailed descriptions of expected consequences as physical or natural units, without monetization or discounting

As stipulated in both the Reagan and Clinton executive orders on regulatory review, an RIA is intended to be a document that aids in agency decisionmaking, not only at the level of the technical experts, but also at the level of agency heads and, if it comes to that, the White House. In addition, as Nat Keohane suggests, the RIA could also inform the public about the consequences of agency actions.

These purposes would be promoted if agencies included in their RIAs detailed descriptions of the concrete consequences of their decisions, presented in physical endpoints or natural units rather than solely in monetized and discounted form, at least for the major benefit categories. A key issue is how much detail can be developed with reasonable scientific confidence and at reasonable cost. If, for example, an environmental rule is expected to reduce premature mortality and adverse health conditions, then a range of details about those expected health outcomes may be of interest to decisionmakers; these details might include the expected nature of the death or adverse health condition, the likely age of the populations affected, the likely timing of the effects, and the socioeconomic status of the populations most affected. In cases where a strong scientific basis supports the development of such estimates at a reasonable cost, they should be provided.

In addition, it would be useful to have baseline information on these natural units wherever possible, or to at least include contextual information that gives the reader some perspective on the significance of the changes. For example, if a regulation is expected to reduce the frequency of asthma attacks in sensitive populations, what is the current attack frequency in those populations?

Baseline information of this sort is useful in at least two ways. It allows for a determination of not only the expected absolute change in outcomes, but also the relative or percentage change.

It's true that this baseline information is not relevant to the economic criterion of maximizing net benefits—only the marginal conditions are. But that applies specifically to monetary measures. Because good things gain in value as they become scarcer, the change relative to the baseline matters, and decisionmakers might want to know whether the regulatory proposal is going to reduce bad outcomes by 1 percent or 10 percent, for example. If a regulation is expected to reduce fish mortality, by how much are fish populations expected to change relative to the baseline? For example, if billions of fish are dying each year, it should matter whether you have billions or trillions to start with. In addition, having baseline information can provide a sense of perspective that can aid in assessing the credibility of the estimated changes in outcomes.

Agencies would provide this information in a summary chart just as they currently provide monetized and discounted benefit estimates. EPA's summary tables for the CAIR and the CAMR are good examples of this practice. Indeed, with respect to RIAS on the regulation of the criteria air pollutants, EPA generally does a good job of reporting expected consequences in natural units.

Where regulatory consequences are routinely captured by economic terminology, agencies should continue to supply information about these consequences in economic terms. An agency proposing a rule that will result in greater use of scrubber technology, for example, could report the estimated price of the scrubbers along with the number and expected location, of the scrubbers. But where regulatory consequences are not ordinarily stated in economic terms, where the "price" of a consequence must be divined by reference to complex revealed or stated preference methodologies, the economic description of these consequences should be supplemented by the description of natural units.

For any of the personnel directly involved in decisionmaking—top-level agency officials and White House staff—and even the general public, description of consequences in natural units could serve as useful aids in evaluating agency decisions. Officials unschooled in economics might be confused by the translation of human lives into dollars and by the process of discounting of future illness and other elements of the CBA. Presumably, many would gain additional insights from a comparison of economic costs and tangible consequences expressed in natural units. If the head of EPA, for example, were asked whether average utility customers ought to be asked to pay a penny a day to save billions of fish—an estimate of the cost of the CWIS rule for the typical household (Ackerman and Heinzerling 2003)—she might find this a much more tractable decision than one that invites her to evaluate the economic machinery that EPA deployed to calculate the precise value of those billions of fish. Prominent display of the natural units information will also be helpful to those comfortable with economic valuation because it makes it easier to understand the benefits calculations and judge their credibility. A further advantage of this approach is that it might create added incentives for the agency to develop quantitative estimates of some physical endpoints not typically quantified in the RIAS, such as noncancer health effects.

7. Ensure greater transparency at all stages of the process

As a number of participating authors have noted, RIAS have become huge, dense documents that are almost impenetrable to all but those with training in the relevant technical fields, especially economics. Even to the well-trained eye, RIAS are often opaque; it can be hard to find, for example, exactly what value the agency has placed on human life or exactly which discount rate it has used, over what time interval.

Because an important purpose of RIAS, beyond their use as aids to decisionmaking, is to communicate to Congress and the broader public about the benefits and costs of federal regulations, greater transparency in the analysis would be highly desirable. Accordingly, we recommend that agencies endeavor to make RIAS more comprehensible by nonexpert audiences. Obviously, the complexity of the analysis in RIAS constrains to some extent the degree of transparency that can be achieved. Even so, three quite straightforward changes in practice could considerably improve the transparency of RIAS.

Wherever possible, agencies should use plain English to describe their analysis. They should avoid technical jargon, or at least amplify it with parallel descriptions in plain English. OMB's Office of Information and Regulatory Affairs already monitors agency rules for plainness of speech; it should monitor RIAS for the same quality.

Agencies should use a similar format, across RIAS, to provide information on the key variables in the economic analysis. They should provide this information in the same location in each RIA. For example, in the portion of the RIA describing the benefits analysis for an environmental rule, the value of a statistical life, value of illness, value of ecosystem effects, discount rate, and time interval for discounting should all be presented in the same order and format across RIAS. One perusing many RIAS could then know exactly where to look in the RIA for information on crucial inputs to the analysis.

The executive summaries of most RIAS focus on the conclusions of the analysis rather than on the methods and assumptions used. With the adoption of a standardized format for summarizing the methods and assumptions, as described above, it might be useful to incorporate the same or similar information into the executive summary.

Several of the other reforms we suggest in this chapter (such as the recommendations that the benefits of regulation be expressed in natural units and that agencies complete a checklist relating to the quality of the RIA) also would enhance transparency.

Treatment of New Scientific Information

8. Update EPA guidance documents for RIAs more frequently to reflect significant developments in the literature

As in the natural sciences, the professional literature on environmental economics is evolving at a quite rapid pace. RIAS typically incorporate a range of analytical and empirical findings from the recent economics literature. Failure to incorporate these new findings into the RIAS can lead to biased estimates of benefits and costs.

Although in principle the concern about updating the RIA guidelines applies to virtually all parameters, the most recent examples involve discounting, the value of a statistical life, and cost analysis. In all of these cases, a similar pattern applied: recent research indicated a departure from past studies, yet the guidelines lagged behind. Fortunately, during the preparation of this volume, EPA has acted in a number of cases to update its approaches. At the same time, it is fair to observe that in the interim several RIAS were produced using the older values, which resulted in various biases in the estimates of benefits and/or costs.

Our purpose here is not to debate the individual issues. Rather, we would emphasize the dynamic nature of the economics literature and the corresponding need for EPA to keep abreast of the changes and, when appropriate, update the guidelines.

9. Reform current practices on nonmonetized benefits in a number of ways

EPA should indicate clearly and up front an enumeration of benefits into at least three categories: those that have been monetized, those that have been quantified but not monetized, and those that have neither been quantified nor monetized. This classification should be summarized in an easy-to-read table in the executive summary of the RIA. In case of substantial disagreement or uncertainty regarding which category an effect of a regulation belongs in, it should be further disaggregated, if possible, until the categorization is no longer ambiguous. Comments on the proposed regulation should be explicitly invited on the definition of major expected effects and their categorization.

Encourage the SAB to provide expedited review for new or innovative analyses presumed to be of high quality, including those unpublished studies that have particular relevance to RIAs. Currently, virtually all studies included in EPA's economic and scientific assessments are those that have been published in peer-reviewed journals or accepted for publication by such journals. Excluded are those studies still undergoing peer review, which can sometimes be a quite lengthy process, as well as those that represent solid research but are not deemed sufficiently novel to warrant publication in peer-reviewed journals.[5]

One possible approach to address this problem would be for EPA to encourage the SAB to establish an expedited review process for studies deemed to be potentially important for agency regulatory decisions. EPA should issue guidance on this expedited review process, covering both the nature of the process and the criteria for selecting studies for review. The goal of this expedited review should not be to lower the quality bar for the acceptability of new research, but rather to recognize the complexities of the peer review process and encourage inclusion in RIAs of high-quality research regardless of its publication status.

Consider whether it is better to include some number or distribution of values in place of the default of zero, either as a new scenario or as part of an uncertainty analysis. Notwithstanding the preceding suggestions for the expedited review of economic and scientific papers relevant to regulatory decisions, many regulations will probably still involve some nonmonetized categories of benefits. There are several reasons for this, some unavoidable and some even desirable. First, there may be a consensus that some effects are relatively small and under any reasonable assumption may not contribute much to total benefits. Second, the quantitative effects may be large enough to matter but not well understood or well estimated, in which case proceeding to a potentially arbitrary valuation step will appear to be meaningless twice over. Third, even when estimated, the quantitative effects may be subject to large and possibly asymmetric errors. Estimating WTP for such effects is likely to give misleading results. Climate change is the canonical example; economic estimates using conventional assumptions may greatly underestimate the likely consequences. Fourth, environmental effects may be understood quantitatively, but the link between the regulation and the change in the effect may not yet be established. Similarly, valuation information may be available, but not in a form that links easily to predicted changes in quantitative effects.

The well-known mismatch between water quality indicators, which measure decrements in water quality in contaminant concentrations, and recreational benefits, measured by increases in days spent in various recreation activities, is a case in point.

It is in the cases (and there are many) where total compliance costs exceed monetized benefits that the disposition of the nonmonetized benefits plays a crucial role in the regulator's decision. This reality poses what can be a difficult choice for regulatory decisionmakers: either enter a zero for benefits that have not been monetized, running the risk that they will be ignored by decisionmakers, or use some arbitrary values, if for no other reason than to prevent them from being ignored. Obviously, no regulatory decision strictly requires monetization of all benefits; we pay decisionmakers to make decisions in the hard cases, after all. But still, any perspective the RIA can provide on the potential magnitude of those benefits will be helpful to decisionmakers. In addition to providing potentially valuable information, better description and quantification of the value of nonmonetized benefits will provide explanation and justification for observing stakeholders.

EPA has usually opted for leaving out nonmonetized benefits. We believe there is something to be said for the other approach, heretical as it may be: the inclusion of nonzero benefit values for some benefit categories where such values are not currently supported by empirical benefit studies. At worst, including nonzero benefits in such cases is harmless as long as it is understood by decisionmakers that they are not supported by benefit studies. At best, they can prevent decisionmakers from disregarding such categories, and they can force all parties, from decisionmakers to analysts to stakeholders, to try to think through what numbers might be reasonable. If enough observers think that the potential benefits in such categories are sufficiently large, it may give an impetus for research to try to provide real estimates.

Nevertheless, simply assigning an arbitrary benefit number is not likely to gain instant acceptance among many observers. It is worth considering whether there are defensible approaches to assigning such numbers. Below are some options that may be worth considering, including some that have in fact been employed, at least informally, to assign benefits to previously nonmonetized effects or, at least, to put the benefits in other categories in perspective.

Imputation of necessary benefits. Calculate the implicit value of the nonmonetized benefits that, when added to other benefits, make the regulation a break-even proposition. Like all of the methods proposed here, this approach invites the decisionmaker to subject the benefits claim to his or her own judgment and experience. Inevitably, this approach assigns a single value to the total package of nonmonetized benefits. If many disparate effects remain nonmonetized, it may not be easy for decisionmakers to decide whether the resulting value is worth investing in. In other words, this top-down approach is wanting in the detail that might allow the decisionmaker to make an informed decision.

Expert elicitation. Convene a panel of recognized experts in economic benefit estimation, risk perception, and the appropriate natural sciences, and solicit their views on several matters, including the link between the regulatory options and the environmental improvement and the link between environmental improvement and WTP. This is more of a bottom-up approach, in principle at least, that allows explicit valuation of the individual components. At the same time, it raises a different set of methodological issues having to do with disaggregation. Are the experts to assign a monetary value to all of the benefits in the aggregate? Should they assign values to distinct benefit categories? Should they assign benefits to unit changes or to the aggregate change resulting from the regulation?

The convening of an expert panel brings another issue to the forefront that is worthy of consideration by EPA and indeed by all students of regulation. Should the opinions of the scientific experts be limited to the physical effects of the regulation? Or should their views on the monetary valuation of those effects, or at least what the trade-offs might be with other relevant effects, be accorded special weight? It is customary to solicit valuation from random samples of adults, an approach that makes sense when the benefits being valued are familiar to the average person, such as the valuation of health effects or recreation experiences. But is this practice justified when ecological changes are at stake and the environmental effects are subtle, hard to observe, and not directly connected to matters that people care about on a day-to-day basis? At the same time, is it reasonable to turn such authority over to an unelected panel of experts who may have personal and professional biases that can skew results?

Balance in Both the Analyses and the Regulatory Process, Including the Treatment of Distributional Consequences

10. Promote evenhanded treatment of decisions to regulate, deregulate, and decline to regulate

We recommend that agencies' decisions not to regulate—as well as their decisions to regulate—be subject to regulatory review when they pass the threshold of EO 12866: that is, when they "have an annual effect on the economy of $100 million or more or adversely affect in a material way the economy, a sector of the economy, productivity, competition, jobs, the environment, public health or safety, or State, local, or tribal governments or communities."[6] Because EO 12866 currently applies only to rules, however, agencies' decisions not to regulate at all do not come within the formal terms of the order. Thus, this recommendation would involve amending the executive order to clarify that agency decisions not to regulate are also subject to regulatory review when they meet the triggering conditions of the order. In the case of deregulation, only a change in practice is required. To keep the process manageable, we would propose that these decisions be subject to regulatory review only when they are formal agency announcements, published in the *Federal Register.* This limitation would ensure that the process of regulatory review would not be set in motion by every agency decision that might possibly have an adverse effect on the environment.

This recommendation would respond to a long-standing criticism of regulatory review: decisions to regulate are subject to CBA, yet decisions to deregulate or not to regulate at all do not undergo this formal examination. This one-sidedness introduces a potential bias against regulation into the process of regulatory review that is unwarranted (Olson 1984).

The history on this issue is instructive. When the U.S. Department of Agriculture (USDA) Forest Service in 2002 reversed a Clinton-era initiative protecting almost 60 million acres of roadless areas in the national forests, it maintained that the rule would not "adversely affect in a material way . . . the environment." The Forest Service noted that an RIA had been prepared for the rule being discarded, but stated that it could not produce a quantitative analysis of its new approach because there was "no experience with implementing the roadless rule, and thus there are no data available" (USDA Forest Service 2005, 25649). When EPA issued its first rule relaxing the requirements for the Clean Air Act's New Source Review program, it did not prepare an RIA because it concluded that the rule would not adversely affect the environment (EPA 2002, 1). When the U.S. Department of Interior proposed trimming back the requirements for consultation with the wildlife agencies under the Endangered Species Act and changed regulatory definitions to make the statute inapplicable to effects resulting from climate change, it noted that the action was a "significant rule" within the meaning of EO 12866, but it did not prepare an RIA (U.S. Department of Interior and U.S. Department of Commerce 2008, 47872). Likewise, when the U.S. Department of Interior proposed easing rules regulating mountaintop mining, it stated that the rule would not have an adverse effect on the environment and it prepared no RIA (U.S. Department of Interior 2004, 1045).

Rather than rest with a conclusory and potentially questionable statement that deregulatory actions have no adverse effect on the environment, agencies should undertake the same process of regulatory review for deregulatory decisions as for regulatory ones when those decisions likely will have a material adverse effect on the environment. In principle, RIAs for deregulatory actions should be relatively easy to conduct because a regulatory RIA already exists. Moreover, a decision to deregulate will not come out of thin air, and the fact that a regulation already exists usually means that there is already some real-world experience with it. This experience provides a basis for analysis that is not available to newly proposed regulations, and agencies should report on this experience and use it to motivate their decisions.

Economic logic supports this recommendation for evenhandedness. Likewise, economic logic supports treating decisions not to regulate at all with the same degree of scrutiny as decisions to regulate. There is no more reason to believe, for example, that EPA's outright refusal, in 2003, to regulate greenhouse gases in any fashion promoted efficiency than to believe that its decision to regulate conventional pollutants in the CAIR promoted inefficiency. If one kind of decision deserves economic scrutiny, so does the other.

11. Reform the federal data collection request process

The Paperwork Reduction Act (PRA) of 1995, as well as OMB regulations issued in its name, impose stringent requirements on data collection from firms and individuals. To conduct a survey with more than nine respondents, any federal agency and any organization conducting a project sponsored by a federal agency must submit the survey instrument for public comment and OMB approval. These restrictions are intended to minimize the recordkeeping and survey burden on private citizens and firms and to prevent undue invasion of privacy. They apply not only to mandatory data collections, such as Internal Revenue Service forms and EPA data requests to support regulatory development, but also to voluntary participation in WTP surveys if those surveys are supported by federal grants and contracts. In addition, the PRA and OMB regulations require that surveys "must be adequately designed and justified, with an opportunity for public comment" (OMB 2005, 51).

Many researchers who do federally supported research on the benefits and costs of regulations, as well as federal agency personnel who are responsible for developing and supporting economically justified regulations, report horror stories about extensive delays in getting surveys approved. Virtually all would agree that the required public comment on surveys is an unwarranted and unwelcome intrusion on research autonomy. Most would further agree that the PRA and OMB's interpretation of its requirements are a little too energetic and could be tweaked to make data collection to support regulation more efficient without compromising the goals of the PRA. Below we offer some possible solutions.

Exempt or relax voluntary surveys from the survey size restrictions in some cases. Arguably, there is little distinction between voluntary and mandatory surveys of firms regulated by an agency. What regulated firm wishes to risk being regarded as "uncooperative" by its regulator? However, that issue does not apply to the main concern raised here, namely, surveys of attitudes of and WTP for public goods by private citizens, because such surveys offer little possibility of coercion. In addition, OMB approval is required for surveys that do not directly support regulation; this requirement should be reviewed.

Limit OMB technical review of some survey instruments. No doubt OMB review has limited poor research designs and weak sampling methods in some cases. However, OMB has also rejected surveys where the issues involve unsettled methodological controversies. In some instances, for example, researchers have been denied the use of controlled web-based surveys because they are not in OMB's view properly randomized. OMB has also rejected the use of cash incentives for completed surveys. Both practices are generally accepted by social science researchers as the only cost-effective ways to recruit an adequate sample and achieve an acceptable response rate.

Eliminate or severely restrict the public comment requirement on WTP surveys, possibly replacing it with a peer review requirement. Most issues of experimental design are quite technical in nature, and self-selected laypersons rarely have much useful to add. Not surprisingly, comments often reflect interest group positions rather than independent professional judgments. However, it might not be inappropriate for OMB to request reviews from qualified professionals, or to invite commentary from all members of relevant scientific disciplines.

Replace OMB review of survey instruments and methodology with peer review by technically qualified persons outside of the federal government. OMB is regarded by many regulatory stakeholders as a nonneutral party, generally hostile to most social regulations. To some extent, these attitudes may be inevitable given OMB's executive and statutory role as regulatory gatekeeper, but it is not clear that the gatekeeping function should extend to survey quality. Indeed, it sits uneasily with OMB's recently acquired responsibilities under the Data Quality Act. It is strange for OMB to be, in effect, limiting the acquisition of information through surveys on the front end of the regulatory process and then criticizing the poor quality of regulatory information later in the process.

12. Consider interactions between the distribution of regulatory costs and benefits

Distributional consequences of regulation are important. At present, however, EPA tends to consider the distribution of regulatory costs and the distribution of benefits independently. It is possible, however, that strong and potentially adverse interactions exist, and these interactions should be considered explicitly in the RIA and during the rulemaking process.

EPA has demonstrated considerable concern about distributional consequences of its regulation, although sometimes not on the issues of most concern to environmental advocates. The agency clearly pays some attention to issues of environmental justice and the identification of disproportionately affected communities, but by statute and executive order it is at least as concerned about the impacts of regulatory costs and other restrictions on various types of industrial facilities. Estimates of plant closings remain an important metric in assessments of economic impacts of regulation, and small plants routinely receive exemptions from the more stringent regulations governing larger plants. But small plants may be older and dirtier than their larger counterparts, and are probably located in the more run-down parts of inner cities or small towns, surrounded by low-income and perhaps minority communities. The location and continuous existence of these plants could therefore exacerbate adverse environmental justice outcomes that in other ways EPA explicitly attempts to avoid. At the same time, people living in these communities are frequently employed by the very plants whose actions may be harmful to their health, so that any action against small or old plants conceivably could increase local unemployment precisely at locations where few alternative jobs exist. Thus, any regulatory response here should be considered carefully, based on credible analysis of the potential for injustice, the potential interactions between regulatory costs and benefits, and disadvantaged communities.

Research-Oriented Recommendations

13. Consider the use of group- as well as individual-respondent methods for calculating WTP

Critics of CBA have argued persistently that when considering public goods, it is more appropriate to value them in a collective context than in the individual-consumer context prescribed by welfare analysis. According to this view, people's decisionmaking calculus about public goods is different from their valuation of private goods because the context is different. Their thinking is supposedly less parochial, more future-oriented, and more altruistic. In addition, critics argue that the context in which WTP is elicited in individual surveys is artificial and inconsistent with how individuals actually make market-based decisions. The issue we want to focus on is this: How might those concerns be addressed by the use of group processes to elicit WTP?

Primarily, advocates of group processes have in mind fully group-determined decisions, reached by some deliberative process followed by the exercise of some kind of voting mechanism. Group valuation methods seem to have risen out of the citizens' jury, a method of illuminating public policy controversies by convening one or more panels of citizens. In the United States, for example, the Hubert Humphrey School at the University of Minnesota has been prominent in its use of citizens' juries to compare pricing policies to other approaches to deal with traffic congestion. Group valuation methods add a valuation step to the citizens' jury concept. See Sagoff (1998) or Spash (2007) for brief reviews of different approaches to group-determined benefit estimates.

To economists, the problem with this sort of group valuation is that it breaks the link between theoretical welfare economics and CBA. In principle, CBA accepts only one method for valuing the outcomes of social regulations or of public investments: the sum of individual valuations, elicited either directly by survey or indirectly by inference from consumer behavior via autonomous market agents. Moreover, observers from numerous backgrounds see potential practical problems related to group-elicited WTP estimates.

A bigger problem is that little consensus exists regarding how to conduct such group elicitations, and many observers fear that any such estimate may reflect more than just the valuation of the public good or service in question. For this reason, most observers would predict that the use of group methods would probably produce higher WTP estimates than would standard methods. List and his colleagues (2004), for example, agree that social approaches can lead to higher WTP values, but not because the values more faithfully reflect true WTP for the public good in question. Rather, group processes can include individuals' willingness to be accommodating to the values of others, as well as their signaling of their environmental and social concerns. These latter effects may be valid, but they could be connected to any public good or to no public good and, according to List et al., can only distort the estimates of WTP for the good in question.

However, perhaps it is possible to have a middle ground in which information and attitudes about the public good in question can be aired in a group setting but coupled with private elicitation of WTP in a manner consistent with welfare theory and current practice. To see how this group interaction might help, consider briefly how WTP surveys are typically conducted now. To elicit individual WTP, the general procedure is to conduct a single 15- to 30-minute personal or telephone interview in which the environmental problem or public good to be valued is described in some detail, and a public policy remedy that will regulate the harm is proposed, as is a method of covering its costs. The payment method is designed to make it clear to the respondent that the respondent would have to pay, and so would everyone else. Thus, most well-designed WTP studies attempt to eliminate concerns about free riding (unless altruism is the focus of the research). After the setup is explained to the respondent's satisfaction, a series of yes–no WTP questions is asked. These data are then aggregated across respondents to get the demand curve.

In other words, respondents come to the survey cold, are presented with the environmental problem and potential remedy having perhaps never thought of it before, and then are asked to absorb a great deal of information and make value decisions with very little time for consideration or introspection and without being able to discuss the matter with friends or colleagues. All this despite the fact that most people do spend time thinking about major decisions, and often consult friends and colleagues for advice or additional perspective.

For use values that are broadly familiar to the public and that have more or less direct counterparts in market activities, such as increased availability of outdoor recreation, improved health, or greater commercial fishing yields, estimation of WTP is relatively uncontroversial. These estimates are most often produced by indirect methods, but the fact that they are familiar to the public means that they are also better suited than other benefit categories to individual survey methods. For more obscure or less empirically supported use values, such as the water-purifying and flood-control benefits of wetlands, or nonuse values such as endangered species and habitat protection, few if any market surrogates are available, and survey methods are the only game in town. Unfortunately, such goods are also the ones for which respondents will most likely have greater difficulties in valuation surveys.

Coupling group information provision with individual WTP elicitation has begun to attract empirical attention. In one interesting empirical study of WTP for the preservation of wildlife habitat of endangered geese in Scotland, for example, McMillan et al. (2002) outline a group informational approach they call the Market Stall.[7] The authors recruit several groups of participants in a focus group-like setting, explaining to attendees the usual survey preliminaries of problem, potential solution, and payment method. A question-and-answer session follows. Researchers then ask the valuation questions in a format in which participants respond without revealing their answers to other participants. Respondents are then excused and invited back one week later for a follow-up discussion. In the meantime, they are encouraged to do their own research and talk to their friends and families. At the follow-up meeting, participants are once again asked if they have any questions, and discussion is encouraged. When no one has anything else to say, WTP is again elicited privately. For comparison purposes, researchers also conduct a more conventional WTP survey without the group discussion.

The results were dramatic. Compared to the Market Stall participants, the conventional survey participants were nearly twice as likely to indicate they would "definitely pay" (DP; 33 percent to 18 percent). The mean WTP of DP respondents was £15.29 in the survey, compared to £3.67 for the Market Stall participants in the first session and £4.49 for the same participants in the second session. The Market Stall estimates also had much smaller standard errors. Learning about this problem in a group session appears to affect WTP dramatically, but probably not in the direction that most would expect. Obviously, one cannot conclude on the basis of one study that group methods will reliably produce lower estimates of WTP, but it does suggest that we might have much to learn from group processes and that some of these lessons are likely to be surprising.

Recently EPA's SAB recommended against the use of group sharing of information in WTP surveys and of group elicitation of WTP. In view of the substantial development of literature on these issues, we suggest that EPA revisit this issue.

14. Investigate WTP to Avoid the Dread Associated with Increased Risk to Oneself or to One's Family

A persistent theme in the debate between proponents and opponents of CBA has been the question of whether the risk perceptions of experts or of laypeople should dominate in public decisionmaking about risk. One lesson from this discussion has been that risk involves more than the probability of material harm. Depending on the circumstances, it can also involve fear, anger, hopelessness, a sense of losing control, and more—myriad emotional and psychological reactions we will gather under the common heading, *dread.*

To the extent that CBA estimates only the WTP to avoid an increased probability of material harm, and ignores the dread associated with that probability, it may be missing an important category of regulatory benefits. Regulation may reduce both the probability of harm and the dread that often accompanies it. There is no theoretical reason for ignoring the latter in CBA if empirical evidence eventually shows a meaningful WTP to avoid dread. An important concern here is not to use WTP from studies of a health effect that is not expected, such as death in an automobile accident, to estimate WTP of a health effect where dread may play an important role. Comparison of valuation studies for various health effects suggest that differences beyond the direct risks of dying may raise WTP by 0 to 100 percent.

Final Observations

Our recommendations for the reform of RIAS cover a range of topics: the quality of the analyses, relevance to agency decisionmaking, transparency, treatment of new scientific information, and the proper balance in both the analyses and the process, including the distributional consequences. The overall message is clear: improve the quality, scientific credibility, and timeliness of RIAS and, at the same time, make them more transparent and relevant to the decisionmaking process.

The natural pushback is to ask how much these improvements will cost. Presently, a small cottage industry is involved in preparing RIAS, both inside and outside of EPA. At an estimated cost of $1 million for each of the 8 to 10 RIAS produced annually, the agency is already committing substantial resources to this effort.[8] Despite a number of cost-reducing proposals among our recommendations, such as a more selective focus on particular topics to be studied in individual RIAS, we recognize that our proposals would probably add to the total costs of developing RIAS. It is also possible that some of our recommendations may be at odds with others. For example, more SAB review might well conflict with the goal of developing a preliminary RIA six months in advance of agency decision meetings.

Recalling that RIAS are generally focused on rules with a minimum of $100 million of annual costs and/or benefits, the potential gains from improved regulatory decisionmaking are large. Unfortunately, the evidence that RIAS actually add net benefits to regulation is limited. Despite one early study demonstrating the gains from RIAS, limited recent data are available on the subject.[9] Nonetheless, based on our review of the RIAS examined in this report, as well as other evidence, it is our judgment that recent RIAS have fallen well short of the mark in generating information and analyses that are truly useful to decisionmakers. We appear to be at a crossroads: either we fix the current system or we accept it without major reform. The recommendations developed here represent our judgment on an agenda for the former effort. We hope to spur further debate on these issues to stimulate constructive change.

■ ■ ■

Notes

1. *See* http://yosemite.epa.gov/ee/epa/eed.nsf/webpages/Guidelines.html.

2. Steinzor (2008), 122.

3. According to EPA (2000), the guidelines "... establish a sound scientific framework for performing economic analyses of environmental regulations and policies. They incorporate recent advances in theoretical and applied work in the field of environmental economics. The Guidelines provide guidance on analyzing the economic impacts of regulations and policies and on assessing the distribution of costs and benefits among various segments of the population, with a particular focus on disadvantaged and vulnerable groups." *See* http://yosemite.epa.gov/ee/epa/eed.nsf/webpages/Guidelines.html.

4. Morgenstern and Landy (1997) also proposed that a NEPA-style scoping exercise be added to the RIA process.

5. An example drawn from the RIAS examined in this volume is the paper by Bell et al. (2004) on ozone mortality. Although the RIA on the CAIR cited the Bell et al. study, it was not included in the agency's benefit calculations because it had not yet been formally accepted for journal publication when the RIA was completed.

6. Executive Order 12866.

7. Strictly speaking, the experiment was to elicit the amount citizens were willing to contribute to compensate farmers for damages to land and crops caused by the protected species on their land. This illustrates a common problem of WTP studies: their connection to a regulation or to a policy outcome is tenuous at best. Based on the description in the paper, the respondent is not told how farmers would respond to the offer of compensation or how the goose populations would respond to the increase in habitat.

8. The estimate of $1 million is from Morgenstern and Landy (1997), based on a dozen RIAS conducted by EPA in the 1980s and 1990s. The Congressional Budget Office (CBO 1997) estimated the cost at about $700,000 apiece, although it highlighted the large variance in costs among different RIAS. Averaging the two estimates and inflating to current dollars yields about $1 million.

9. Morgenstern and Landy (1997) found that in a group of a dozen RIAS conducted in the 1980s–1990s, the increase in net benefits of the rules attributable to the RIAS greatly exceeded the costs of the actual studies. For a contrary view on the usefulness of RIAS, *see* Hahn and Tetlock (2008).

References

Ackerman, Frank, and Lisa Heinzerling. 2003. *Priceless: On Knowing the Price of Everything and the Value of Nothing.* New York: The New Press.

Bell, M.L., A. McDermott, S.L. Zeger, J.M. Samet, and F. Dominici. 2004. Ozone and Short-term Mortality in 95 U.S. Urban Communities, 1987–2000. *Journal of the American Medical Association* 292:2372–2378.

Congressional Budget Office (CBO). 1997. *Regulatory Impact Analysis: Costs at Selected Agencies and Implications for the Legislative Process.* Washington, DC: CBO

Executive Order (EO) 12866. 1993. *Federal Register* 58:51735, October 4.

Hahn, Robert W. and Patrick Dudley. 2007. How Well Does Government Do Cost-Benefit Analysis? *Review of Environmental Economics and Policy.* 1(2): 192–211.

Hahn, Robert W. and Paul C. Tetlock. 2008. Has Economic Analysis Improved Regulatory Decisions? *Journal of Economic Perspectives.* 22:1, 67–84.

List, John A., Robert P. Berrens, Alok K. Bohara, and Joe Kerkvliet. 2004. Examining the Role of Social Isolation on Stated Preferences. *American Economic Review* 94(3): 741–752.

McMillan, Douglas C., Lorna Philip, Nick Hanley, and Begona Alvarez-Farizo. 2002. Valuing the Nonmarket Benefits of Wild Goose Conservation: A Comparison of Interview and Group-Based Approaches. *Ecological Economics* 43:49–59.

Morgenstern, Richard, and Mark Landy. 1997. Chapter 15 (Conclusions), in *Economic Analyses at EPA: Assessing Regulatory Impact,* edited by Richard Morgenstern. Washington, DC: RFF Press.

Olson, Erik D. 1984. The Quiet Shift of Power: Office of Management and Budget Supervision of Environmental Protection Agency Rulemaking under Executive Order 12291. *Virginia Journal of Natural Resources Law* 4:1.

Office of Management and Budget (OMB). 2005. *Report to Congress on the Benefits and Costs of Federal Regulations.* Washington, DC: Office of Information and Regulatory Affairs, OMB.

Robinson, Lisa A., James K. Hammitt, Joseph E. Aldy, Alan Krupnick, and Jennifer Baxter. Forthcoming in 2009. Valuing the Risk of Death from Terrorist Attacks. RFF Discussion Paper. www.rff.org.

Sagoff, Mark. 1998. Aggregation and Deliberation in Valuing Environmental Public Goods: A Look beyond Contingent Pricing. *Ecological Economics* 24: 213–230.

Spash, Clive L. 2007. Deliberative Monetary Valuation (DMV): Issues in Combining Economic and Political Processes to Value Environmental Change. *Ecological Economics* 63: 690–699.

Steinzor, Rena I. 2008. *Mother Earth and Uncle Sam: How Pollution and Hollow Government Hurt our Kids,* Austin, TX: University of Texas Press.

U.S. Department of Agriculture (USDA) Forest Service. 2005. Special Areas, State Petitions for Inventoried Roadless Area Management, Final Rule. *Federal Register* 70:25654, May 15.

U.S. Department of Interior. 2004. Office of Surface Mining Reclamation and Enforcement, Surface Coal Mining and Reclamation Operations, Excess Spoil, Stream Buffer Zones, Diversions, Proposed Rule. *Federal Register* 69:1035–1048, January 7.

U.S. Department of Interior, Fish and Wildlife Service, and U.S. Department of Commerce, National Marine Fisheries Service. 2008. Interagency Cooperation Under the Endangered Species Act, Proposed Rule. *Federal Register* 73:47868, August 15.

U.S. Environmental Protection Agency (EPA). 2000. *Guidelines for Preparing Economic Analyses,* Office of the Adminstrator (EPA 240-R-00-003), Washington, DC. http://yosemite.epa.gov/ee/epa/eed.nsf/web pages/Guide lines.html (accessed January 31, 2009).

———. 2002. New Source Review (NSR) Improvements, Supplemental Analysis of the Environmental Impact of the 2002 Final NSR Improvement Rules, November 21. www.epa. gov/NSR/documents/nsr-analysis.pdf (accessed January 11, 2008).

■ ■ ■

CONTRIBUTORS

David A. Evans. *Economist, Center for Environmental Economics, U.S. Environmental Protection Agency*

David Evans' research interests include normative and positive analyses of regulatory design and the application of stated preference methods. At EPA, he performs and evaluates the economic analyses that support decisionmaking on federal air quality regulations. Evans was formerly a research assistant and then research associate at Resources for the Future. He holds a Ph.D. from the University of Maryland, a master's degree from the University of Illinois, and an undergraduate degree from Hiram College.

Scott Farrow. *Professor and Chair, Department of Economics, and Faculty Fellow, Center for Urban Environmental Research and Education, UMBC (University of Maryland, Baltimore County)*

Scott Farrow's research interests include risk-based economic approaches to government program evaluation, including decisionmaking under uncertainty. Previously, he was chief economist at the U.S. Government Accountability Office, where he led agency development with respect to risk management and economic performance, among other duties. Prior to that, he was director and principal research economist at the Center for the Study and Improvement of Regulation in the Department of Engineering and Public Policy and the H. John Heinz III School of Public Policy and Management at Carnegie Mellon University. Farrow holds a Ph.D. in economics and a master's degree in industrial organization and environmental economics from Washington State University, and an undergraduate degree in economics from Whitman College.

Winston Harrington. *Senior Fellow, Resources for the Future*

Winston Harrington's work encompasses urban transportation, motor vehicles and air quality, and problems of estimating the costs of environmental policy. He has worked extensively on the economics of enforcing environmental regulations, the health benefits derived from improved air quality, the costs of waterborne disease outbreaks, endangered species policy, federal rulemaking procedures, and the economics of outdoor recreation. Harrington has served as a consultant to state and federal governments, the World Bank, and the Harvard Institute for International Development. He also is on the adjunct faculty at Georgetown University. Harrington holds a Ph.D. in city and regional planning from the University of North Carolina at Chapel Hill and a master's degree in mathematics from Cornell University.

Lisa Heinzerling. *Professor of Law, Georgetown University Law Center (on leave of absence)*

Lisa Heinzerling received an A.B. from Princeton University and a J.D. from the University of Chicago Law School, where she was editor-in-chief of the *Law Review.* She clerked for Judge Richard A. Posner on the United States Court of Appeals for the Seventh Circuit and for Justice William J. Brennan, Jr. on the United States Supreme Court. Later, Heinzerling served as an assistant attorney general in Massachusetts, specializing in environmental law, before becoming a faculty member at Georgetown. She has been a visiting professor at the law schools at both Yale and Harvard, and has published extensively, including a book with Frank Ackerman, *Priceless: On Knowing the Price of Everything and the Value of Nothing* (The New Press, 2004). Heinzerling is currently on leave of absence from Georgetown and serving as senior climate policy counsel at the Environmental Protection Agency.

Nathaniel O. Keohane. *Director of Economic Policy and Analysis, Climate and Air Program, Environmental Defense Fund (EDF)*

Nat Keohane oversees EDF's analytical work on the economics of climate policy, and helps to develop and advocate the organization's policy positions on global warming. His academic research has focused on the design and performance of market-based environmental policies. From 2001 to 2007, Keohane was an assistant and then associate professor of economics at the Yale School of Management. He has published articles on environmental economics in academic journals including the *Journal of Public Economics*, the *RAND Journal of Economics*, the *Journal of Environmental Economics and Management*, and the *Harvard Environmental Law Review*. Keohane is also the co-author of *Markets and the Environment* (Island Press, 2007) and co-editor of *Economics of Environmental Law* (Edward Elgar, 2009).

Alan Krupnick. *Senior Fellow and Director of Research, Resources for the Future*

Alan Krupnick's research focuses on analyzing environmental issues, in particular, the benefits, costs, and design of air pollution policies, both in the United States and in developing countries. He also studies the valuation of health and ecological improvements and, more recently, the ancillary benefits of climate policy and urban transportation and development problems. Krupnick has served as a consultant to state governments, federal agencies, private corporations, the Canadian government, the European Union, the World Health Organization, and the World Bank. Krupnick also served as senior economist on the President's Council of Economic Advisers, advising the Clinton administration on environmental and natural resource policy issues. He is a regular member of expert committees from the National Academy of Sciences and has served on a Royal Society of Canada committee analyzing ambient air quality standard setting in Canada. Krupnick holds a Ph.D. and a master's degree in economics from the University of Maryland.

Douglas A. Kysar. *Professor of Law, Yale Law School*

Douglas Kysar's teaching and research areas include tort law, international environmental law, sustainable development, products liability, and risk regulation. He received his bachelor's degree from Indiana University in 1995 and his J.D. from Harvard Law School in 1998, where he served on the student board of advisors. He has published widely on competing policymaking paradigms for the regulation of environmental, health, and safety risks, examining in particular certain underappreciated moral and political assumptions that underlay invocation of cost–benefit analysis and the precautionary principle within environmental policymaking contexts. He has recently completed a book on these subjects, *Regulating from Nowhere: Environmental Law and the Search for Objectivity* (Yale University Press, forthcoming).

Richard D. Morgenstern. *Senior Fellow, Resources for the Future*

Richard Morgenstern's work centers on the economic analysis of environmental issues with an emphasis on the costs, benefits, evaluation, and design of environmental policies, especially economic incentive measures. His analysis also focuses on climate change, including the design of cost-effective policies to reduce emissions in the United States and abroad. Immediately prior to joining RFF, Morgenstern was senior economic counselor to the undersecretary for global affairs at the U.S. Department of State. Previously, he served at the U.S. Environmental Protection Agency, where he acted as deputy administrator (1993); assistant administrator for policy, planning, and evaluation (1991–1993); and director of the Office of Policy Analysis (1983–1995). He has served on expert committees of the National Academy of Sciences and as a consultant to various organizations. Morgenstern holds a Ph.D. in economics from the University of Michigan.

Catherine A. O'Neill. *Associate Professor of Law, Seattle University School of Law*

Catherine O'Neill focuses her research on issues of justice in environmental law and policy, particularly issues affecting fish and other resources relied upon by American Indian tribes, and minority and low-income communities. She also works on methods of risk assessment and management as well as on approaches to regulatory analysis. She is a member scholar with the Center for Progressive Reform and has testified before Congress on regulations governing mercury emissions from coal-fired power plants. O'Neill has also served as a pro bono consultant to the attorneys for the National Congress of American Indians and other tribes in litigation challenging mercury regulations. She has published numerous scholarly articles in such journals as the *Stanford Environmental Law Journal, Ecology Law Quarterly,* and the *Vermont Law Review.* O'Neill was a Ford Foundation graduate fellow at Harvard Law School. She received her J.D. from the University of Chicago School of Law.

Wendy E. Wagner. *Joe A. Worsham Centennial Professor, School of Law, University of Texas at Austin*

In addition to being on the faculty at UT Austin, Wendy Wagner also teaches one semester per year at Case Western Law School. She serves on the National Research Council's Committee on Reducing Stormwater Discharge Contributions to Water Pollution and is a member scholar of the Center for Progressive Reform. Previously, she served as an honors attorney in the Enforcement Division of the Department of Justice's Environment and Natural Resources Division, and then as pollution control coordinator with the Department of Agriculture's Office of the General Counsel. Wagner holds a J.D. and a master's degree in environmental studies from Yale University, where she was senior editor of the *Yale Law Journal* and managing editor of the *Yale Journal of Regulation.*

■ ■ ■

INDEX

Note: Page numbers in *italics* indicate figures and tables. Page numbers followed by an 'n' indicate notes.

For Product Safety Concerns and Information please contact our EU
representative GPSR@taylorandfrancis.com
Taylor & Francis Verlag GmbH, Kaufingerstraße 24, 80331 München, Germany

www.ingramcontent.com/pod-product-compliance
Lightning Source LLC
LaVergne TN
LVHW080651011026
849160LV00014B/1388

* 9 7 8 1 9 3 3 1 1 5 7 5 7 *